AF358651

Advances in Cardiovascular
Magnetic Resonance

Advances in Cardiovascular Magnetic Resonance

Guest Editors

Minjie Lu
Arlene Sirajuddin

Basel • Beijing • Wuhan • Barcelona • Belgrade • Novi Sad • Cluj • Manchester

Guest Editors

Minjie Lu
Department of Magnetic
Resonance Imaging
Fuwai Hospital
Chinese Academy of Medical
Sciences & Peking Union
Medical College
Beijing
China

Arlene Sirajuddin
Radiology and Imaging
Sciences
Clinical Center
National Institutes of Health
(NIH)
Bethesda
MD
USA

Editorial Office
MDPI AG
Grosspeteranlage 5
4052 Basel, Switzerland

This is a reprint of the Special Issue, published open access by the journal *Diagnostics* (ISSN 2075-4418), freely accessible at: https://www.mdpi.com/journal/diagnostics/special_issues/Cardiac_MRI.

For citation purposes, cite each article independently as indicated on the article page online and as indicated below:

Lastname, A.A.; Lastname, B.B. Article Title. *Journal Name* **Year**, *Volume Number*, Page Range.

ISBN 978-3-7258-2967-5 (Hbk)
ISBN 978-3-7258-2968-2 (PDF)
https://doi.org/10.3390/books978-3-7258-2968-2

Contents

Editorial

The Multi-Faceted Utility of Cardiovascular Magnetic Resonance Imaging: Editorial on Special Issue "Advances in Cardiovascular Magnetic Resonance"

Minjie Lu [1,2,*] **and Arlene Sirajuddin** [3,*]

1. Department of Magnetic Resonance Imaging, Fuwai Hospital, Chinese Academy of Medical Sciences & Peking Union Medical College, Beijing 100037, China
2. Key Laboratory of Cardiovascular Imaging, Chinese Academy of Medical Sciences, Beijing 100037, China
3. Department of Health and Human Services, National Heart, Lung and Blood Institute, National Institutes of Health, Bethesda, MD 20892, USA
* Correspondence: coolkan@163.com (M.L.); arlene.sirajuddin@nih.gov (A.S.); Tel.: +86-010-883-969-41 (M.L.)

Citation: Lu, M.; Sirajuddin, A. The Multi-Faceted Utility of Cardiovascular Magnetic Resonance Imaging: Editorial on Special Issue "Advances in Cardiovascular Magnetic Resonance". *Diagnostics* **2023**, *13*, 3501. https://doi.org/10.3390/diagnostics13233501

Received: 14 November 2023
Accepted: 17 November 2023
Published: 22 November 2023

Cardiovascular magnetic resonance (CMR) imaging has emerged as a versatile tool for evaluating and managing a variety of cardiovascular diseases. In this Special Issue, "Advances in Cardiovascular Magnetic Resonance", a total of 14 articles (11 original research articles, 1 case report, and 2 comprehensive reviews) demonstrated various advances in the clinical applications of CMR imaging. These studies highlighted the efficacy of CMR imaging in characterizing idiopathic dilated cardiomyopathy (IDCM), diagnosing Loeffler endocarditis, evaluating cardiac function in males with acquired immune deficiency syndrome (AIDS), predicting left ventricular remodeling (LVR) following primary percutaneous coronary intervention (PPCI), and assessing the risk of major adverse cardiovascular events (MACE) in patients with left ventricular noncompaction (LVNC). Additionally, CMR imaging techniques, such as tissue-tagging (TT-CMR) and feature-tracking CMR (FT-CMR), showed promising results in determining cardiac deformation and functional dynamic geometry parameters.

1. Ischemic Heart Disease

The burden of ischemic heart disease remains significant, both in terms of its prevalence and its impact on individuals and society. According to the World Health Organization (WHO), ischemic heart disease is the leading cause of death worldwide, and it accounted for more than 8 million deaths in 2019 [1].

Ma et al. focused on exploring predictive parameters for LVR following PPCI in patients with acute anterior myocardial infarction (AAMI) using CMR imaging [2]. This study found that both global and regional CMR parameters were valuable in predicting LVR in AAMI patients following PPCI, with the local parameters of the infarct zones found to be superior to those of the global ones.

Palumbo et al. evaluated the role of stress perfusion cardiac magnetic resonance (spCMR) in predicting the risk of major cardiac events in patients with long-standing chronic coronary syndrome (CCS) and ischemia [3]. They included 35 patients who underwent coronary CT angiography (CCTA) and additional adenosine spCMR and the primary outcomes measured were heart failure and all major cardiac events. They concluded that spCMR modeling, including perfusion and strain anomalies, could be a powerful tool in predicting the risk of major cardiac events in patients with long-standing CCS, even in the absence of conventional imaging predictors.

2. IDCM

IDCM is a type of myocardial disease of unknown origin, with many cases thought to be genetic in nature. This condition can lead to heart failure, arrhythmias, and sudden cardiac death. Due to the lack of a clear cause, diagnosis and management of IDCM can be challenging.

The study by Tsabedze et al. investigated the use of CMR imaging in the characterization of IDCM in a cohort of patients in southern Africa who were suspected of having a genetic cause of their cardiomyopathy [4]. The study found that late gadolinium enhancement (LGE) was present in the majority of the participants and that mid-wall LGE enhancement was the most common pattern observed. The study also found that patients with LGE on CMR imaging had a lower risk of death than those without. These findings highlight the potential of CMR imaging for improving not only the diagnosis but also the management of IDCM in sub-Saharan Africa.

Zhang et al. investigated the expression of microRNAs (miRNAs) in myocardial tissues of hypertrophic cardiomyopathy (HCM) patients and found that some of the miRNAs had significant correlations with cardiac function and myocardial fibrosis, indicating their potential as biomarkers for left ventricular hypertrophy, fibrosis, and remodeling [5]. These findings suggest that miRNA levels in myocardial tissues could serve as potential biomarkers for HCM, potentially aiding in the early diagnosis and treatment of this disease.

Huang evaluated the role of LGE in predicting MACE in patients with LVNC and found that a specific ring-like pattern of LGE, free-wall, or mid-wall LGE, and LGE extent greater than 7.5% was associated with an increased risk of MACE [6]. This study suggests that risk stratification based on LGE may have value in predicting MACE in patients with LVNC.

Revnic et al. focused on the use of CMR imaging to detect myocardial replacement fibrosis in patients with nonischemic dilated cardiomyopathy (NIDCM) and sought to evaluate the association between collagen turnover biomarkers and replacement myocardial scarring by CMR and test their ability to predict outcomes in conjunction with LGE in patients with NIDCM [7]. They found that galectin-3 (Gal3), procollagen type I carboxy-terminal pro-peptide (PICP), and N-terminal pro-peptide of procollagen type III (PIIINP) were significantly increased in LGE+ individuals and were directly correlated with LGE mass. These circulating collagen turnover biomarkers could significantly predict cardiovascular outcomes, and their joint use with LGE could improve outcome prediction in patients with NIDCM. Overall, this study highlights the potential of using biomarkers in conjunction with CMR to detect myocardial fibrosis and predict outcomes in patients with NIDCM.

Brendel et al. evaluated the diagnostic performance of dark-blood LGE compared to that of conventional bright-blood LGE in detecting non-ischemic myocardial scarring [8]. The study included 343 patients with suspected non-ischemic cardiomyopathy who underwent both dark-blood and bright-blood LGE imaging. The results showed that dark-blood LGE had a sensitivity of 99%, a specificity of 99%, and an accuracy of 99% for detecting non-ischemic scarring, with no significant difference in scar size compared to bright-blood LGE. The study concludes that dark-blood LGE imaging is non-inferior to bright-blood LGE imaging in detecting non-ischemic scarring, suggesting it may be an equivalent method for detecting both ischemic and non-ischemic scars.

3. Inflammatory Cardiomyopathy

Inflammatory cardiomyopathy is a myocardial disease caused by inflammation that can result from several factors, including viral infections, autoimmune diseases, and drug reactions. Early diagnosis and proper management are critical to prevent severe complications and enhance patient outcomes.

Lupu et al. reported a case of Loeffler endocarditis in a patient presenting with heart failure with preserved ejection fraction (HFpEF) [9]. This case highlights the importance of early diagnosis and prompt management of Loeffler endocarditis in order to prevent serious complications and improve patient outcomes. CMR imaging was able to detect eosinophil and lymphocyte infiltration of the endomyocardium, as well as the formation of thrombus and fibrosis, leading to the diagnosis of Loeffler endocarditis.

Hou et al. evaluated the cardiac function of antiretroviral therapy-treated (ART-treated) males with AIDS using CMR imaging and found that those with a short disease duration may not develop obvious cardiac dysfunction as evaluated by routine CMR [10]. The findings suggest that routine CMR may not be necessary for ART-treated males with AIDS with a short disease duration, but it is important to consider individual patient characteristics and other factors when determining follow-up intervals.

4. MR Contrast Related and Extra-Cardiac Disease

In this issue, we feature two special research articles. The first study examines the relationship between MR contrast and nephrogenic systemic fibrosis (NSF). The second explores myocardial impairment associated with extra-cardiac disorders.

Gallo-Bernal et al. provide a comprehensive review of the existing evidence on the use of gadolinium-based contrast agents in cardiac imaging and the risk of NSF in patients with chronic kidney disease [11]. They highlight the importance of understanding the clinical characteristics and risk factors of NSF in order to prevent and recognize it as well as summarize the pathophysiology, clinical manifestations, diagnosis, and prevention of NSF related to the use of gadolinium-based contrast agents.

Yang et al. focused on the use of CMR imaging to examine the temporal changes in the cardiac cycle of patients with pulmonary hypertension (PH), a condition that alters the biventricular shape and temporal phases of the cardiac cycle [12]. They found that only the right ventricular ejection fraction (RVEF) was decreased in the ventricular function of the interventricular septal (IVS) non-displacement (IVS_{ND}) group, and no temporal change in the cardiac cycle was found. In contrast, a prolonged isovolumetric relaxation time (IRT) and shortened filling time (FT) in both ventricles, along with biventricular dysfunction, were detected in the IVS displacement (IVS_D) group. The IRT of the right ventricle (IRTRV) and the FT of the right ventricle (FTRV) in PH patients were associated with pulmonary vascular resistance, right cardiac index, and IVS curvature, and the IRTRV was also associated with the RVEF in a multivariate regression analysis. The researchers concluded that the temporal changes in the cardiac cycle were related to IVS displacement and mainly impacted the diastolic period of the two ventricles in the PH patients. Both IRT and FT changes may provide useful pathophysiological information on the progression of PH. The study suggests that CMR imaging can be a useful tool for understanding cardiac dysfunction in PH patients and for risk stratification of PH patients.

5. CMR Emerging Techniques

CMR has evolved significantly over the years, with the development of emerging techniques that have expanded its role in clinical practice. Among them, myocardial strain imaging is one of the most important new technologies in recent years, which provides information on myocardial deformation and has the potential to detect subtle changes in myocardial function that may not be apparent with conventional imaging methods.

The work written by Zlibut et al. provides a comprehensive overview of the available data on the role of CMR in evaluating myocardial strain and biomechanics—highlighting its potential as a valuable tool for early detection, diagnosis, and management of cardio-vascular diseases [13]. The authors highlight two specific CMR techniques, tissue-tagging (TT-CMR) and feature-tracking CMR (FT-CMR), that have been shown to accurately deter-mine deformation parameters and functional dynamic geometry parameters. The authors point out that these techniques have been studied extensively in ischemic heart disease and primary myocardial illnesses and have shown utility in prognostic prediction in various cardiovascular patients. They also discuss the recent emergence of fast strain-encoded imaging CMR-derived myocardial strain as a potentially superior method for measuring myocardial strain due to its accuracy and reduced acquisition time. However, it should be acknowledged that more studies need to be carried out to establish its clinical impact.

Michler et al. compared two methods, the conventional contour surface method (KfM) and the pixel-based evaluation method (PbM), for determining left ventricular function parameters in CMR imaging [14]. The KfM includes the papillary muscle as part of the left ventricular volume, leading to a systematic error in the calculation of the left ventricular ejection fraction (LVEF). The PbM, on the other hand, excludes the papillary muscle volume and provides more accurate results. The study analyzed 191 CMR image data sets and found that the PbM showed a negative difference for end-diastolic volume (EDV), a negative difference for end-systolic volume (ESV), and a positive difference for LVEF compared to the KfM. There was no difference in stroke volume (SV). The PbM also had a mean papillary muscle volume of 14.2 mL and took an average of 2:02 min for evaluation. The authors conclude that the PbM is an easy and fast method for determining left ventricular cardiac function and provides comparable results to the established KfM while omitting the papillary muscles. This can have a significant influence on therapy decisions, as the LVEF may be 6% higher with the PbM.

A study conducted by Ueda et al. explored the use of FT-CMR and self-gated magnetic resonance cine imaging to evaluate cardiac function in a young mouse model of Duchenne muscular dystrophy (mdx) [15]. The results show that the LVEF was significantly lower in the mdx group compared to that in the control group at both time points. Strain analysis also revealed significantly lower strain values in mdx mice, with exception of the longitudinal strain of the four-chamber view. This study concludes that strain analysis with feature tracking and self-gated magnetic resonance cine imaging is useful for assessing cardiac function in young mdx mice.

6. Summary

This Special Issue of "Advances in Cardiovascular Magnetic Resonance" highlights the role of CMR imaging in improving the diagnosis, management, and prognosis of various cardiovascular diseases. From its ability to detect LGE in patients with IDCM to its role in diagnosing Loeffler endocarditis as well as its potential for predicting LVR following PPCI in AAMI patients, CMR imaging is proving to be a valuable diagnostic tool.

Finally, the comprehensive review of the available data on the role of CMR in evaluating myocardial strain and biomechanics underscores the potential of these techniques in providing a more complete assessment of cardiac function and mechanics, as well as in detecting subtle changes in myocardial strain and deformation in various cardiovascular conditions.

As the field of cardiovascular imaging continues to advance, CMR imaging is poised to play an increasingly pivotal role in detecting, diagnosing, and managing cardiovascular diseases.

Conflicts of Interest: The authors declare no conflict of interest.

References

1. World Health Organization. The Top 10 Causes of Death. 2020. Available online: https://www.who.int/news-room/fact-sheets/detail/the-top-10-causes-of-death (accessed on 10 November 2023).
2. Ma, W.; Li, X.; Gao, C.; Gao, Y.; Liu, Y.; Kang, S.; Pan, J. Predictive Value of Cardiac Magnetic Resonance for Left Ventricular Remodeling of Patients with Acute Anterior Myocardial Infarction. *Diagnostics* **2022**, *12*, 2780. [CrossRef]
3. Palumbo, P.; Cannizzaro, E.; Di Cesare, A.; Bruno, F.; Arrigoni, F.; Splendiani, A.; Barile, A.; Masciocchi, C.; Di Cesare, E. Stress Perfusion Cardiac Magnetic Resonance in Long-Standing Non-Infarcted Chronic Coronary Syndrome with Preserved Systolic Function. *Diagnostics* **2022**, *12*, 786. [CrossRef]
4. Tsabedze, N.; du Plessis, A.; Mpanya, D.; Vorster, A.; Wells, Q.; Scholtz, L.; Manga, P. Cardiovascular Magnetic Resonance Imaging Findings in Africans with Idiopathic Dilated Cardiomyopathy. *Diagnostics* **2023**, *13*, 617. [CrossRef]
5. Zhang, C.; Zhang, H.; Zhao, L.; Wei, Z.; Lai, Y.; Ma, X. Differential Expression of microRNAs in Hypertrophied Myocardium and Their Relationship to Late Gadolinium Enhancement, Left Ventricular Hypertrophy and Remodeling in Hypertrophic Cardiomyopathy. *Diagnostics* **2022**, *12*, 1978. [CrossRef]
6. Huang, W.; Sun, R.; Liu, W.; Xu, R.; Zhou, Z.; Bai, W.; Hou, R.; Xu, H.; Guo, Y.; Yu, L.; et al. Prognostic Value of Late Gadolinium Enhancement in Left Ventricular Noncompaction: A Multicenter Study. *Diagnostics* **2022**, *12*, 2457. [CrossRef]
7. Revnic, R.; Cojan-Minzat, B.O.; Zlibut, A.; Orzan, R.I.; Agoston, R.; Muresan, I.D.; Horvat, D.; Cionca, C.; Chis, B.; Agoston-Coldea, L. The Role of Circulating Collagen Turnover Biomarkers and Late Gadolinium Enhancement in Patients with Non-Ischemic Dilated Cardiomyopathy. *Diagnostics* **2022**, *12*, 1435. [CrossRef]

8. Brendel, J.M.; Holtackers, R.J.; Geisel, J.N.; Kübler, J.; Hagen, F.; Gawaz, M.; Nikolaou, K.; Greulich, S.; Krumm, P. Dark-Blood Late Gadolinium Enhancement MRI Is Noninferior to Bright-Blood LGE in Non-Ischemic Cardiomyopathies. *Diagnostics* **2023**, *13*, 1634. [CrossRef]
9. Lupu, S.; Pop, M.; Mitre, A. Loeffler Endocarditis Causing Heart Failure with Preserved Ejection Fraction (HFpEF): Characteristic Images and Diagnostic Pathway. *Diagnostics* **2022**, *12*, 2157. [CrossRef]
10. Hou, K.; Fu, H.; Xiong, W.; Gao, Y.; Xie, L.; He, J.; Feng, X.; Zeng, T.; Cai, L.; Xiong, L.; et al. Clinical Application of Cardiac Magnetic Resonance in ART-Treated AIDS Males with Short Disease Duration. *Diagnostics* **2022**, *12*, 2417. [CrossRef]
11. Gallo-Bernal, S.; Patino-Jaramillo, N.; Calixto, C.A.; Higuera, S.A.; Forero, J.F.; Lara Fernandes, J.; Gongora, C.; Gee, M.S.; Ghoshhajra, B.; Medina, H.M. Nephrogenic Systemic Fibrosis in Patients with Chronic Kidney Disease after the Use of Gadolinium-Based Contrast Agents: A Review for the Cardiovascular Imager. *Diagnostics* **2022**, *12*, 1816. [CrossRef]
12. Yang, F.; Ren, W.; Wang, D.; Yan, Y.; Deng, Y.L.; Yang, Z.W.; Yu, T.L.; Li, D.; Zhang, Z. The Variation in the Diastolic Period with Interventricular Septal Displacement and Its Relation to the Right Ventricular Function in Pulmonary Hypertension: A Preliminary Cardiac Magnetic Resonance Study. *Diagnostics* **2022**, *12*, 1970. [CrossRef]
13. Zlibut, A.; Cojocaru, C.; Onciul, S.; Agoston-Coldea, L. Cardiac Magnetic Resonance Imaging in Appraising Myocardial Strain and Biomechanics: A Current Overview. *Diagnostics* **2023**, *13*, 553. [CrossRef]
14. Michler, K.; Hessman, C.; Prümmer, M.; Achenbach, S.; Uder, M.; Janka, R. Cardiac MRI: An Alternative Method to Determine the Left Ventricular Function. *Diagnostics* **2023**, *13*, 1437. [CrossRef]
15. Ueda, J.; Saito, S. Evaluation of Cardiac Function in Young Mdx Mice Using MRI with Feature Tracking and Self-Gated Magnetic Resonance Cine Imaging. *Diagnostics* **2023**, *13*, 1472. [CrossRef]

Article

Dark-Blood Late Gadolinium Enhancement MRI Is Noninferior to Bright-Blood LGE in Non-Ischemic Cardiomyopathies

Jan M. Brendel [1], Robert J. Holtackers [2,3], Jan N. Geisel [1], Jens Kübler [1], Florian Hagen [1], Meinrad Gawaz [4], Konstantin Nikolaou [1], Simon Greulich [4,*,†] and Patrick Krumm [1,†]

[1] Department of Radiology, Diagnostic and Interventional Radiology, University of Tübingen Hoppe-Seyler-Straße 3, 72076 Tübingen, Germany

[2] Department of Radiology and Nuclear Medicine, Maastricht University Medical Centre, 6229 HX Maastricht, The Netherlands

[3] Cardiovascular Research Institute Maastricht (CARIM), Maastricht University, 6229 ER Maastricht, The Netherlands

[4] Department of Internal Medicine III, Cardiology and Angiology, University of Tübingen Otfried-Müller-Straße 10, 72076 Tübingen, Germany

* Correspondence: simon.greulich@med.uni-tuebingen.de

† These authors contributed equally to this work.

Abstract: **(1) Background and Objectives:** Dark-blood late gadolinium enhancement has been shown to be a reliable cardiac magnetic resonance (CMR) method for assessing viability and depicting myocardial scarring in ischemic cardiomyopathy. The aim of this study was to evaluate dark-blood LGE imaging compared with conventional bright-blood LGE for the detection of myocardial scarring in non-ischemic cardiomyopathies. **(2) Materials and Methods:** Patients with suspected non-ischemic cardiomyopathy were prospectively enrolled in this single-centre study from January 2020 to March 2023. All patients underwent 1.5 T CMR with both dark-blood and conventional bright-blood LGE imaging. Corresponding short-axis stacks of both techniques were analysed for the presence, distribution, pattern, and localisation of LGE, as well as the quantitative scar size (%). **(3) Results:** 343 patients (age 44 ± 17 years; 124 women) with suspected non-ischemic cardiomyopathy were examined. LGE was detected in 123 of 343 cases (36%) with excellent inter-reader agreement (κ 0.97–0.99) for both LGE techniques. Dark-blood LGE showed a sensitivity of 99% (CI 98–100), specificity of 99% (CI 98–100), and an accuracy of 99% (CI 99–100) for the detection of non-ischemic scarring. No significant difference in total scar size (%) was observed. Dark-blood imaging with mean 5.35 ± 4.32% enhanced volume of total myocardial volume, bright-blood with 5.24 ± 4.28%, $p = 0.84$. **(4) Conclusions:** Dark-blood LGE imaging is non-inferior to conventional bright-blood LGE imaging in detecting non-ischemic scarring. Therefore, dark-blood LGE imaging may become an equivalent method for the detection of both ischemic and non-ischemic scars.

Keywords: magnetic resonance imaging; heart; contrast media; gadolinium; LGE; cardiomyopathies; dark blood; bright blood

Citation: Brendel, J.M.; Holtackers, R.J.; Geisel, J.N.; Kübler, J.; Hagen, F.; Gawaz, M.; Nikolaou, K.; Greulich, S.; Krumm, P. Dark-Blood Late Gadolinium Enhancement MRI Is Noninferior to Bright-Blood LGE in Non-Ischemic Cardiomyopathies. *Diagnostics* **2023**, *13*, 1634. https://doi.org/10.3390/diagnostics13091634

Academic Editors: Minjie Lu and Arlene Sirajuddin

Received: 4 April 2023
Revised: 29 April 2023
Accepted: 3 May 2023
Published: 5 May 2023

1. Introduction

Late gadolinium enhancement (LGE) is the most established imaging modality for myocardial tissue characterisation in cardiac magnetic resonance (CMR). It is routinely used to assess and quantify myocardial fibrosis, irrespective of an ischemic or non-ischemic origin [1–4]. The ability to detect areas of scarring within the myocardium using LGE is critical for the diagnosis of patients with cardiomyopathy. Scarring or fibrosis compromises the structural integrity of the myocardium, predisposing to dysfunction leading to heart failure, arrhythmias, and even sudden cardiac death. Therefore, in addition to its great diagnostic value, LGE has been shown in several studies to have a high predictive value, identifying patients at high risk for adverse cardiac events and death [5–10]. Various

LGE techniques have been developed for the purpose of distinct scar assessment and its delineation from healthy myocardium. Over time, numerous LGE techniques have been introduced, each with the goal of improving myocardial scar detection. In recent years, the so-called dark-blood LGE technique has been proposed to further improve the contrast between subendocardial ischemic scar tissue and the ventricular blood pool [11]. This technique has shown promising results in improving the accuracy and specificity of LGE. Previous studies compared dark-blood LGE with conventional bright-blood LGE in the detection of ischemic scarring, suggesting that dark-blood LGE allows for better delineation and quantification, thus improving the diagnosis of ischemic cardiomyopathy [12,13]. Compared to ischemic scars, non-ischemic scars show different forms of distribution within the myocardium, including linear or patchy types of enhancement. In addition, non-ischemic scars are located mid-wall or subepicardial, typically sparing subendocardial regions supplied by a specific coronary artery. Although some studies have reported on the detection of non-ischemic areas of fibrosis [3,14], there is a lack of data evaluating the diagnostic performance of dark-blood LGE versus conventional bright-blood LGE in visualising non-ischemic scars in a direct head-to-head comparison. To date, there is no clear recommendation for the potential use of dark-blood LGE in cardiac MRI for non-ischemic cardiomyopathy.

We hypothesised that the detection of non-ischemic scars may not differ between the two LGE techniques. Therefore, the aim of this non-inferiority cardiac MRI study was to evaluate the diagnostic performance of dark-blood LGE compared with conventional bright-blood LGE in the detection of non-ischemic myocardial scarring.

2. Materials and Methods

2.1. Study Population

In this single-centre study (Tübingen University Hospital, Tübingen, Germany), we prospectively enrolled patients referred for cardiac MRI (CMR) between January 2020 and March 2023 for clinical suspicion of non-ischemic cardiomyopathy. Patients with incomplete LGE data sets, insufficient image quality or an ischemic LGE pattern were excluded. The study was approved by the Institutional Ethical Review Board, and all patients provided written, informed consent to participate in the study.

2.2. Cardiac MRI Image Acquisition

All patients underwent both conventional bright-blood late gadolinium enhancement and dark-blood late gadolinium enhancement as part of the cardiac MRI imaging protocol using a 1.5 T scanner (MAGNETOM Aera, SIEMENS Healthcare, Erlangen, Germany) within the same scan. In detail, ten minutes after intravenous injection of 0.15 mmol/kg gadobutrol (Gadovist, Bayer Healthcare, Leverkusen, Germany) [15], both LGE techniques were applied according to current recommendations [16]. Dark-blood LGE was performed at 10 min after contrast injection, immediately followed by bright-blood LGE (starting at 15 min after contrast injection). For both LGE methods, an inversion time (TI) scout scan was performed to select the optimal TI. For bright-blood LGE, the TI was set to null viable myocardium of the left ventricle; whereas for dark-blood LGE, the TI was set to null the left ventricular blood pool signal. Sequence parameters for bright-blood LGE were: readout type steady state free precession, echo time 1.24 ms, flip angle 45°, acquired resolution 1.33×1.33 mm^2; phase sensitive inversion recovery (PSIR) steady state free precession, TE 1.26 ms, flip angle 90°, acquired resolution 1.48×1.48 mm^2. For dark-blood LGE, 10–15 2D 8-mm short-axis slices and one 4-chamber slice were acquired; for conventional bright-blood LGE, a 3D data set was acquired and reconstructed in identical slices. The mechanism of the used dark-blood LGE method without additional magnetisation preparation (blood nulled PSIR LGE) has been described in detail previously [17].

2.3. Cardiac MRI Image Analysis

Cardiac magnetic resonance image analysis was performed in three readings by readers with different levels of experience: reading 1 by J.M.B. (6 years of cardiac MRI experience), reading 2 by J.N.G. (1 year of cardiac MRI experience), reading 3 in consensus by P.K. (13 years of cardiac MRI experience) and S.G. (22 years of cardiac MRI experience). Analysis was performed according to the recommendations of the Society for Cardiovascular Magnetic Resonance (SCMR) [18,19] using a dedicated commercially available software package (cvi42 version 5.14, Circle Cardiovascular Imaging Inc, Calgary, AB, Canada). Readers were blinded to the clinical data. In all short-axis LGE slices, endocardial and epicardial contours of the left ventricle were manually delineated, followed by an automated calculation of total scar size using a threshold of 2 SD above remote myocardium [18]. LGE distribution (linear or patchy) and pattern (subendocardial, transmural, mid-wall, subepicardial) were noted [5]. LGE localisation was assessed segmentally according to the adapted American Heart Association 16-segment model, excluding the apical segment 17 [20]. Confidence in the presence or absence of scarring was assessed using a 4-point scale (1 = nondiagnostic exam, 2 = low confidence, 3 = moderate confidence, 4 = high confidence). The CMR reporting and diagnosis of "myocarditis", "non-ischemic cardiomyopathy", or "normal" was made in accordance with current SCMR recommendations and ESC guidelines [21–23].

2.4. Statistical Analysis

A priori power calculation was performed, and sample size estimation was based on Tango [24] (alpha = 0.05; power = 0.90; proportion discordant 0.10), the number to be included was n = 343 patients. Data distribution was assessed using histograms and measures of skewness and kurtosis. Myocardial scar size (normally distributed) was compared using a paired samples t-test (JMP, version 16.2, SAS Institute Inc., Heidelberg, Germany). Differences in myocardial scar size measurements between conventional bright-blood and dark-blood LGE were assessed by Bland–Altman analysis (MedCalc, Version 18.1, MedCalc Software Ltd., Ostend, Belgium). Inter-reader variability of LGE scar size measurements (%) was assessed using intraclass correlations. The diagnostic performance of dark-blood LGE in non-ischemic cardiomyopathy was assessed using the MedCalc diagnostic test evaluation calculator (version 20.112, MedCalc Software Ltd., Ostend, Belgium), considering bright-blood LGE as the reference standard. A two-sided McNemar test was performed to test the marginal homogeneity of both LGE techniques in the dichotomous diagnosis of LGE-positive patients (LGE-positive vs. -negative). Cohen's κ statistic was used to assess inter-reader agreement for the presence of LGE. Wilcoxon signed rank test for paired samples was used to compare the readers' confidence scores in assessing the presence or absence of scarring in dark-blood and bright-blood images. Continuous data are presented as mean ± standard deviation. Categorical data are expressed as frequencies (%). p-values < 0.05 were considered to indicate a significant difference.

3. Results

3.1. Patient Characteristics

A total of 420 consecutive patients underwent 1.5 T cardiac magnetic resonance (CMR) for clinically suspected non-ischemic cardiomyopathy. A total of 25 datasets were incomplete due to missing acquisition of LGE (n = 13), the bright-blood datasets (n = 10) or the dark-blood datasets (n = 2). Figure 1. 23 cases were excluded because of insufficient image quality due to artifacts: fold-over (n = 8), trigger (n = 6), MR conditional implantable cardiac devices (n = 5), ghosting (n = 2), and motion (n = 2). N = 4 datasets were excluded due to incomplete coverage of the left ventricle. Evaluating the remaining 368 complete and diagnostic datasets, we further excluded cases with ischemic LGE patterns (n = 25). Finally, bright-blood and dark-blood LGE data sets from 343 patients (age 44 ± 17 years; age range 18 to 82 years) were used for comparative analysis. The study population consisted of 124/343 (36%) women and 219/343 men (64%).

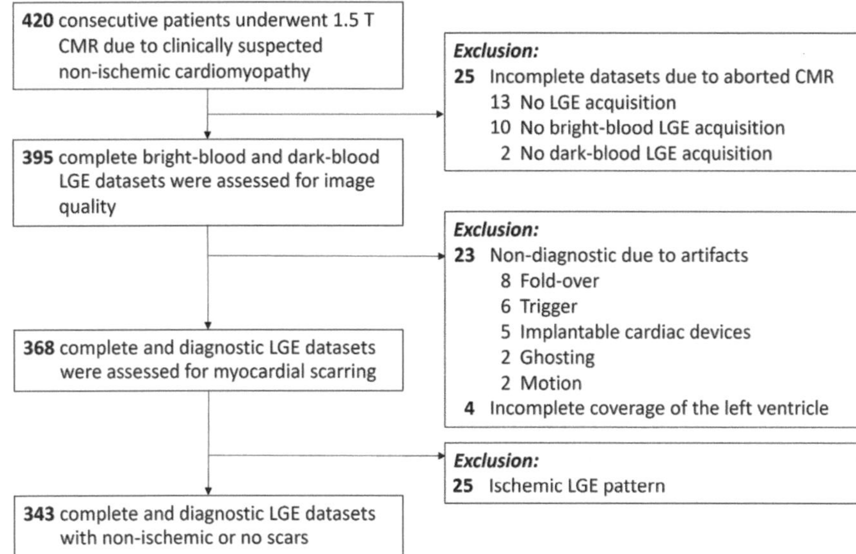

Figure 1. Flowchart showing patient enrolment and reasons for exclusion from the study. Patients referred for cardiac MRI because of clinical suspicion of non-ischemic cardiomyopathy were prospectively enrolled at a single centre (University Hospital of Tübingen) between January 2020 and March 2023. Patients with incomplete LGE data sets, insufficient image quality, or an ischemic LGE pattern were excluded. *CMR* = Cardiac magnetic resonance. *LGE* = Late gadolinium enhancement.

Baseline characteristics of the study population are shown in Table 1. The most common cardiac MRI diagnosis was myocarditis (*n* = 125/343, 36%), followed by patients without abnormalities on CMR (*n* = 109/343, 32%), and patients with different forms of non-ischemic cardiomyopathy (*n* = 109/343, 32%). In detail: *n* = 31 with hypertrophic cardiomyopathy, *n* = 29 with dilated cardiomyopathy, *n* = 3 with arrhythmogenic cardiomyopathy, *n* = 3 with non-compaction cardiomyopathy, *n* = 3 with tako-tsubo cardiomyopathy, *n* = 2 amyloidosis, *n* = 1 with peripartum cardiomyopathy, and *n* = 37 not further classified non-ischemic cardiomyopathy.

Table 1. Baseline Characteristics of Study Population (*n* = 343).

Parameter	Study Populatio
Age (years)	44 ± 17
Age total range (years)	18–82
Female	124/343 (36%)
Male	219/343 (64%)
Final diagnosis by cardiac MRI	
Myocarditis	125/343 (36%)
Non-ischemic cardiomyopathy	109/343 (32%)
Normal	109/343 (32%)

Table summarises the core baseline characteristics of the study population. Values are presented as mean ± standard deviation or numerator/denominator (frequency %).

3.2. Evaluation of Left Ventricular LGE Frequency, Pattern, and Localisation

Late gadolinium enhancement was present in 123 of 343 cases (36%). Regarding the distribution of LGE, linear LGE was found in 72 of 123 cases (59%), and 70 of 123 cases (57%) showed patchy LGE. All scars were of a non-ischemic type with a mid-wall LGE pattern (87 of 123 cases, 71%) or a subepicardial LGE pattern (63 of 123 cases, 51%). LGE was predominantly located in the basal inferolateral wall. The inferior right ventricular insertion points showed LGE more frequently than the anterior right ventricular insertion points. The localisation of LGE with segmental counts is depicted in Figure 2.

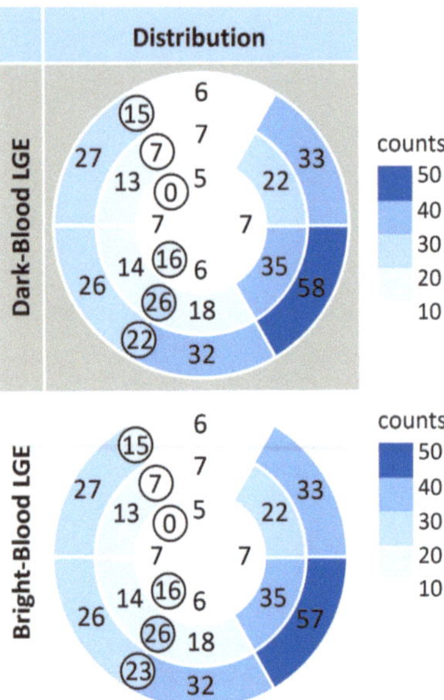

Figure 2. Bullseye plots represent the distribution of late gadolinium enhancement (LGE) depicted by the dark-blood LGE technique (top row) and the bright-blood LGE technique (bottom row). Numbers are the counts of LGE appearances per AHA—segment and at the right ventricular insertion points (circled). The blue colouring serves to better visualise the number of segmental counts (dark blue = many counts, light blue/grey = few counts).

3.3. Comparison of Total Scar Size

Total scar size did not differ significantly between the two LGE techniques: mean $5.35 \pm 4.32\%$ enhanced volume for dark-blood LGE and $5.24 \pm 4.28\%$ for bright-blood LGE, $p = 0.84$. No systematic bias for dark-blood LGE was found in the measurement of myocardial scar size, Figure 3. The LGE methods showed a mean difference of 0.1% ($p = 0.57$), with slightly higher values measured with dark-blood LGE than with bright-blood LGE. The limits of agreement were +1.5% (+1.96 standard deviations) and −1.3% (−1.96 standard deviations).

3.4. Reader Agreement and Diagnostic Confidence

Excellent inter-reader agreement was observed for the assessment of the presence of late gadolinium enhancement: $\kappa_{\text{dark-blood LGE}} = 0.99$ (read 1 vs. 2), 0.97 (read 2 vs. 3), and 0.98 (read 1 vs. 3); and $\kappa_{\text{bright-blood LGE}} = 0.98$ (read 1 vs. 2), 0.97 (read 2 vs. 3), and 0.97 (read 1 vs. 3). Regarding the confidence level for scar detection, there were no significant differences between dark-blood LGE (3.79 ± 0.41) and bright-blood LGE (3.77 ± 0.46), $p = 0.60$. Regarding the confidence level for the absence of scarring, no significant differences were found between dark-blood LGE (3.88 ± 0.40) and bright-blood LGE (3.91 ± 0.34), $p = 0.57$. When assessing the total quantitative scar size, excellent inter-reader ICC coefficients were observed: 0.93 for dark-blood LGE and 0.94 for bright-blood LGE.

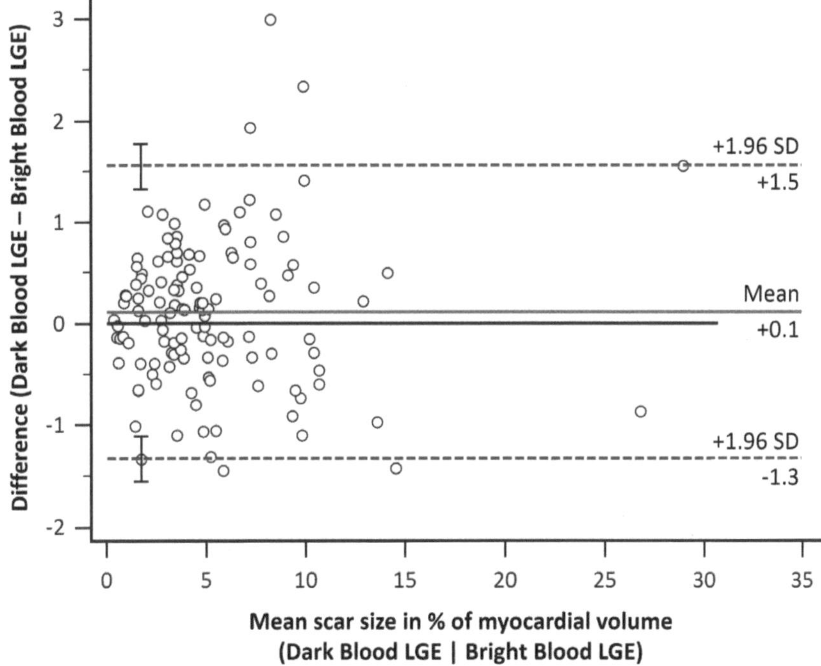

Figure 3. Bland–Altman plot of myocardial scar size as portrayed by dark-blood late gadolinium enhancement (LGE) and bright-blood LGE. No significant bias (solid red line) was found between the two methods. The limits of agreement (%, ±1.96 standard deviations) are indicated as dashed blue lines.

3.5. Diagnostic Performance of Dark-Blood LGE

In the dichotomous per-patient assessment of LGE (yes/no), 361 concordant ratings for the presence of LGE in both techniques were given by the readers; and in 661 evaluations, the absence of myocardial scarring was rated concordantly. This adds to a total of 1022/1029 concordant ratings (99%) for the presence or absence of scarring. The numbers of positive and negative LGE counts for both the dark-blood LGE technique and the bright-blood LGE technique are presented in Table 2.

Table 2. Detection of non-ischemic scarring using dark-blood LGE and bright-blood LGE.

		Bright-Blood LGE		
		Positive	Negative	Σ
Dark-blood LGE	Positive	361	5	366
	Negative	2	661	663
	Σ	363	666	1029

Contingency table depicts dichotomous per-patient evaluations for the presence (positive) or the absence (negative) of enhancement in both the dark-blood technique and the bright-blood technique. *LGE* = late gadolinium enhancement.

In five cases, one of the readers indicated late gadolinium enhancement (scar) on the dark-blood images but no LGE on the bright-blood images. In two cases, one of the readers detected LGE on the bright-blood images but no LGE on the dark-blood images. All patients with discordant assessments demonstrated only small focal LGE lesions with minimal scar volume (median 1.0% [IQR, 0.6–2.1] for dark-blood LGE, and median 1.6%

[IQR, 0.7–2.2] for bright-blood LGE), resulting in low-to-moderate confidence scores for both presence and absence of a scar by all readers.

Cardiac MRI examples of both LGE techniques displaying typical non-ischemic, subepicardial linear LGE are shown in Figure 4.

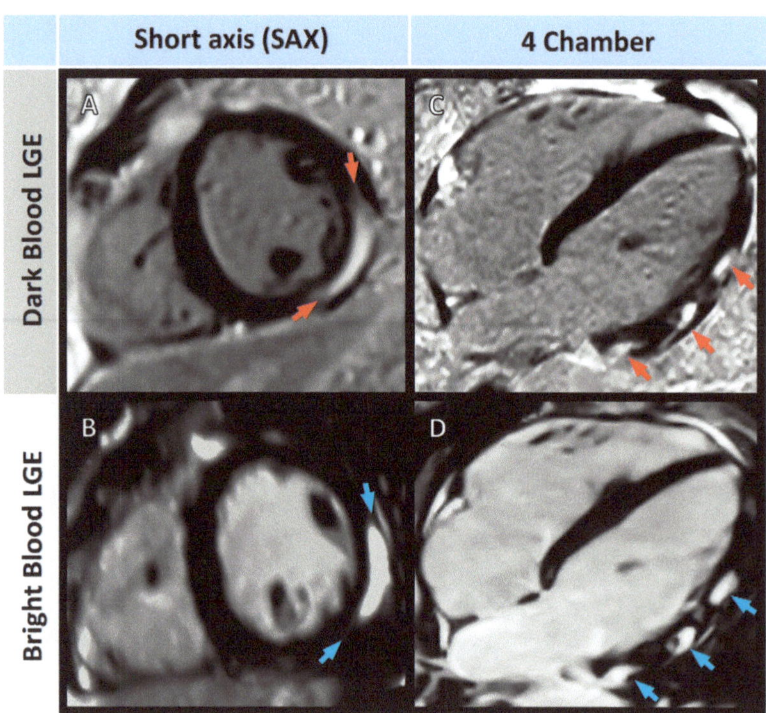

Figure 4. Short-axis dark-blood (top row) and bright-blood (bottom row) images of non-ischemic subepicardial linear LGE in a 19-year-old man with myocarditis. Red and blue arrows indicate the areas of enhancement in (**A,B**) short-axis (SAX) views, and in (**C,D**) four-chamber views.

Considering conventional bright-blood LGE as the reference standard, dark-blood LGE showed a sensitivity of 99% (confidence interval, CI 98–100), a specificity of 99% (CI 98–100), and an accuracy of 99% (CI 99–100), Table 3. The positive predictive value was 99% (CI 97–99), and the negative predictive value was 100% (CI 99–100). The positive likelihood ratio was 132 (CI 55–317), and the negative likelihood ratio was 0.01 (CI 0.00–0.02). McNemar's test revealed no significant marginal inhomogeneity between the methods ($p = 0.45$).

Table 3. Diagnostic performance of dark-blood late gadolinium enhancement in non-ischemic cardiomyopathy.

	Sensitivity	Specificity	Positive Likelihood Ratio	Negative Likelihood Ratio	Positive Predictive Value	Negative Predictive Value	Accuracy
Dark-blood LGE	99 % (CI 98–100)	99 % (CI 98–100)	132 (CI 55–317)	0.01 (CI 0.00–0.02)	99 % (CI 97–99)	100 % (CI 99–100)	99 % (CI 99–100)

Table depicts the diagnostic test statistics for the detection of non–ischemic scarring by dark-blood late gadolinium enhancement (LGE). Bright-blood LGE was considered as reference standard. *CI* = Confidence interval.

4. Discussion

This study prospectively evaluated dark-blood late gadolinium enhancement in a head-to-head comparison with conventional bright-blood late gadolinium enhancement imaging for the visualisation of non-ischemic scarring. Our findings suggest that dark-blood late gadolinium enhancement allows for accurate scar detection not only in ischemic scars but also in non-ischemic cardiomyopathies, demonstrating an excellent sensitivity of 99% and an accuracy of 99%.

Accurate detection of scarring by late gadolinium enhancement (LGE) is of paramount importance for cardiac magnetic resonance-based (CMR-based) diagnosis of both non-ischemic and inflammatory cardiomyopathy, as LGE is known to have both high diagnostic and prognostic value. As previously demonstrated, the presence of LGE in non-ischemic cardiomyopathy is associated with a poor prognosis for heart failure and hospitalisation [25]. It portends an increased risk of major adverse cardiac events including sudden cardiac death [8,26]. In addition, not only the presence but also the extent and localisation of LGE areas is predictive. In particular, mid-wall and (antero)septal LGE is associated with increased mortality [27,28]. As a consequence, recent CMR recommendations consider LGE imaging as an indispensable tool for both diagnosis and risk prediction in patients with non-ischemic cardiomyopathies [21,29].

The advantage of dark-blood LGE is optimised to reduce blood pool signal and optimise contrast to the blood pool [30], which may be particularly helpful in cases with possible overlapping or non-specific enhancement patterns on conventional LGE imaging. While the contrast between healthy myocardium and scarring is inherently high in LGE imaging [31], the dark-blood technique is advantageous for detecting small and subendocardial scars without sacrificing contrast or acquisition time [11]. PSIR LGE imaging is generally available on standard CMR scanners from all vendors and scanner types, and the dark-blood-style nulling of the blood pool for PSIR imaging can be easily implemented on any scanner. In addition, dark-blood LGE imaging generally does not require any additional contrast agent or hardware modifications, making it a feasible and cost-effective option for patients with cardiomyopathy.

The proposed technique can be applied on any CMR scanner without requiring costly sequence updates or extensive technical training, as only the inversion time in the PSIR LGE sequence is set differently [17]. Dark-blood LGE imaging can be applied to both 2D and 3D LGE imaging [32].

Dark-blood late gadolinium enhancement has been shown to be more sensitive than conventional bright-blood late gadolinium enhancement in detecting ischemic scarring [12,33]. The improved scar-to-blood contrast improves the delineation of even small subendocardial scars, facilitating the detection of unrecognised myocardial infarction [12]. In this context, the method has recently been histopathologically correlated and confirmed in an animal model with induced myocardial infarction [13]. Unrecognised myocardial infarction is a common finding in patients with coronary artery disease and typically occurs in the posterolateral wall [34]. The question arises whether dark-blood LGE can be used in a standardised comprehensive cardiac MRI protocol to detect ischemic and non-ischemic myocardial scarring and fibrosis, even in equivocal cases. To date, dark-blood LGE has typically been used in cardiac MRI scans dedicated to patients with coronary artery disease and ischemic cardiomyopathy, with no clear recommendation for non-ischemic cardiomyopathy. In general, LGE imaging is optimised to provide contrast between healthy myocardium and focal fibrosis or scarring. However, any LGE technique runs the risk of inadvertently underestimating diffuse fibrosis when there is no healthy myocardium present to serve as a contrast to pathological enhancement [35].

To our knowledge, this is the first study to implement dark-blood late gadolinium enhancement in an exclusively non-ischemic cardiac MRI cohort. The results of our study demonstrate that dark-blood LGE is not inferior to bright-blood LGE in the detection of non-ischemic scarring, which typically spares the sub-endocardium [36]. With an overall concordance of 99% for the presence or absence of scarring by both LGE techniques, dark-blood LGE demonstrated excellent sensitivity, specificity, and accuracy for the detection of non-ischemic scarring in this study. Dark-blood LGE appears to be a reliable diagnostic method for evaluating LGE in both ischemic and non-ischemic cardiomyopathies in daily clinical practice.

As an outlook, dark-blood LGE may become an equivalent acquisition method in LGE imaging for a routine cardiac MRI protocol, considering the proven superiority of dark-blood LGE in the detection of ischemic scars and its non-inferiority in the delineation of non-ischemic scars, as demonstrated in this study. Small undetected myocardial infarction scars or non-ischemic myocardial scars may be better delineated with dark-blood PSIR LGE.

This study has several limitations. First, histologic confirmation of the presence of myocardial scarring was not obtained. Histopathologic correlation has recently been performed in an animal model for ischemic scars [13]; however, it appears difficult to induce non-ischemic scarring in an animal model to allow a direct comparison between LGE technique and histology. Second, data from only one scanner were included, allowing a direct head-to-head comparison between both LGE techniques within the same cardiac MRI examination. Further studies, preferably in a multicentre setting and with an even larger patient population, identical sequence protocols, and scanners from different vendors and different field strengths (1.5 and 3 T), may further investigate the performance of dark-blood LGE (vs. bright-blood LGE) in non-ischemic cardiomyopathies. Third, although the optimal threshold for semi-automated scar quantification of ischemic scarring using dark-blood LGE has recently been investigated using histopathology as a reference standard [37], an optimal threshold for quantification of non-ischemic scarring using dark-blood LGE has not been investigated and is currently unknown [38].

Key Points:

1. A prospective single-centre study enrolled patients referred for cardiac MRI for clinical suspicion of non-ischemic cardiomyopathy between January 2020 and March 2023.
2. Dark-blood late gadolinium enhancement showed excellent sensitivity (99%, CI 98–100) and accuracy (99%, CI 99–100) for detecting non-ischemic scarring compared with bright-blood late gadolinium enhancement as the reference standard.
3. Measurements of total scar size did not differ between dark-blood late gadolinium enhancement and bright-blood late gadolinium enhancement.

5. Conclusions

Dark-blood LGE has been shown to be non-inferior to conventional bright-blood LGE in the evaluation of non-ischemic scarring, which may have potential implications for future cardiac MRI protocols in the evaluation of unknown cardiomyopathy. As an outlook, dark-blood LGE may become an equivalent acquisition method in LGE imaging for a routine cardiac MRI protocol in the coming years, considering the proven superiority of dark-blood LGE in the detection of ischemic scarring and its non-inferiority in the delineation of non-ischemic scarring, as demonstrated in this study.

Author Contributions: Conceptualisation, J.M.B., R.J.H., S.G. and P.K.; methodology, J.M.B., J.N.G.; software, J.M.B., J.N.G.; validation, J.M.B., S.G. and P.K.; formal analysis, J.M.B.; investigation, J.M.B., S.G. and P.K.; resources, K.N. and M.G.; data curation, J.M.B. and J.N.G.; writing—original draft preparation, J.M.B.; writing—review and editing, S.G., R.J.H., F.H., J.K. and P.K.; visualisation, J.M.B.; supervision, P.K.; project administration, K.N.; funding acquisition, M.G. All authors have read and agreed to the published version of the manuscript.

Funding: This project was supported by the Deutsche Forschungsgemeinschaft (DFG, German Research Foundation)—project number 374031971–TRR 240.

Institutional Review Board Statement: The Institutional Review Board of the University of Tuebingen approved this study (514/2015BO2, 31 January 2019).

Informed Consent Statement: Written, informed consent was obtained from all subjects involved in this study.

Data Availability Statement: The datasets analysed in our study are available from the corresponding author on reasonable request.

Acknowledgments: We acknowledge support by Open Access Publishing Fund of the University of Tuebingen. We would like to thank Gunnar Blumenstock of the Institute for Clinical Epidemiology and Applied Biometry of the University of Tuebingen for his assistance with statistical analysis.

Conflicts of Interest: The authors declare no conflict of interest.

Abbreviations

CI = Confidence interval, CMR = Cardiac magnetic resonance, LGE = Late gadolinium enhancement.

References

1. Pop, M.; Ghugre, N.R.; Ramanan, V.; Morikawa, L.; Stanisz, G.; Dick, A.J.; Wright, G. A Quantification of fibrosis in infarcted swine hearts by ex vivo late gadolinium-enhancement and diffusion-weighted MRI methods. *Phys. Med. Biol.* **2013**, *58*, 5009–5028. [CrossRef] [PubMed]
2. Greulich, S.; Arai, A.E.; Sechtem, U.; Mahrholdt, H. Recent advances in cardiac magnetic resonance. *F1000Research* **2016**, *5*, 1–8. [CrossRef] [PubMed]
3. Bohl, S.; Wassmuth, R.; Abdel-Aty, H.; Rudolph, A.; Messroghli, D.; Dietz, R.; Schulz-Menger, J. Delayed enhancement cardiac magnetic resonance imaging reveals typical patterns of myocardial injury in patients with various forms of non-ischemic heart disease. *Int. J. Cardiovasc. Imaging* **2008**, *24*, 597–607. [CrossRef]
4. Greulich, S.; Mayr, A.; Kitterer, D.; Latus, J.; Henes, J.; Vecchio, F.; Kaesemann, P.; Patrascu, A.; Greiser, A.; Groeninger, S.; et al. Advanced myocardial tissue characterisation by a multi-component CMR protocol in patients with rheumatoid arthritis. *Eur. Radiol.* **2017**, *27*, 4639–4649. [CrossRef] [PubMed]
5. Gräni, C.; Eichhorn, C.; Bière, L.; Murthy, V.L.; Agarwal, V.; Kaneko, K.; Cuddy, S.; Aghayev, A.; Steigner, M.; Blankstein, R.; et al. Prognostic Value of Cardiac Magnetic Resonance Tissue Characterization in Risk Stratifying Patients With Suspected Myocarditis. *J. Am. Coll. Cardiol.* **2017**, *70*, 1964–1976. [CrossRef]
6. Gräni, C.; Eichhorn, C.; Bière, L.; Kaneko, K.; Murthy, V.L.; Agarwal, V.; Aghayev, A.; Steigner, M.; Blankstein, R.; Jerosch-Herold, M.; et al. Comparison of myocardial fibrosis quantification methods by cardiovascular magnetic resonance imaging for risk stratification of patients with suspected myocarditis. *J. Cardiovasc. Magn. Reson.* **2019**, *21*, 1–11. [CrossRef]
7. Mahrholdt, H.; Wagner, A.; Deluigi, C.C.; Kispert, E.; Hager, S.; Meinhardt, G.; Vogelsberg, H.; Fritz, P.; Dippon, J.; Bock, C.T.; et al. Presentation, patterns of myocardial damage, and clinical course of viral myocarditis. *Circulation* **2006**, *114*, 1581–1590. [CrossRef]
8. Greulich, S.; Seitz, A.; Müller, K.A.L.; Grün, S.; Ong, P.; Ebadi, N.; Kreisselmeier, K.P.; Seizer, P.; Bekeredjian, R.; Zwadlo, C.; et al. Predictors of mortality in patients with biopsy-proven viral myocarditis: 10-year outcome data. *J. Am. Heart Assoc.* **2020**, *9*, e015351. [CrossRef]
9. Halliday, B.P.; Baksi, A.J.; Gulati, A.; Ali, A.; Newsome, S.; Izgi, C.; Arzanauskaite, M.; Lota, A.; Tayal, U.; Vassiliou, V.S.; et al. Outcome in Dilated Cardiomyopathy Related to the Extent, Location, and Pattern of Late Gadolinium Enhancement. *J. Am. Coll. Cardiol.* **2019**, *12*, 1645–1655. [CrossRef]
10. Iles, L.; Pfluger, H.; Lefkovits, L.; Butler, M.J.; Kistler, P.M.; Kaye, D.M.; Taylor, A.J. Myocardial fibrosis predicts appropriate device therapy in patients with implantable cardioverter-defibrillators for primary prevention of sudden cardiac death. *J. Am. Coll. Cardiol.* **2011**, *57*, 821–828. [CrossRef]
11. Holtackers, R.J.; Van De Heyning, C.M.; Chiribiri, A.; Wildberger, J.E.; Botnar, R.M.; Kooi, M.E. Dark-blood late gadolinium enhancement cardiovascular magnetic resonance for improved detection of subendocardial scar: A review of current techniques. *J. Cardiovasc. Magn. Reson.* **2021**, *23*, 1–18. [CrossRef] [PubMed]
12. Holtackers, R.J.; Van De Heyning, C.M.; Nazir, M.S.; Rashid, I.; Ntalas, I.; Rahman, H.; Botnar, R.M.; Chiribiri, A. Clinical value of dark-blood late gadolinium enhancement cardiovascular magnetic resonance without additional magnetization preparation. *J. Cardiovasc. Magn. Reson.* **2019**, *21*, 44. [CrossRef] [PubMed]
13. Holtackers, R.J.; Gommers, S.; Heckman, L.I.B.; Van De Heyning, C.M.; Chiribiri, A.; Prinzen, F.W. Histopathological Validation of Dark-Blood Late Gadolinium Enhancement MRI Without Additional Magnetization Preparation. *J. Magn. Reson. Imaging* **2022**, *55*, 190–197. [CrossRef]
14. Francis, R.; Kellman, P.; Kotecha, T.; Baggiano, A.; Norrington, K.; Martinez-Naharro, A.; Nordin, S.; Knight, D.S.; Rakhit, R.D.; Lockie, T.; et al. Prospective comparison of novel dark blood late gadolinium enhancement with conventional bright blood imaging for the detection of scar. *J. Cardiovasc. Magn. Reson.* **2017**, *19*, 1–12. [CrossRef] [PubMed]
15. Monti, C.B.; Codari, M.; Cozzi, A.; Alì, M.; Saggiante, L.; Sardanelli, F.; Secchi, F. Image quality of late gadolinium enhancement in cardiac magnetic resonance with different doses of contrast material in patients with chronic myocardial infarction. *Eur. Radiol. Exp.* **2020**, *4*, 1–9. [CrossRef] [PubMed]
16. Kramer, C.M.; Barkhausen, J.; Bucciarelli-Ducci, C.; Flamm, S.D.; Kim, R.J.; Nagel, E. Standardized cardiovascular magnetic resonance imaging (CMR) protocols: 2020 update. *J. Cardiovasc. Magn. Reson.* **2020**, *22*, 1–18. [CrossRef] [PubMed]

17. Holtackers, R.J.; Chiribiri, A.; Schneider, T.; Higgins, D.M.; Botnar, R.M. Dark-blood late gadolinium enhancement without additional magnetization preparation. *J. Cardiovasc. Magn. Reson.* **2017**, *19*, 1–10. [CrossRef] [PubMed]

18. Schulz-Menger, J.; Bluemke, D.A.; Bremerich, J.; Flamm, S.D.; Fogel, M.A.; Friedrich, M.G.; Kim, R.J.; von Knobelsdorff-Brenkenhoff, F.; Kramer, C.M.; Pennell, D.J.; et al. Standardized image interpretation and post-processing in cardiovascular magnetic resonance—2020 update. *J. Cardiovasc. Magn. Reson.* **2020**, *22*, 1–22. [CrossRef] [PubMed]

19. Bunck, A.C.; Baeβler, B.; Ritter, C.; Kröger, J.R.; Persigehl, T.; Pinto Santos, D.; Steinmetz, M.; Niehaus, A.; Bamberg, F.; Beer, M.; et al. Structured Reporting in Cross-Sectional Imaging of the Heart: Reporting Templates for CMR Imaging of Cardiomyopathies (Myocarditis, Dilated Cardiomyopathy, Hypertrophic Cardiomyopathy, Arrhythmogenic Right Ventricular Cardiomyopathy and Siderosis). *Fortschr. Röntgenstr.* **2020**, *192*, 27–37. [CrossRef]

20. Liu, B.; Dardeer, A.M.; Moody, W.E.; Hayer, M.K.; Baig, S.; Price, A.M.; Leyva, F.; Edwards, N.C.; Steeds, R.P. Reference ranges for three-dimensional feature tracking cardiac magnetic resonance: Comparison with two-dimensional methodology and relevance of age and gender. *Int. J. Cardiovasc. Imaging* **2018**, *34*, 761–775. [CrossRef]

21. Messroghli, D.R.; Moon, J.C.; Ferreira, V.M.; Grosse-Wortmann, L.; He, T.; Kellman, P.; Mascherbauer, J.; Nezafat, R.; Salerno, M.; Schelbert, E.B.; et al. Clinical recommendations for cardiovascular magnetic resonance mapping of T1, T2, T2* and extracellular volume: A consensus statement by the Society for Cardiovascular Magnetic Resonance (SCMR) endorsed by the European Association for Cardiovascular Imagi. *J. Cardiovasc. Magn. Reson.* **2017**, *19*, 75. [CrossRef]

22. McDonagh, T.A.; Metra, M.; Adamo, M.; Gardner, R.S.; Baumbach, A.; Böhm, M.; Burri, H.; Butler, J.; Čelutkienė, J.; Chioncel, O.; et al. 2021 ESC Guidelines for the diagnosis and treatment of acute and chronic heart failure. *Eur. Heart J.* **2021**, *42*, 3599–3726. [CrossRef] [PubMed]

23. Hundley, W.G.; Bluemke, D.A.; Bogaert, J.; Flamm, S.D.; Fontana, M.; Friedrich, M.G.; Grosse-Wortmann, L.; Karamitsos, T.D.; Kramer, C.M.; Kwong, R.Y.; et al. Society for Cardiovascular Magnetic Resonance (SCMR) guidelines for reporting cardiovascular magnetic resonance examinations. *J. Cardiovasc. Magn. Reson.* **2022**, *24*, 1–26. [CrossRef] [PubMed]

24. Tango, T. Equivalence test and confidence interval for the difference in proportions for the paired-sample design. *Stat. Med.* **1998**, *17*, 891–908. [CrossRef]

25. Kuruvilla, S.; Adenaw, N.; Katwal, A.B.; Lipinski, M.J.; Kramer, C.M.; Salerno, M. Late gadolinium enhancement on cardiac magnetic resonance predicts adverse cardiovascular outcomes in nonischemic cardiomyopathy: A systematic review and meta-analysis. *Circ. Cardiovasc. Imaging* **2014**, *7*, 250–257. [CrossRef]

26. Grün, S.; Schumm, J.; Greulich, S.; Wagner, A.; Schneider, S.; Bruder, O.; Kispert, E.M.; Hill, S.; Ong, P.; Klingel, K.; et al. Long-term follow-up of biopsy-proven viral myocarditis: Predictors of mortality and incomplete recovery. *J. Am. Coll. Cardiol.* **2012**, *59*, 1604–1615. [CrossRef]

27. Aquaro, G.D.; Perfetti, M.; Camastra, G.; Monti, L.; Dellegrottaglie, S.; Moro, C.; Pepe, A.; Todiere, G.; Lanzillo, C.; Scatteia, A.; et al. Cardiac MR With Late Gadolinium Enhancement in Acute Myocarditis with Preserved Systolic Function: ITAMY Study. *J. Am. Coll. Cardiol.* **2017**, *70*, 1977–1987. [CrossRef]

28. Arzanauskaite, M.; Newsome, S.; Vassiliou, V.; Alpendurada, F.; Pennell, D.; Guha, K.; Cowie, M.; Gregson, J.; Lota, A.; Vazir, A.; et al. Association between midwall late gadolinium enhancement and sudden cardiac death in patients with dilated cardiomyopathy and mild and moderate left ventricular systolic dysfunction. *Circulation* **2017**, *135*, 2106–2115. [CrossRef]

29. Eichhorn, C.; Greulich, S.; Bucciarelli-Ducci, C.; Sznitman, R.; Kwong, R.Y.; Gräni, C. Multiparametric Cardiovascular Magnetic Resonance Approach in Diagnosing, Monitoring, and Prognostication of Myocarditis. *JACC Cardiovasc. Imaging* **2022**, *15*, 1325–1338. [CrossRef]

30. Holtackers, R.J.; Emrich, T.; Botnar, R.M.; Kooi, M.E.; Wildberger, J.E.; Kreitner, K.F. Late Gadolinium Enhancement Cardiac Magnetic Resonance Imaging: From Basic Concepts to Emerging Methods. *RoFo Fortschr. Auf Dem Geb. Der Rontgenstrahlen Und Der Bildgeb. Verfahr.* **2022**, *194*, 491–504. [CrossRef]

31. Jenista, E.R.; Wendell, D.C.; Azevedo, C.F.; Klem, I.; Judd, R.M.; Kim, R.J.; Kim, H.W. Revisiting how we perform late gadolinium enhancement CMR: Insights gleaned over 25 years of clinical practice. *J. Cardiovasc. Magn. Reson.* **2023**, *25*, 18. [CrossRef] [PubMed]

32. Holtackers, R.J.; Gommers, S.; Van De Heyning, C.M.; Mihl, C.; Smink, J.; Higgins, D.M.; Wildberger, J.E.; ter Bekke, R.M.A. Steadily Increasing Inversion Time Improves Blood Suppression for Free-Breathing 3D Late Gadolinium Enhancement MRI with Optimized Dark-Blood Contrast. *Invest. Radiol.* **2021**, *56*, 335–340. [CrossRef]

33. Krumm, P.; Greulich, S.; Nikolaou, K. Editorial for "Histopathological Validation of Dark-Blood Late Gadolinium Enhancement Cardiovascular Magnetic Resonance Without Additional Magnetization Preparation. " *J. Magn. Reson. Imaging* **2022**, *55*, 198–199. [CrossRef] [PubMed]

34. Krumm, P.; Zitzelsberger, T.; Weinmann, M.; Mangold, S.; Rath, D.; Nikolaou, K.; Gawaz, M.; Kramer, U.; Klumpp, B.D. Cardiac MRI left ventricular global function index and quantitative late gadolinium enhancement in unrecognized myocardial infarction. *Eur. J. Radiol.* **2017**, *92*, 11–16. [CrossRef] [PubMed]

35. Krumm, P.; Mueller, K.A.L.; Klingel, K.; Kramer, U.; Horger, M.S.; Zitzelsberger, T.; Kandolf, R.; Gawaz, M.; Nikolaou, K.; Klumpp, B.D.; et al. Cardiovascular magnetic resonance patterns of biopsy proven cardiac involvement in systemic sclerosis. *J. Cardiovasc. Magn. Reson.* **2016**, *18*, 70. [CrossRef]

36. Shanbhag, S.M.; Greve, A.M.; Aspelund, T.; Schelbert, E.B.; Cao, J.J.; Danielsen, R.; Porgeirsson, G.; Sigurðsson, S.; Eiríksdóttir, G.; Harris, T.B.; et al. Prevalence and prognosis of ischaemic and non-ischaemic myocardial fibrosis in older adults. *Eur. Heart J.* **2019**, *40*, 529–538. [CrossRef] [PubMed]
37. Nies, H.M.J.M.; Gommers, S.; Bijvoet, G.P.; Heckman, L.I.B.; Prinzen, F.W.; Vogel, G.; Van De Heyning, C.M.; Chiribiri, A.; Wildberger, J.E.; Mihl, C.; et al. Histopathological validation of semi-automated myocardial scar quantification techniques for dark-blood late gadolinium enhancement magnetic resonance imaging. *Eur. Hear. J. Cardiovasc. Imaging* **2022**, *31*, 1–9. [CrossRef]
38. van der Velde, N.; Hassing, H.C.; Bakker, B.J.; Wielopolski, P.A.; Lebel, R.M.; Janich, M.A.; Kardys, I.; Budde, R.P.J.; Hirsch, A. Improvement of late gadolinium enhancement image quality using a deep learning–based reconstruction algorithm and its influence on myocardial scar quantification. *Eur. Radiol.* **2021**, *31*, 3846–3855. [CrossRef]

diagnostics

Article

Evaluation of Cardiac Function in Young Mdx Mice Using MRI with Feature Tracking and Self-Gated Magnetic Resonance Cine Imaging

Junpei Ueda [1] and Shigeyoshi Saito [1,2,*]

[1] Department of Medical Physics and Engineering, Division of Health Sciences, Osaka University Graduate School of Medicine, Suita 560-0871, Osaka, Japan
[2] Department of Advanced Medical Technologies, National Cardiovascular and Cerebral Research Center, Suita 564-8565, Osaka, Japan
* Correspondence: saito@sahs.med.osaka-u.ac.jp

Abstract: This study aimed to evaluate cardiac function in a young mouse model of Duchenne muscular dystrophy (mdx) using cardiac magnetic resonance imaging (MRI) with feature tracking and self-gated magnetic resonance cine imaging. Cardiac function was evaluated in mdx and control mice (C57BL/6JJmsSlc mice) at 8 and 12 weeks of age. Preclinical 7-T MRI was used to capture short-axis, longitudinal two-chamber view and longitudinal four-chamber view cine images of mdx and control mice. Strain values were measured and evaluated from cine images acquired using the feature tracking method. The left ventricular ejection fraction was significantly less ($p < 0.01$ each) in the mdx group at both 8 (control, 56.6 ± 2.3% mdx, 47.2 ± 7.4%) and 12 weeks (control, 53.9 ± 3.3% mdx, 44.1 ± 2.7%). In the strain analysis, all strain value peaks were significantly less in mdx mice, except for the longitudinal strain of the four-chamber view at both 8 and 12 weeks of age. Strain analysis with feature tracking and self-gated magnetic resonance cine imaging is useful for assessing cardiac function in young mdx mice.

Keywords: cardiac function; young mdx mice; feature tracking; self-gated magnetic resonance cine imaging

Citation: Ueda, J.; Saito, S. Evaluation of Cardiac Function in Young Mdx Mice Using MRI with Feature Tracking and Self-Gated Magnetic Resonance Cine Imaging. *Diagnostics* **2023**, *13*, 1472. https://doi.org/10.3390/ diagnostics13081472

Academic Editors: Minjie Lu and Arlene Sirajuddin

Received: 1 March 2023
Revised: 31 March 2023
Accepted: 18 April 2023
Published: 19 April 2023

1. Introduction

Duchenne muscular dystrophy (DMD) is an X-linked severe progressive muscle wasting disease caused by a deficiency in dystrophin protein [1–3]. A dystrophin deficiency leads to dilated cardiomyopathy (DCM), which may occur during adolescence [4,5]. DCM is a myocardial disease characterized by left ventricular or biventricular diastolic and systolic dysfunction in the absence of sufficient pressure/volume loading or coronary artery disease [6,7]. The C57BL/10-mdx (mdx) mouse model of human muscular dystrophy carries a mutation in exon 23 of the Xp21 region of the genome, resulting in the loss of dystrophin protein expression [8]. The DMD research has been conducted using the mdx model, which has a mutation in the dystrophin gene itself like DMD patients [9,10]. DCM is considered to be the main cause of death in DMD, and the mechanism of DCM is not yet fully elucidated [11]. Therefore, it is very important to elucidate the early pathogenesis of DCM by young mdx mice.

Cardiovascular magnetic resonance (CMR) imaging is used to diagnose various cardiac diseases, and its usefulness in DCM has been reported [12–14]. The evaluation of cardiac function using CMR in mice is as reproducible as it is in humans [15] and has also been used to evaluate the pathogenesis of DCM [16,17]. In CMR, the evaluation of cardiac function can be quantified, including the left ventricular ejection fraction and strain analysis. In recent years, feature-tracking methods have attracted attention in addition to conventional methods, such as tagging and harmonic phase methods in strain analysis. The feature-tracking method can calculate strain values from standard cine CMR images without additional CMR images [18,19]. The cardiac function of mdx mice has been evaluated with CMR [20,21]; however, these studies evaluated older mice or calculated strain values without using the feature tracking method.

This study aimed to assess cardiac function in young mdx mice, including strain assessment using the feature tracking method, and to observe changes in cardiac function in mdx mice.

2. Materials and Methods

2.1. Animal Preparation

All experimental protocols were approved by the Research Ethics Committee of Osaka University. All experimental procedures involving animals and their care were performed in accordance with the University Guidelines for Animal Experimentation and the National Institutes of Health Guide for the Care and Use of Laboratory Animals. Animal experiments were performed on male C57BL/6JJmsSlc mice (control mice) 8–12 weeks old purchased from Japan SLC (Hamamatsu, Japan) and C57BL/10ScSn-Dmdmdx/JicJcl mice (mdx mice) purchased from CLEA Japan, Inc. (Tokyo, Japan). All mice were housed in a controlled vivarium environment (24 °C; 12:12 h light:dark cycle) and fed a standard pellet diet with water ad libitum. MRI experiments were performed at 2 time points at 8 weeks of age (10 control mice and 10 mdx mice) and 12 weeks of age (10 control mice and 10 mdx mice).

2.2. Magnetic Resonance Imaging

MR images of animal hearts were acquired using a horizontal 7-T scanner (PharmaScan 70/16 US; Bruker Biospin; Ettlingen, Germany) equipped with a volume coil with an inner diameter of 30 mm. For MRI, the mice were positioned with their mouth in a stereotaxic frame to prevent movement in a prone position during acquisition. The mice were maintained at a body temperature of 36.5 °C, with water flow regulated, and continuously monitored using a physiological monitoring system (SA Instruments Inc., Stony Brook, NY, USA). All MRI experiments on mice were performed under general anesthesia with 1.0–2.0% isoflurane in an air–oxygen mixture (Abbott Laboratories; Abbott Park, IL, USA) administered through a mask covering the nose and mouth of the mice.

Short-axis, long-axis two-chamber, and long-axis four-chamber images were obtained using fast low-angle shots (FLASH) with a self-gated magnetic resonance cine imaging system (). For short-axis images, the following parameters were used: repetition time (TR)/echo time (TE) = 44.5/2.5 ms, flip angle = 25°, movie frames = 14 frames per cardiac cycle, field of view (FOV) = 25.6 × 25.6 mm, acquisition time = 23:43 min, in-plane resolution per pixel = 133 μm, matrix = 192 × 192, number of excitations (NEX) = 300, oversampling = 1, and five concomitant slices covering the whole heart from the apex to the base. For the long-axis two- and four-chamber views, the parameters were TR/TE = 6.5/3.1 ms, flip angle = 10°, movie frames = 14 frames/s, FOV = 25.6 mm × 25.6 mm, acquisition time = 3:52 min, in-plane resolution per pixel = 133 μm, matrix = 192 × 192, NEX = 1, and oversampling = 250. The total scan time per animal was approximately 40 min.

2.3. MRI Data Analysis

The cardiac MR images were analyzed using cvi42 software (Circle Cardiovascular Imaging, Calgary, AB, Canada). The borders of the epicardium and endocardium were outlined manually on the short-axis images, two- chamber long-axis images and four-chamber long-axis images at both end-diastolic phase and end-systolic phase. The left ventricular end-diastolic volume (LVEDV), left ventricular end-systolic volume (LVESV), left ventricular stroke volume (LVSV), left ventricular ejection fraction (LVEF), and left ventricular mass (LVM) were calculated from cine images. All values were calculated automatically by Cvi42. In addition, strain analysis was performed using feature tracking. For myocardial strain analysis, the global radial strain (RS), global circumferential strain (CS), and longitudinal strain (LS) values were calculated. LS was analyzed from the 2- and 4-chamber long-axis views. The peak value of each strain was used for statistical evaluation.

2.4. Statistical Analysis

The LVESV, LVEDV, LVSV, LVEF, LVM, and each strain value calculated from the MR images are presented as the mean ± standard deviation. All statistical analyses were performed using Prism, version 9 (GraphPad Software; San Diego, CA, USA). Differences were compared using a one-way analysis of variance and Tukey's multiple comparison test. Statistical significance was set at $p < 0.05$ (* $p < 0.05$, ** $p < 0.01$, *** $p < 0.001$).

2.5. Histological Analysis and Immunostaining

After completion of the MRI at 12 weeks of age, the mice were sacrificed, and their hearts were removed and fixed in formalin. Fixed hearts were sliced in the direction of the short axis of the left ventricle. Heart specimens were embedded in paraffin and sectioned at 5 μm; some sections were stained with hematoxylin and eosin (H&E), and some sections were immunostained. The stained tissues were observed using an optical microscope (Keyence Corporation; Osaka, Japan). H&E staining was performed by immersing samples in a hematoxylin solution (5 min) and alcohol–eosin staining solution (3 min). H&E-stained tissues were dehydrated six times with 100% alcohol and then permeabilized using xylene. Immunostaining was performed using the enzyme–antibody method. Dystrophin rabbit polyclonal antibodies (12715-1-P; Proteintech Group, Inc., Rosemont, IL, USA) were used as the primary antibody; the samples were incubated for 1 h. After washing three times for 5 min each, the samples were incubated with the EnVision+ horseradish peroxidase-conjugated anti-rabbit secondary antibodies (K4003; Agilent Technologies, Inc., Santa Clara, CA, USA) for 30 min.

3. Results

3.1. Animals' Characteristics

The characteristics of the animals are summarized in Table 1. No significant difference in body weight or respiratory rate was found between the control and mdx groups of the same age. However, both the 8- and 12-week-old groups showed significant differences between the groups in heart rate.

Table 1. Animal characteristics.

	8 Week		12 Week	
	Control ($n = 10$)	**Mdx ($n = 10$)**	**Control ($n = 10$)**	**Mdx ($n = 10$)**
Body weight (g)	25.6 ± 1.3	26.2 ± 1.2	28.3 ± 1.1	29.3 ± 1.6
Heart rate (bpm)	321 ± 32	275 ± 23 **	334 ± 35	277 ± 22 **
Respiratory rate (brpm)	73.4 ± 3.4	65.5 ± 6.6	76.2 ± 4.6	70 ± 9.4

** $p < 0.01$ between mdx mice and age-matched controls.

3.2. Visual Evaluation of Cine Images of Short Axis, Two-Chamber, and Four-Chamber View

In short-axis cine images and the two-chamber view cine images, contraction and expansion of the entire myocardium were observed in the control group at 8 and 12 weeks of age. For LVEDV, no significant difference was observed between the control and mdx groups (Figures 1B,D and 2B,D). However, the myocardial contraction was weaker in the mdx group than in the control group (Figures 1F,H and 2F,H). In the four-chamber view cine images, contraction and expansion of the entire myocardium were observed in the control group at 8 and 12 weeks of age. However, no differences in shrinkage were observed in four-chamber views (Figure 3F,H).

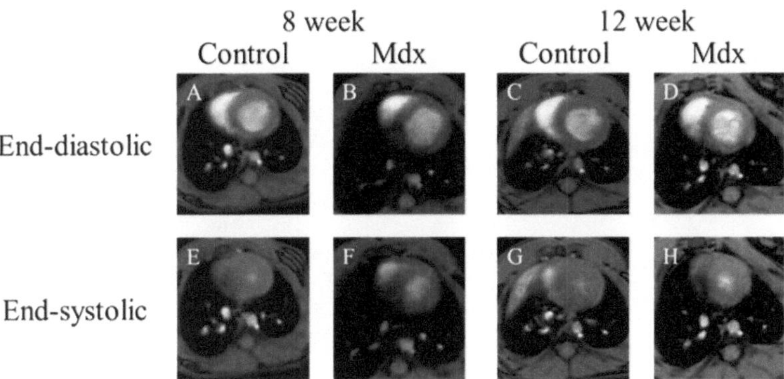

Figure 1. Representative short-axis view cine magnetic resonance images: (**A**,**E**) 8-week-old control mouse, (**B**,**F**) 8-week-old mdx mouse, (**C**,**G**) 12-week-old control, (**D**,**H**) 12-week-old mdx mouse. (**A**–**D**) The end-diastolic phase of a mouse heart; (**E**–**H**) the end-systolic phase of a mouse heart.

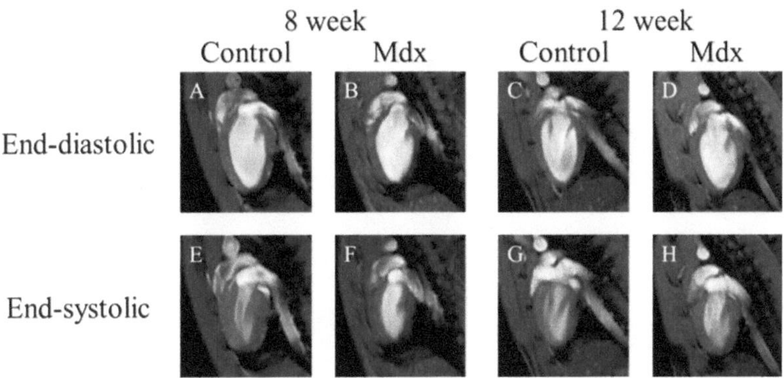

Figure 2. Representative two-chamber view cine magnetic resonance images: (**A**,**E**) 8-week-old control mouse, (**B**,**F**) 8-week-old mdx mouse, (**C**,**G**) 12-week-old control, (**D**,**H**) 12-week-old mdx mouse. (**A**–**D**) The end-diastolic phase of a mouse heart; (**E**–**H**) the end-systolic phase of a mouse heart.

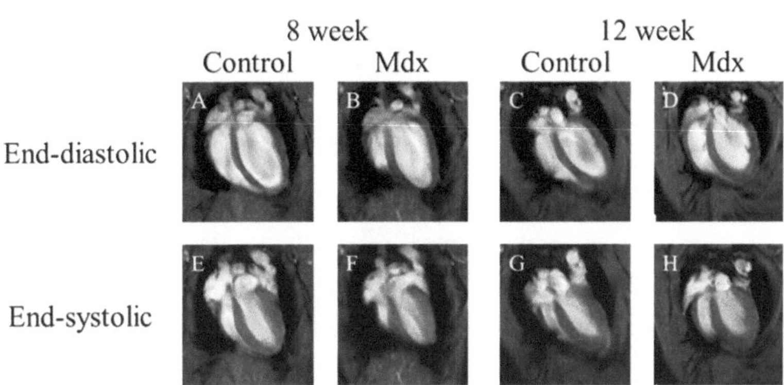

Figure 3. Representative four-chamber view cine magnetic resonance images: (**A**,**E**) 8-week-old control mouse, (**B**,**F**) 8-week-old mdx mouse, (**C**,**G**) 12-week-old control, (**D**,**H**) 12-week-old mdx mouse. (**A**–**D**) The end-diastolic phase of a mouse heart; (**E**–**H**) the end-systolic phase of a mouse heart.

3.3. Cardiac Function in 8- and 12-Week-Old Mdx Mice vs. Age-Matched Controls

Regarding the LVEDV (Figure 4A), no significant difference was found between the control and mdx groups at 8 weeks old (control, 52.6 ± 4.9 µL; mdx, 52.5 ± 7.9 µL) and 12 weeks old (control, 55.9 ± 5.1 µL; mdx, 60.3 ± 9.1 µL). Regarding the LVESV (Figure 4B), no significant difference was found between the control and mdx groups at 8 weeks old (control, 22.8 ± 2.6 µL; mdx, 28.1 ± 7.0 µL). However, at 12 weeks old, LVESV was significantly less in the mdx group compared to the control group (control, 25.8 ± 3.6 µL; mdx, 33.9 ± 5.6 µL, $p < 0.01$). Regarding the LVSV (Figure 4C), a significant difference was identified between the control and mdx groups at 8 weeks old (control, 29.9 ± 2.8 µL; mdx, 24.3 ± 2.8 µL, $p < 0.01$). However, at 12 weeks old, no significant difference was found between the control and mdx groups (control, 30.6 ± 3.1 µL; mdx, 26.6 ± 4.2 µL). Regarding the LVEF (Figure 4D), a significant difference was identified between the control and mdx groups at 8 weeks (control, 56.6 ± 2.3%; mdx, 47.2 ± 7.4%, $p < 0.01$) and 12 weeks (control, 53.9 ± 3.3%; mdx, 44.1 ± 2.7%, $p < 0.01$). A significant difference was also identified between the 8- and 12-week-old mice in the mdx group ($p < 0.05$). Regarding the LVM (Figure 4E), no significant difference was found between the control and mdx groups at 8 weeks old (control, 36.3 ± 3.3 mg; mdx, 35.1 ± 3.8 mg) and 12 weeks old (control, 36.7 ± 4.3 mg; mdx, 39.0 ± 4.0 mg).

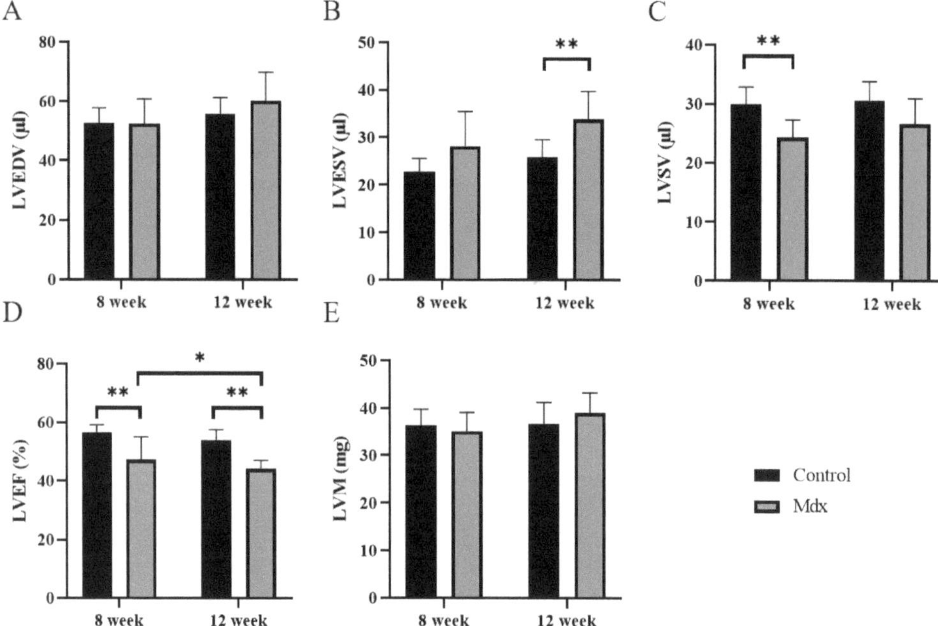

Figure 4. Graphs quantifying LVEDV (**A**), LVESV (**B**), LVSV (**C**), LVEF (**D**), and LVM (**E**) in the 8-week-old control mice, 8-week-old mdx mice, 12-week-old control mice, and 12-week-old mdx mice. LVEDV, left ventricular end-diastolic volume; LVESV, left ventricular end-systolic volume; LVSV, left ventricular stroke volume; LVEF, left ventricular ejection fraction; LVM, left ventricular mass. * $p < 0.05$, ** $p < 0.01$, respectively.

3.4. Strain Analysis by Feature-Tracking Method in 8- and 12-Week-Old Mdx Mice vs. Age-Matched Controls

In the strain analysis of the 2ch-LS and short-axis images, the peak strain values were lower in the mdx group compared with the control group. The areas of reduction in strain values were diffuse, and no regularity was apparent (Figures 5–7). According to statistical analysis of strain values, the global radial strain of the left ventricle was significantly less in the mdx group compared to the control group at 8 weeks old (control, 22.2 ± 1.7%; mdx, 19.0 ± 2.0%, $p < 0.05$) and 12 weeks old (control, 20.9 ± 2.2%; mdx, 17.2 ± 2.7%, $p < 0.01$; Figure 8A). The global circumferential strain of the left ventricle was significantly less in the mdx group compared to the control group at 8 weeks old (control, −14.6 ± 0.8%; mdx, −13.2 ± 0.9%, $p < 0.05$) and 12 weeks old (control, −13.9 ± 1.0%; mdx, −12.1 ± 1.4%, $p < 0.01$; Figure 8B). The two-chamber-view longitudinal strain of the left ventricle (2ch-LVLS) was significantly less in the mdx group compared with the control group at 8 weeks old (control, −14.1 ± 1.7%; mdx, −11.4 ± 2.0%, $p < 0.05$) and 12 weeks old (control, −13.5 ± 1.1%; mdx, −11.3 ± 1.9%, $p < 0.05$; Figure 8C). Regarding the strain values of the global four-chamber-view longitudinal strain of the left ventricle (4ch-LVLS), no significant difference was found between the control and mdx groups at 8 weeks (control, −14.5 ± 2.4%; mdx, −15.0 ± 1.8%) and 12 weeks (control, −15.4 ± 2.4%; mdx, −13.4 ± 1.6%; Figure 8D).

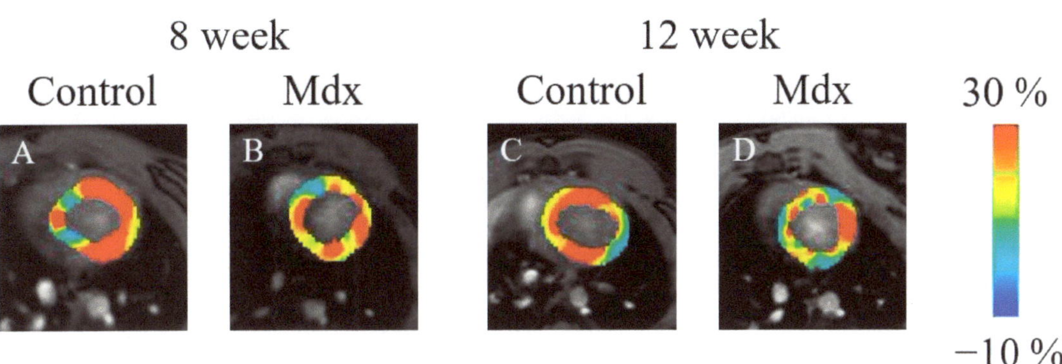

Figure 5. Radial strain-encoded functional magnetic resonance imaging of the end-systolic left ventricle: (**A**) 8-week-old control mouse, (**B**) 8-week-old mdx mouse, (**C**) 12-week-old control mouse, (**D**) 12-week-old mdx mouse. The color bar shows the scale of the strain based on the end-diastolic left ventricle, with maximum contraction shown in red and minimum contraction in blue.

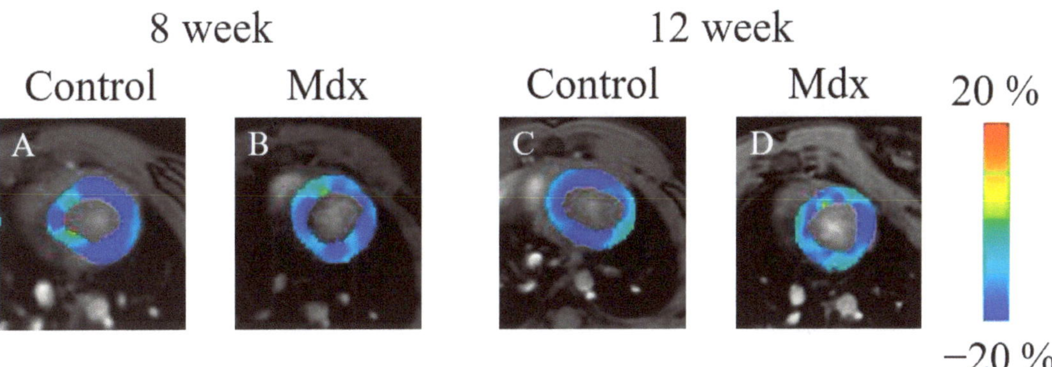

Figure 6. Circumferential strain-encoded functional magnetic resonance imaging of the end-systolic left ventricle: (**A**) 8-week-old control mouse, (**B**) 8-week-old mdx mouse, (**C**) 12-week-old control mouse, (**D**) 12-week-old mdx mouse. The color bar shows the scale of the strain based on the end-diastolic left ventricle, with maximum contraction shown in blue and minimum contraction in red.

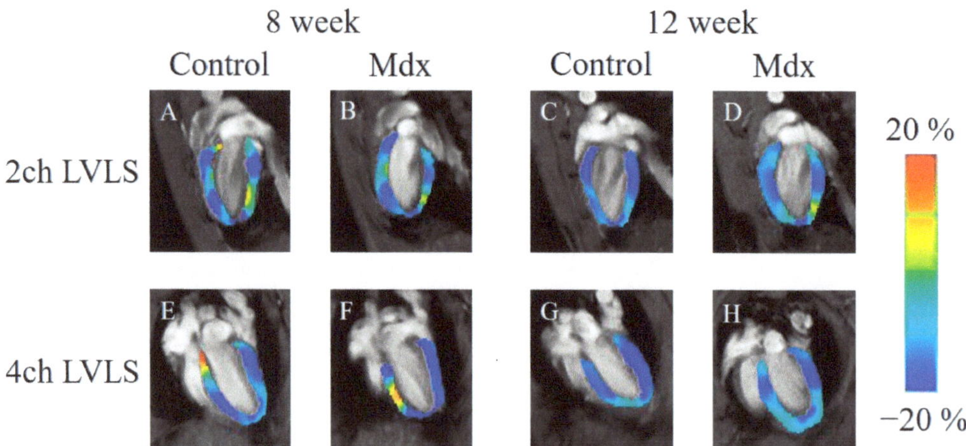

Figure 7. Longitudinal strain-encoded functional magnetic resonance imaging of the end-systolic left ventricle: (**A**,**E**) 8-week-old control mouse, (**B**,**F**) 8-week-old mdx mouse, (**C**,**G**) 12-week-old control mouse, (**D**,**H**) 12-week-old mdx mouse. The color bar shows the scale of the strain based on the end-diastolic left ventricle, with maximum contraction shown in blue and minimum contraction in red. 2ch LVLS: two-chamber-view longitudinal strain of left ventricle, 4ch LVLS: four-chamber-view longitudinal strain of left ventricle.

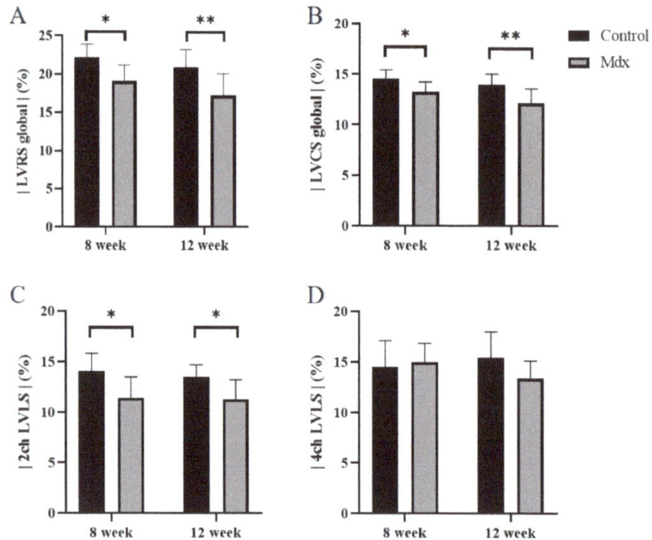

Figure 8. Graphs quantifying the strain analysis in the 8-week-old control mice, 8-week-old mdx mice, 12-week-old control mice, and 12-week-old mdx mice. (**A**) Global radial strain of left ventricle (LvRS global), (**B**) global circumferential strain of left ventricle (LVCS global), (**C**) two-chamber-view longitudinal strain of left ventricle (2ch-LVLS), (**D**) four-chamber-view longitudinal strain of left ventricle (4ch-LVLS). * $p < 0.05$, ** $p < 0.01$, respectively.

3.5. H&E Staining and Immunostaining of Myocardium Tissue

In the H&E staining results, no differences were observed between the mdx and control groups (Figure 9). However, staining with antibodies against dystrophin showed differences between the mdx and control groups (Figure 10). Staining of the intercellular matrix was observed in the control group (Figure 10A,B) but not in the mdx group (Figure 10C,D).

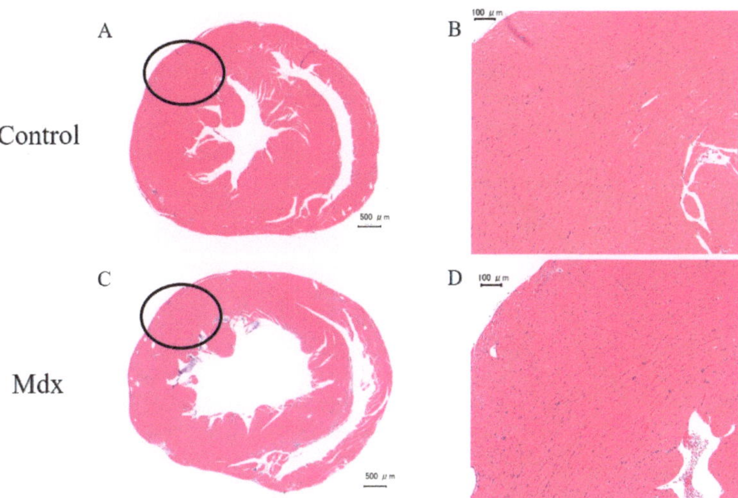

Figure 9. Heart sections stained with hematoxylin and eosin from (**A**,**B**) control and (**C**,**D**) mdx mice. (**B**,**D**) Magnified images of the black-circle regions. Scale bars represent 500 μm (**A**,**C**) and 100 μm (**B**,**D**).

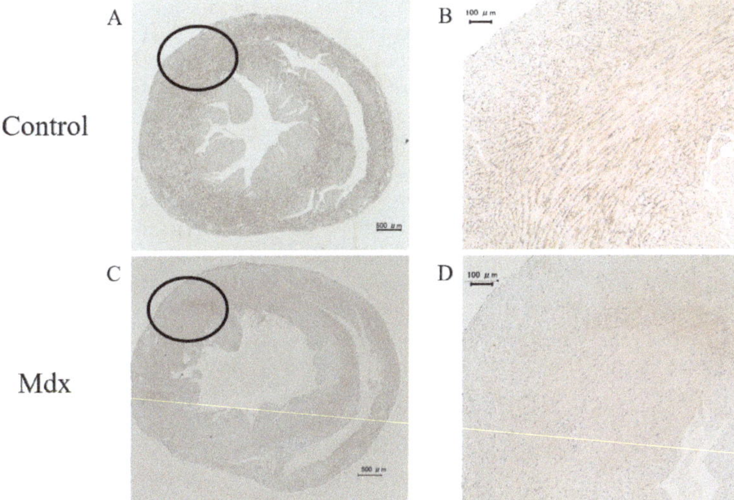

Figure 10. Heart sections stained with antibodies against dystrophin from (**A**,**B**) control and (**C**,**D**) mdx mice. (**B**,**D**) Magnified images of the black-circle regions. Scale bars represent 500 μm (**A**,**C**) and 100 μm (**B**,**D**).

4. Discussion

In this study, myocardial cine images of mdx mice were acquired using 7-T MRI at 8 and 12 weeks of age to evaluate cardiac function and myocardial strain values. No significant difference was found in the left ventricular end-diastolic volume at 8 and 12 weeks of age. However, both at 8 and 12 weeks of age, the left ventricular ejection fraction showed a decreasing trend in the mdx group compared with the control group. In strain analysis, all peak strain values showed a decreasing trend, except for 4ch-LS at 8 and 12 weeks of age. Previous studies using CMR have reported lower cardiac function in older mdx mice than we found in our study [20,21]. However, these previous studies did not use the feature tracking, ECG, and self-gated magnetic resonance cine imaging used in our study. The present study's novel strain analysis, ECG, and respiratory-gated methods suggest that young mdx mice have less cardiac function.

Compared to the conventional tagging method, the feature-tracking method does not require additional imaging and can be performed from cine images for strain analysis. Self-gated magnetic resonance cine imaging allows ECG and respiration gating without the need to wear monitoring equipment. Therefore, in addition to the advantage of being able to detect a decline in cardiac function that could not be detected by conventional methods, these novel methods also have the significant advantage of reducing examination time.

Previous studies have reported no change in cardiac function until older age [21–24]. At the age of 8 months, the contractile and diastolic functions of the left ventricular myocardium decrease, thereby decreasing the left ventricular ejection fraction [20]. However, a previous study showed that CMR taken at a high temporal resolution showed reduced cardiac function in 3-month-old mdx mice [25]. Although the temporal resolution differs from this previous study, our study and other studies [20,26] suggest a decline in cardiac function in young mdx mice. Consistent with a previous study [20], immunostaining showed no staining of the extracellular matrix of cardiomyocytes in mdx mice compared to the control group (Figure 10). Hence, immunostaining showed a difference in cardiomyocytes between the mdx and control groups, and CMR also detected less cardiac function in mdx group. However, the CMR results did not show ventricular enlargement, which is a symptom of dilated cardiomyopathy. In addition, no areas of necrosis were found in the cardiomyocytes of mdx mice in H&E staining unlike previous studies [20]. Thus, our study suggests that mdx mice have impaired cardiac function at a stage prior to DCM diagnosis.

This study had limitations. First, by evaluating cardiac function in 8- and 12-week-old mdx mice, this study demonstrated this methodology's utility for detecting changes in cardiac function in young mdx mice. However, cardiac function after 12 weeks of age has not been evaluated. Evaluation of cardiac function after 12 weeks of age would provide more details on the transition of cardiac function in mdx mice during aging. Analysis of cardiac function after 12 weeks may reveal some changes in LVEDV, which did not differ significantly in this study. A more detailed cardiac function transition may allow better investigation of cardiac function treatment in mdx mice [27,28], including different genotypes [29,30]. Second, to demonstrate the usefulness of the present method, we evaluated the cardiac function of mdx mice using feature tracking and self-gated magnetic resonance cine imaging and compared them with previous studies. By comparing this study with conventional ECG synchronization and tagging methods, it may be possible to demonstrate the usefulness of the method used in this study in more detail. Third, it is considered that there is an effect of sex difference on the decrease in cardiac function caused by DMD [31]. Since only male mice were used in this study, it is necessary to examine the effects of sex differences on the results in the future. To make evaluating cardiac function in mdx mice more accurate, further studies are needed to evaluate longer-term cardiac function and assess cardiac function using various methods.

5. Conclusions

This study showed that strain analysis with feature tracking and self-gated magnetic resonance cine imaging is useful for assessing cardiac function in young mdx mice. Using these methods, we demonstrated that young mdx mice decline in cardiac function.

Author Contributions: Conceptualization, J.U. and S.S.; methodology, J.U. and S.S.; software, J.U. and S.S.; investigation, J.U. and S.S.; data curation, J.U. and S.S., writing—original draft preparation, J.U. and S.S.; writing—review and editing, J.U. and S.S.; supervision, S.S.; project administration, S.S.; funding acquisition, S.S. All authors have read and agreed to the published version of the manuscript.

Funding: This work was the result of using research equipment shared in the MEXT Project for promoting public utilization of advanced research infrastructure (Program for Advanced Research Equipment Platforms MRI Platform), Grant Number JPMXS0450400022 and JPMXS0450400023.

Institutional Review Board Statement: All experimental protocols were approved by the Research Ethics Committee of Osaka University. All experimental procedures involving animals and their care were performed in accordance with the University Guidelines for Animal Experimentation and the National Institutes of Health Guide for the Care and Use of Laboratory Animals. The study was conducted according to the guidelines of the Declaration of Helsinki and approved by the Research Ethics Committee of Osaka University (R02-05-0, 20 November 2019).

Informed Consent Statement: Not applicable.

Data Availability Statement: The data presented in this study are available on request from the corresponding author.

Conflicts of Interest: The authors declare no conflict of interest.

References

1. Jelinkova, S.; Sleiman, Y.; Fojtík, P.; Aimond, F.; Finan, A.; Hugon, G.; Scheuermann, V.; Beckerová, D.; Cazorla, O.; Vincenti, M.; et al. Dystrophin Deficiency Causes Progressive Depletion of Cardiovascular Progenitor Cells in the Heart. *Int. J. Mol. Sci.* **2021**, *22*, 5025. [CrossRef] [PubMed]
2. Biggar, W.D.; Klamut, H.J.; Demacio, P.C.; Stevens, D.J.; Ray, P.N. Duchenne muscular dystrophy: Current knowledge, treatment, and future prospects. *Clin. Orthop. Relat. Res.* **2002**, *401*, 88–106. [CrossRef] [PubMed]
3. Hoffman, E.P.; Kunkel, L.M. Dystrophin abnormalities in Duchenne/Becker muscular dystrophy. *Neuron* **1989**, *2*, 1019–1029. [CrossRef]
4. Hoogerwaard, E.M.; van der Wouw, P.A.; Wilde, A.A.; Bakker, E.; Ippel, P.F.; Oosterwijk, J.C.; Majoor-Krakauer, D.F.; van Essen, A.J.; Leschot, N.J.; de Visser, M. Cardiac involvement in carriers of Duchenne and Becker muscular dystrophy. *Neuromuscul. Disord.* **1999**, *9*, 347–351. [CrossRef] [PubMed]
5. Schade van Westrum, S.M.; Hoogerwaard, E.M.; Dekker, L.; Standaar, T.S.; Bakker, E.; Ippel, P.F.; Oosterwijk, J.C.; Majoor-Krakauer, D.F.; van Essen, A.J.; Leschot, N.J.; et al. Cardiac abnormalities in a follow-up study on carriers of Duchenne and Becker muscular dystrophy. *Neurology* **2011**, *77*, 62–66. [CrossRef]
6. Merlo, M.; Cannatà, A.; Gobbo, M.; Stolfo, D.; Elliott, P.M.; Sinagra, G. Evolving concepts in dilated cardiomyopathy. *Eur. J. Heart Fail.* **2018**, *20*, 228–239. [CrossRef]
7. Elliott, P.; Andersson, B.; Arbustini, E.; Bilinska, Z.; Cecchi, F.; Charron, P.; Dubourg, O.; Kühl, U.; Maisch, B.; McKenna, W.J.; et al. Classification of the cardiomyopathies: A position statement from the European Society Of Cardiology Working Group on Myocardial and Pericardial Diseases. *Eur. Heart J.* **2008**, *29*, 270–276. [CrossRef]
8. Sicinski, P.; Geng, Y.; Ryder-Cook, A.S.; Barnard, E.A.; Darlison, M.G.; Barnard, P.J. The molecular basis of muscular dystrophy in the mdx mouse: A point mutation. *Science* **1989**, *244*, 1578–1580. [CrossRef]
9. Nowak, D.; Kozlowska, H.; Gielecki, J.S.; Rowinski, J.; Zurada, A.; Goralczyk, K.; Bozilow, W. Cardiomyopathy in the mouse model of Duchenne muscular dystrophy caused by disordered secretion of vascular endothelial growth factor. *Med. Sci. Monit.* **2011**, *17*, BR332–BR338. [CrossRef]
10. Lorin, C.; Gueffier, M.; Bois, P.; Faivre, J.-F.; Cognard, C.; Sebille, S. Ultrastructural and functional alterations of EC coupling elements in mdx cardiomyocytes: An analysis from membrane surface to depth. *Cell. Biochem. Biophys.* **2013**, *66*, 723–736. [CrossRef]
11. Law, M.L.; Cohen, H.; Martin, A.A.; Angulski, A.B.B.; Metzger, J.M. Dysregulation of Calcium Handling in Duchenne Muscular Dystrophy-Associated Dilated Cardiomyopathy: Mechanisms and Experimental Therapeutic Strategies. *J. Clin. Med.* **2020**, *9*, 520. [CrossRef]
12. Singla, N.; Mehra, S.; Garga, U.C. Diagnostic Role of Cardiovascular Magnetic Resonance Imaging in Dilated Cardiomyopathy. *Indian J. Radiol. Imaging* **2021**, *31*, 116–123. [CrossRef]
13. Ziółkowska, L.; Śpiewak, M.; Małek, L.; Boruc, A.; Kawalec, W. The usefulness of cardiovascular magnetic resonance imaging in children with myocardial diseases. *Kardiol. Pol.* **2015**, *73*, 419–428. [CrossRef]
14. Pirruccello, J.P.; Bick, A.; Wang, M.; Chaffin, M.; Friedman, S.; Yao, J.; Guo, X.; Venkatesh, B.A.; Taylor, K.D.; Post, W.S.; et al. Analysis of cardiac magnetic resonance imaging in 36,000 individuals yields genetic insights into dilated cardiomyopathy. *Nat. Commun.* **2020**, *11*, 2254. [CrossRef]
15. Manka, R.; Jahnke, C.; Hucko, T.; Dietrich, T.; Gebker, R.; Schnackenburg, B.; Graf, K.; Paetsch, I. Reproducibility of small animal cine and scar cardiac magnetic resonance imaging using a clinical 3.0 tesla system. *BMC Med. Imaging* **2013**, *13*, 44. [CrossRef]
16. Saito, S.; Tanoue, M.; Masuda, K.; Mori, Y.; Nakatani, S.; Yoshioka, Y.; Murase, K. Longitudinal observations of progressive cardiac dysfunction in a cardiomyopathic animal model by self-gated cine imaging based on 11.7-T magnetic resonance imaging. *Sci. Rep.* **2017**, *7*, 9106. [CrossRef]

17. Vanhoutte, L.; Guilbaud, C.; Martherus, R.; Bouzin, C.; Gallez, B.; Dessy, C.; Balligand, J.-L.; Moniotte, S.; Feron, O. MRI Assessment of Cardiomyopathy Induced by β1-Adrenoreceptor Autoantibodies and Protection Through β3-Adrenoreceptor Overexpression. *Sci. Rep.* **2017**, *7*, 43951. [CrossRef]

18. Amzulescu, M.S.; De Craene, M.; Langet, H.; Pasquet, A.; Vancraeynest, D.; Pouleur, A.C.; Vanoverschelde, J.L.; Gerber, B.L. Myocardial strain imaging: Review of general principles, validation, and sources of discrepancies. *Eur. Heart J. Cardiovasc. Imaging* **2019**, *20*, 605–619. [CrossRef]

19. Claus, P.; Omar, A.M.S.; Pedrizzetti, G.; Sengupta, P.P.; Nagel, E. Tissue Tracking Technology for Assessing Cardiac Mechanics: Principles, Normal Values, and Clinical Applications. *JACC Cardiovasc. Imaging* **2015**, *8*, 1444–1460. [CrossRef]

20. Zhang, W.; ten Hove, M.; Schneider, J.E.; Stuckey, D.J.; Sebag-Montefiore, L.; Bia, B.L.; Radda, G.K.; Davies, K.E.; Neubauer, S.; Clarke, K. Abnormal cardiac morphology, function and energy metabolism in the dystrophic mdx mouse: An MRI and MRS study. *J. Mol. Cell. Cardiol.* **2008**, *45*, 754–760. [CrossRef]

21. Li, W.; Liu, W.; Zhong, J.; Yu, X. Early manifestation of alteration in cardiac function in dystrophin deficient mdx mouse using 3D CMR tagging. *J. Cardiovasc. Magn. Reson.* **2009**, *11*, 40. [CrossRef] [PubMed]

22. Au, C.G.; Butler, T.L.; Sherwood, M.C.; Egan, J.R.; North, K.N.; Winlaw, D.S. Increased connective tissue growth factor associated with cardiac fibrosis in the mdx mouse model of dystrophic cardiomyopathy. *Int. J. Exp. Pathol.* **2011**, *92*, 57–65. [CrossRef]

23. Crisp, A.; Yin, H.; Goyenvalle, A.; Betts, C.; Moulton, H.M.; Seow, Y.; Babbs, A.; Merritt, T.; Saleh, A.F.; Gait, M.J.; et al. Diaphragm rescue alone prevents heart dysfunction in dystrophic mice. *Hum. Mol. Genet.* **2011**, *20*, 413–421. [CrossRef] [PubMed]

24. Van Erp, C.; Loch, D.; Laws, N.; Trebbin, A.; Hoey, A.J. Timeline of cardiac dystrophy in 3-18-month-old MDX mice. *Muscle Nerve* **2010**, *42*, 504–513. [CrossRef]

25. Stuckey, D.J.; Carr, C.A.; Camelliti, P.; Tyler, D.J.; Davies, K.E.; Clarke, K. In vivo MRI characterization of progressive cardiac dysfunction in the mdx mouse model of muscular dystrophy. *PLoS ONE* **2012**, *7*, e28569. [CrossRef] [PubMed]

26. Arahata, K.; Ishiura, S.; Ishiguro, T.; Tsukahara, T.; Suhara, Y.; Eguchi, C.; Ishihara, T.; Nonaka, I.; Ozawa, E.; Sugita, H. Immunostaining of skeletal and cardiac muscle surface membrane with antibody against Duchenne muscular dystrophy peptide. *Nature* **1988**, *333*, 861–863. [CrossRef]

27. Potter, R.A.; Griffin, D.A.; Heller, K.N.; Peterson, E.L.; Clark, E.K.; Mendell, J.R.; Rodino-Klapac, L.R. Dose-Escalation Study of Systemically Delivered rAAVrh74.MHCK7.micro-dystrophin in the mdx Mouse Model of Duchenne Muscular Dystrophy. *Hum. Gene Ther.* **2021**, *32*, 375–389. [CrossRef]

28. Salmaninejad, A.; Jafari Abarghan, Y.; Bozorg Qomi, S.; Bayat, H.; Yousefi, M.; Azhdari, S.; Talebi, S.; Mojarrad, M. Common therapeutic advances for Duchenne muscular dystrophy (DMD). *Int. J. Neurosci.* **2021**, *131*, 370–389. [CrossRef]

29. Vohra, R.; Batra, A.; Forbes, S.C.; Vandenborne, K.; Walter, G.A. Magnetic Resonance Monitoring of Disease Progression in mdx Mice on Different Genetic Backgrounds. *Am. J. Pathol.* **2017**, *187*, 2060–2070. [CrossRef]

30. Verhaart, I.E.; van Duijn, R.J.; den Adel, B.; Roest, A.A.; Verschuuren, J.J.; Aartsma-Rus, A.; van der Weerd, L. Assessment of cardiac function in three mouse dystrophinopathies by magnetic resonance imaging. *Neuromuscul. Disord.* **2012**, *22*, 418–426. [CrossRef]

31. Bostick, B.; Yue, Y.; Duan, D. Gender influences cardiac function in the mdx model of Duchenne cardiomyopathy. *Muscle Nerve* **2010**, *42*, 600–603. [CrossRef]

diagnostics

Article

Cardiac MRI: An Alternative Method to Determine the Left Ventricular Function

Kerstin Michler [1,*], Christopher Hessman [2], Marcus Prümmer [3], Stephan Achenbach [4], Michael Uder [2] and Rolf Janka [2]

1 Faculty of Medicine, Friedrich-Alexander-University Erlangen-Nürnberg (FAU), 91054 Erlangen, Germany
2 Institute of Radiology, University Hospital Erlangen, Friedrich-Alexander-Universität Erlangen-Nürnberg, 91054 Erlangen, Germany
3 Chimaera GmbH, 91054 Erlangen, Germany
4 Department of Medicine 2–Cardiology and Angiology, Friedrich-Alexander-University Erlangen-Nürnberg (FAU), Universitätsklinikum, 91054 Erlangen, Germany
* Correspondence: kerstin.michler@fau.de

Abstract: (1) Background: With the conventional contour surface method (KfM) for the evaluation of cardiac function parameters, the papillary muscle is considered to be part of the left ventricular volume. This systematic error can be avoided with a relatively easy-to-implement pixel-based evaluation method (PbM). The objective of this thesis is to compare the KfM and the PbM with regard to their difference due to papillary muscle volume exclusion. (2) Material and Methods: In the retrospective study, 191 cardiac-MR image data sets (126 male, 65 female; median age 51 years; age distribution 20–75 years) were analysed. The left ventricular function parameters: end-systolic volume (ESV), end-diastolic volume (EDV), ejection fraction (EF) and stroke volume (SV) were determined using classical KfW (syngo.via and cvi42 = gold standard) and PbM. Papillary muscle volume was calculated and segmented automatically via cvi42. The time required for evaluation with the PbM was collected. (3) Results: The size of EDV was 177 mL (69–444.5 mL) [average, [minimum–maximum]], ESV was 87 mL (20–361.4 mL), SV was 88 mL and EF was 50% (13–80%) in the pixel-based evaluation. The corresponding values with cvi42 were EDV 193 mL (89–476 mL), ESV 101 mL (34–411 mL), SV 90 mL and EF 45% (12–73%) and syngo.via: EDV 188 mL (74–447 mL), ESV 99 mL (29–358 mL), SV 89 mL (27–176 mL) and EF 47% (13–84%). The comparison between the PbM and KfM showed a negative difference for end-diastolic volume, a negative difference for end-systolic volume and a positive difference for ejection fraction. No difference was seen in stroke volume. The mean papillary muscle volume was calculated to be 14.2 mL. The evaluation with PbM took an average of 2:02 min. (4) Conclusion: PbM is easy and fast to perform for the determination of left ventricular cardiac function. It provides comparable results to the established disc/contour area method in terms of stroke volume and measures "true" left ventricular cardiac function while omitting the papillary muscles. This results in an average 6% higher ejection fraction, which can have a significant influence on therapy decisions.

Keywords: cardiac MRI; left ventricular function; ejection fraction; MRI; cine MRI

Citation: Michler, K.; Hessman, C.; Prümmer, M.; Achenbach, S.; Uder, M.; Janka, R. Cardiac MRI: An Alternative Method to Determine the Left Ventricular Function. *Diagnostics* **2023**, *13*, 1437. https://doi.org/10.3390/diagnostics13081437

Academic Editors: Minjie Lu and Arlene Sirajuddin

Received: 24 February 2023
Revised: 9 April 2023
Accepted: 14 April 2023
Published: 17 April 2023

1. Introduction

The cardiac function parameters end-diastolic volume (EDV) and end-systolic volume (ESV) can be measured using MRI in addition to echocardiography [1–4], with MRI being superior to transthoracic echocardiography [5–7]. For this purpose, SSFP (steady state free precision) cine sequences are scanned as short axis stacks and evaluated with the contour surface method (KfM). In this method, the areas enclosed by endocardium on each acquired slice plane are determined during end-diastole and end-systole and multiplied by the slice thickness (cf. Simpson—method [8,9]). The sum of these slice volumes yields the ESV and the EDV. Secondarily, the stroke volume (SV) and the ejection fraction (EF) can be calculated from these.

By definition, the papillary muscles are part of the determined areas and thus included in the calculated blood volume, which means that the blood volume is incorrectly overstated in end-diastole and end-systole. This systematic error has no influence on the stroke volume, but the EF is incorrectly calculated as too small. In cases where the EF is a decision criterion for therapy, such as in aortic valve replacement [10] or in the indication for implantation of an intracardiac pacemaker (ICD) [11], an accurate method for determining the EF would be desirable.

The image contrast in the cine-SSFP sequence between the blood and myocardium is sufficient to make a signal-based decision as to whether a blood or myocardial/papillary muscle voxel is being displayed. This should allow for accurate volume information (EDV and ESV) given a known voxel volume.

We implemented this approach in combination with an automatic shape recognition of the heart contour in short-axis slices.

The purpose of this study is to compare a pixel-based evaluation of cardiac function parameters with the classical slice method regarding the difference due to the exclusion of the papillary muscle volume.

2. Materials and Methods

The ethics committee of the University Hospital approved the study. All procedures performed in studies with human participants complied with the ethical standards of the institutional research committee and the 1964 Helsinki Declaration and its subsequent amendments or comparable ethical standards. The need for informed consent was waived by the Ethics Committee.

2.1. Patients Studied

Cardiac MRI examinations of patients aged 18 years and older between 1 January 2016 and 11 January 2016 were included via our hospital's Radiology Information System (RIS) for retrospective data collection. Patients with congenital heart defects or cardiac anomalies were excluded. From these patients, 245 were randomly selected alphabetically by first name, and image quality was assessed in the cine short-axis stack. All patients with well-defined cardiac contours in the short-axis cine stack were included in the study [8].

2.2. Examination Technique
2.2.1. MR Parameters

All MR examinations were measured on a Magnetom Aera 1.5 T scanner (Siemens Healthineers GmbH, Erlangen, Germany) using an SSFP-CINE sequence with retrospective ECG gating before contrast administration. A body-phased array coil in combination with the spine coil served as the receiver coil. The scan parameters were TR 42.4 ms, TE 1.1 ms and flip angle 55°. The FOV size was 340 × 276 mm, the matrix was 192 × 109 and the slice thickness was 8 mm. This resulted in a voxel size of 1.8 mm × 2.5 mm × 8.0 mm (=36 mm^3). From the cine sequence, 25 phases were calculated. Depending on the size of the heart, 10–14 layers were measured with a gap of 10% in expiration.

2.2.2. Volume Determination with the Pixel-Based Method (PbM)

The semi-automatic volume determination of the left ventricular heart volume with the pixel-based method (PbM) was evaluated with a software plug-in for the program OsiriX, which was developed by the authors. The software used can be freely purchased from Chimaera GmbH (Erlangen, Germany, www.chimaera.de, accessed on 2 January 2023).

The semi-automatic procedure is based on a brush tool that allows the user to "roughly" colour the target region, such as the left ventricle. In doing so, the algorithm analyses the local image environment of each mouse position and interactively calculates a segmentation mask within the brush size set by the user. The algorithm performs a local intensity analysis based on the minimum, maximum, mean and variance values of the intensity. Using these calculated values, threshold-based region boundaries are determined. Based on each cursor position, local region growth is performed, limited by the determined region boundaries. A morphological "closing" operation additionally closes potential non-detected areas caused by noise in the calculated mask. Optionally, the segmentation tool allows the setting of fixed minima and maxima as intensity thresholds that can be considered in the region's growth. The software has not been developed specifically for the evaluation of cardiac volumes and can therefore be used for any volume determination if there is sufficient contrast to the surrounding tissue.

For volume determination with the PbM, the cardiac base is first determined in the short-axis layer stack of the end-systole (=smallest subjectively determined area circle in the middle third of the heart) and end-diastole (=first image of the cine sequence). For this purpose, we defined the layer in which the myocardial ring is at least 50% closed as the cardiac base layer [9] (Figure 1).

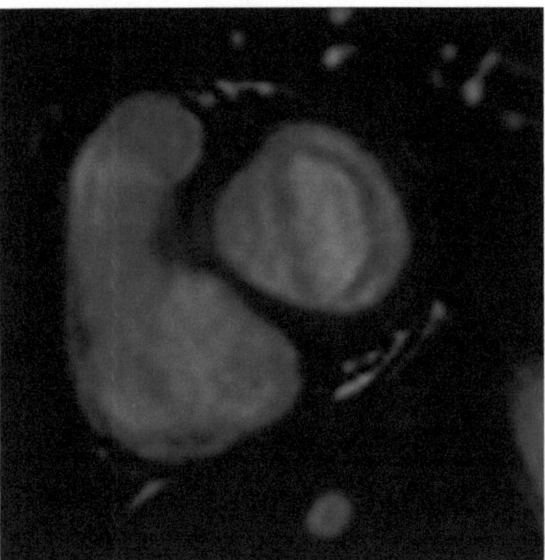

Figure 1. Determination of the most basal layer of the left ventricle at the transition to the left atrium. The myocardial ring is at least 50% complete.

Starting from this layer, the cavity is marked with the brush tool in each individual layer of the end-systole and end-diastole. For pixel-wise marking, the signal intensity range is selected so that the representation of blood falls within this range. This allows for areas of intraluminal blood to be colour-coded and separated from the myocardium and other surrounding tissues. The resulting areas (Figure 2) can then be added together with the slice thickness to form a volume according to Simpson's rule [9].

The pixel-based method for evaluating cardiac volumes is a new method where there are no experienced evaluators yet. To enable the most accurate evaluation possible, all data sets were initially evaluated by a non-board certified radiologist. Subsequently, all data sets were checked by a board certified radiologist and modified if necessary.

2.2.3. Volume Determination According to the Contour Surface Method (=KfM) with SyngoVia

All data sets were analysed with syngo.via, version 20A (Siemens Healthineers GmbH, Erlangen, Germany) according to the procedure of Hammon et al. [12]. Here, the software automatically recognizes the cardiac apex and the cardiac base based on the long-axis slices. Along the endocardium and epicardium, the circles are automatically drawn in all heart phases and the volume is calculated. The largest calculated volume is defined as end-diastole and the smallest calculated volume as end-systole. After the automatic segmentation, the evaluator checks the correctness of the heart base and heart apex in end-systole and end-diastole and corrects them if necessary. The circles are then manually checked along the endocardium and epicardium and corrected if necessary. Here, the parameter heart mass is used as an internal control. Since the cardiac mass does not change during the cardiac cycle, the same value should be obtained in end-systole and end-diastole (Figure 3).

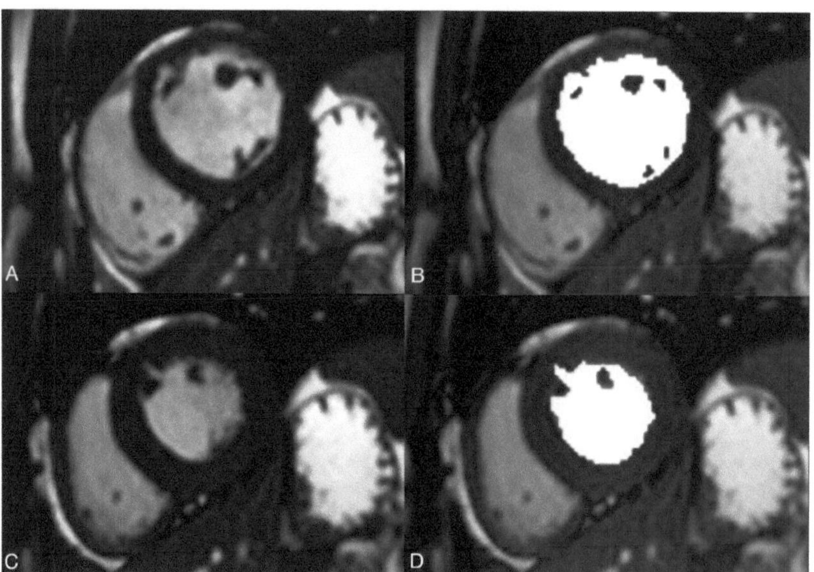

Figure 2. Evaluation of an exemplary layer in diastole (**A,B**) and systole (**C,D**) with the pixel-based method.

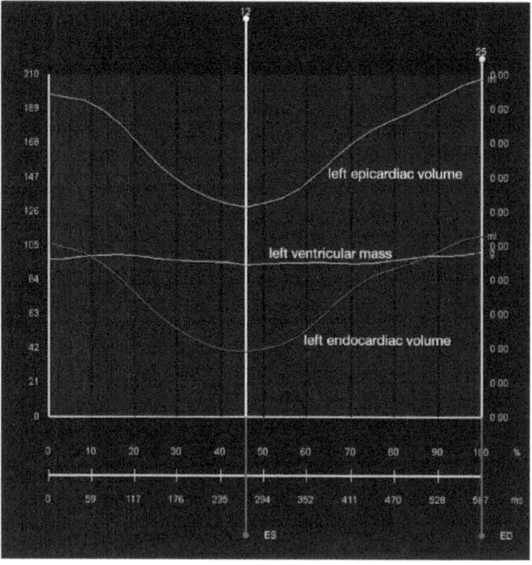

Figure 3. Evaluation of epicardial and endocardial volume and cardiac mass over the entire cardiac cycle. Ideally, the mass line is horizontal over the entire cardiac cycle.

All data obtained via this method was internally validated by a board-certified radiologist at our university hospital department.

2.2.4. Volume Determination according to the Contour Area Method (=KfM) with cvi42

All data sets were automatically evaluated using the contour surface method with cvi42 version 5.12 (Circle Cardiovascular Imaging, Montreal, Canada). For this purpose, the Cine short-axis stack and the Cine 2 Ch view were loaded into the program and automatically evaluated with the AI (artificial intelligence) function. The program first automatically determined the systole and diastole and calculated the volumes using the contour area method. The base of the heart was automatically determined in systole and diastole at 2 Ch View. In addition, the papillary muscle volume was segmented automatically. This fact leads to a measured papillary volume with the program cvi42.

2.2.5. Relationship between the Volume Area Method and the Pixel-Based Method

The difference between the two methods is assumed to be the volume of the papillary muscle. This results in the following mathematical correlations:

$$ESV_{KfM} - PM = ESV_{PbM} \tag{1}$$

$$EDV_{KfM} - PM = EDV_{PbM} \tag{2}$$

Equations (1) and (2) show that the stroke volume must be equal for both methods. Thus, the following applies:

$$SV = ESV_{KfM} - EDV_{KfM} = ESV_{PbM} + PM - (EDV_{PbM} + PM) = ESV_{PbM} - EDV_{PbM} \tag{3}$$

This does not apply to the ejection fraction (*EF*)

$$EF_{PbM} = \frac{EDV_{PbM} - ESV_{PbM}}{EDV_{PbM}} = \frac{\left(EDV_{KfM} - PM\right) - \left(ESV_{KfM} - PM\right)}{EDV_{KfM} - PM} = \frac{SV_{KfM}}{EDV_{KfM} - PM} \tag{4}$$

or

$$EF_{KfM} = \frac{EDV_{KfM} - ESV_{KfM}}{EDV_{KfM}} = \frac{(EDV_{PbM} + PM) - (ESV_{PbM} + PM)}{EDV_{PbM} + PM} = \frac{SV_{PbM}}{EDV_{PbM} + PM} \tag{5}$$

From the Formulas (1), (2), (4) and (5), one can now determine the papillary muscle volume in four ways:

$$PM1 = \frac{EDV_{KfM} - SV_{KfM}}{EF_{PbM}} \text{ based on (4)} \tag{6}$$

$$PM2 = \frac{SV_{PbM}}{EF_{KfM} - EDV_{PbM}} \text{ based on (5)} \tag{7}$$

$$PM3 = EDV_{KfM} - EDV_{PbM} \text{ based on (2)} \tag{8}$$

$$PM4 = ESV_{KfM} - ESV_{PbM} \text{ based on (1)} \tag{9}$$

We compared the results of the pixel-based method (*PbM*) with the three results of the contour surface method (SyngoVia, cvi42 with papillary muscle, cvi42 without papillary muscle).

The analysis was performed with the programme R (version 3.3.1; open source). All continuous variables (*EDV*, *ESV*, *SV* and *PM*) are given as an average with standard deviation. The volume of the papillary muscles (*PM*) was calculated separately for each of the two contour area methods using the four formulae mentioned above (6)–(9) (e.g., $PM = \frac{SV_{Pbm}}{EF_{KfM}} - EDV_{PbM}$). All results are provided as mean values (minimum; maximum).

The null hypothesis was that papillary muscle volume has no effect on stroke volume or ejection fraction. The null hypothesis was tested using the TOST test for paired samples with a Cohen's d value of 0.3 and a significance level of 5%.

3. Results

A search in our radiology information system yielded 406 adult patients with an MRI cardiac examination between 1 January 2016 and 11 January 2016. For randomization, these were sorted alphabetically by first name and the first 245 were considered in more detail. After reviewing the 245 records, 54 of the 245 patients were excluded due to blurred cardiac contours, e.g., due to respiratory artefacts or arrhythmic heartbeats [9]. Thus, a total of 191 data sets with a gender distribution of 126 (66.3%) male and 65 (33.7%) female participants with a median age of 51 years, and an age distribution at the time of study of 20–75 years could be included in the study.

After the first evaluation of the 191 data sets with the PbM, the stroke volumes (EDV—ESV) were calculated and compared with the stroke volumes of the KfM. In 15 cases, the difference was greater than 15%. After re-evaluation of these data sets with the KfM by an independent investigator who was blinded to the results of the pixel-based method and the first evaluation of the contour area method, five cases remained with a difference in beat volume greater than 15%. In these cases, the pixel-based method (PbM) was re-evaluated. Errors in the program operation, e.g., loading of an incomplete short axis stack when individual layers were repeated due to breathing artefacts—and thus the evaluation of too few layers—could be found as the cause. After a new independent evaluation of these five data sets, the difference in SV was not greater than 15% in any case.

After correcting the datasets, the size of EDV in the pixel-based evaluation was 177 mL (64 mL) [average, (standard deviation)], ESV was 87 mL (62.8 mL), SV was 90 mL (25.5 mL) and EF was 54% (16%). The corresponding values with the gold standard by the cvi42 program were EDV 193 mL (67 mL), ESV 101 mL (66.4 mL), SV 89 mL (25 mL) and EF 50% (14.5%).

The KfM results were EDV 189 mL (66 mL), ESV 100.5 mL (64.26 mL), SV 89 mL (25.4 mL) and EF 50.4% (14.7%). The test for equality of stroke volume between the PbM and the KfM (each with cvi42 and syngo.via) showed equivalence ($p < 0.001$ for TOST upper and TOST lower). However, equivalence cannot be assumed for the comparison of the ejection fractions (TOST lower: $p < 0.001$, TOST upper $p > 0.001$). The EF and SV of the two contour area methods are equivalent ($p < 0.001$ for TOST upper and TOST lower). There was a negative difference for end-diastolic volume and end-systolic volume and a positive difference for ejection fraction. There was no difference in stroke volume (Table 1). The test for equality of EDV and ESV between both KfM methods (cvi42 and syngo.via) showed no equivalence (TOST lower: $p < 0.001$, TOST upper $p > 0.001$). For each part, the upper limit was not significant. Comparing EDV and ESV between the PbM and the KfM method (each with cvi42 and syngo.via) showed no equivalence as well (TOST lower: $p < 0.001$, TOST upper $p > 0.001$).

Table 1. Mean values: EDV = end-diastolic volume, ESV = end-systolic volume, SV = stroke volume, EF = ejection fraction, PMV = papillary muscle volume, PbM = pixel-based evaluation method, KfM = contour surface method, cvi42 = evaluation method with cvi42 program (gold standard), * calculated value, ** mean papillary muscle volume given by the program cvi42, *** values in this line are statistically equivalent according to Tost-Test.

	Average PbM	Average KfM	Average cvi42
EDV [mL]	177	188	193
ESV [mL]	87	99	101
SV [mL]	88 ***	89 ***	90 ***
EF [%]	50	47 ***	45 ***
PMV [mL]	—	14.2 *	5.5 **

Figure 4 shows the average left ventricular volumes as a bar diagram. Only, all three values of the stroke volume and the ejection fraction of the clinical finding and gold standard are statistically equivalent.

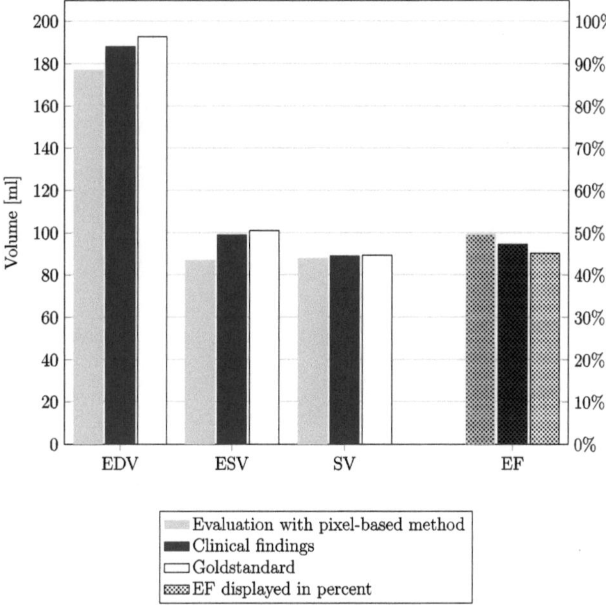

Figure 4. Mean values of the parameters EDV (=end-diastolic volume), ESV (=end-systolic volume), SV (=stroke volume) and EF (=ejection fraction) of all datasets (*n* = 208) using the pixel-based method (grey columns), the conventional contour surface method (black columns) and the gold standard (white columns). *y*-axis: volume [mL].

Using the above formula, an average papillary muscle volume of 14.2 mL (minimum 12.4 mL; maximum 16.3 mL) could be calculated (Table 2).

Table 2. Calculated, average papillary muscle volumes.

Formula	(6)	(7)	(8)	(9)	Mean
KfM syngo.via [mL]	13.4	15.4	12.4	13.5	13.6
KfM cvi42 [mL]	15.7	16.3	13.9	13.3	14.8
Mean [mL]	14.5	15.8	13.1	13.4	14.2

The automatic segmentation of the papillary muscles using the program cvi42 resulted in an average end-diastolic volume of 5.1 mL and an end-systolic volume of 6.0 mL. The mean of both the end-diastolic and end-systolic volume is exemplary shown in Table 1 (average cvi42: papillary muscle volume).

4. Discussion

The measurement of left ventricular function parameters is a common investigation with clinical decision-making relevance. In comparative studies between volume determination using 2D echocardiography and cine MRI, MRI has been shown to be the more accurate method [5–7]. The evaluated pixel-based method for ESV and EDV determination is simple and quick to perform and provides an unbiased ESV and EDV as it does not include the papillary muscle volume. The on average 6% higher ejection fraction in PbM compared to KfM can have a significant impact on treatment decisions.

The common practice for volume determination from MR datasets is the contour area method, where the end-diastolic and end-systolic volumes each include the volume of the papillary muscles. In our patient population, the end-diastolic volume (EDV) was on average 6% higher and the end-systolic volume (ESV) was on average 14% higher with the contour area method than with the pixel-based method. The difference can be explained by the fact that in diastole the endocardium is still easily recognizable, but in systole the papillary muscles and endocardium cannot be separated or can only be separated with difficulty (Figure 5), making it more difficult to draw a line along the endocardium.

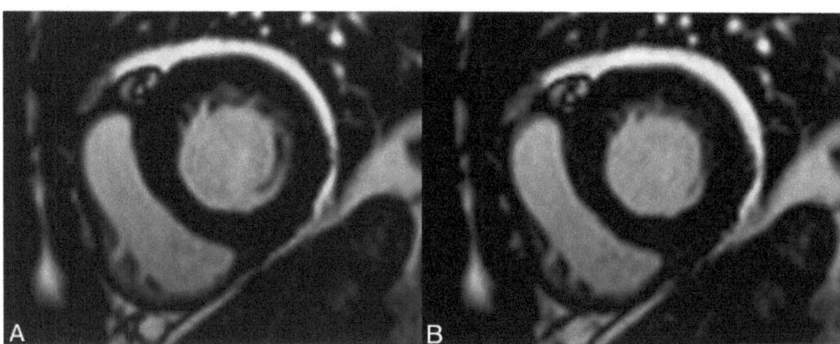

Figure 5. Good visual differentiation of the papillary muscle and the endocardium in the presystolic phase (**A**) and lack of border between papillary muscle and endocardium in systole (**B**).

One approach to solving this problem is to determine the mass of the heart, which is always the same regardless of the cardiac cycle. If the cardiac mass is measured differently in end-diastole and end-systole and the epicardial contour is drawn correctly, a correction of the endocardial cardiac contours is recommended [13,14]. Our data show that despite correction for mass, the difference between the two methods is greater for the ESV than for the EDV. This phenomenon is confirmed by Bailly et al. [15], in whose study the coefficient of variability is larger for the ESV than for the EDV [16,17].

Interestingly, despite a small difference in the average EDV (Δ 5 mL) and ESV (Δ 2 mL), no statistical equality can be assumed when comparing both KfM methods (cvi42 and syngo.via) (cf. Table 1). In particular, for end-diastolic, this could be due to the different proportions of papillary muscle volume within the inner circle.

An important functional parameter for the heart is the ejection fraction $(EF) = \frac{SV}{EDV}$. According to Equation (4) $EF_{PbM} = \frac{SV_{KfM}}{EDV_{KfM}-PM}$, the EF is erroneously calculated too low using the contour method, and the deviation depends on the size of the papillary muscles. In our study, the deviation was 6% on average. Riffel et al. [18] compared the cardiac function parameters with and without segmenting out the papillary muscles and confirmed our results. They also found that the stroke volume was higher and EF lower with the contour area method. However, the difference in EF was smaller in Riffel et al., which may have been due to the fact that only healthy subjects were studied.

The ejection fraction is a cardiological parameter that influences the decision on drug or interventional/surgical therapy. For example, an EF < 50% [10] is decisive for the indication of an artificial heart valve for aortic valve replacement. The EF is also a relevant parameter in the therapy decision trees for treatment concepts of other heart valve diseases, e.g., mitral valve insufficiency [19].

In patients with heart failure, the ejection fraction is not only a diagnostic parameter but also an important functional parameter for therapy decisions concerning both drug therapy and the implantation of a left ventricular assist device [20]. In dilatative cardiomyopathy, implantation of an ICD has indicated if the chronic left ventricular ejection fraction is less than 35% [11]. Furthermore, EF ≤ 40% is included in the CHADS-VASC score as a criterion for heart failure. Thus, it indirectly contributes to the consideration of stroke risk and the resulting decision to use oral anticoagulation for patients with atrial fibrillation [21,22].

Many evaluators of cardiac volumes refer to the standard values published in 2015 in JCMR by Kawel-Böhm et al. [23]. The "standard values" given here for men and women up to and from 60 years of age were determined with the KfM, but the papillary muscles were segmented separately and not evaluated as part of the LV volume. Thus, referencing the measurement results with the KfM to these standard values [23] is severely limited. Likewise, a re-evaluation of the application to current norm values is necessary before the clinically practical use of PbM. Further studies are needed, especially to set new reference values. We would like to emphasize that the gender ratio in our study is asymmetrical (66% male and 34% female). This aspect should be considered in future studies, especially for the setting of new reference values.

The difference between the two methods is the volume of the papillary muscles. According to the formula for calculating the volume, the volume in our study is 14.2 mL on average. The automatic segmentation of the papillary muscles resulted in a value of 5.1 mL end-diastolic and 6.0 mL end-systolic and is clearly smaller than determined via the pixel-based method. The reason for this is that the cvi42 program only segments the papillary muscle located in the lumen, but not the part that is directly adjacent to the myocardium. This also explains the difference between systolic and diastolic volumes of the automatically determined papillary muscle. This results in a larger proportion of the papillary muscle volume for smaller left ventricular volumes. As expected, PbM results in lower end-diastolic and end-systolic volumes compared to KfM. Consecutively, the ejection fraction becomes larger with PbM. The measurement of end-diastolic and end-systolic volume with inclusion of the papillary muscles should therefore be questioned.

The pixel-based method solves the "papillary muscle problem" in a technically simple way by automatic signal detection within predefined limits in combination with contour detection. It is not a "specialist" for cardiac evaluation and can determine many other volumes where there are good signal or density differences to the surrounding tissue. Despite using a development version of the software without automation algorithms, the heart volume of the left ventricle could be determined in about 2 min. A complete automation of the method is conceivable in the future.

In the future, more and more artificial intelligence (AI) will be used in every aspect of our daily lives, including medical applications. He et al. compared the left ventricular ejection fraction echocardiographically using a sonographer vs. artificial intelligence [24]. They were able to show that AI is not inferior to the measurement of the left ventricular ejection fraction with echocardiography compared to a sonographer. Future applications could therefore include AI systems to assist examiners in imaging problems.

A limitation of our study is the unknown actual stroke volume. With an ECG-triggered flow measurement in the ascending aorta, which is not routinely measured in our department, this could have been determined, as flow measurement provides more reproducible results than volumetry [25,26]. Another limitation is that the volume of anatomical structures with bizarre morphology, such as the papillary muscles can only be determined with estimates based on various assumptions. A solution would be direct measurements in body donors and comparison with post-mortem MRI, as already successfully measured by Bertozzi et al. [27].

Before establishing it as a standard, another limitation is the exclusion of patients with moderate image quality and consequently insufficiently identifiable heart contours. This seemed to be in most cases with atrial fibrillation while detecting a higher heart rate during the assessment.

5. Conclusions

PbM is easy and quick to perform for the determination of left ventricular heart function. In terms of stroke volume, it provides comparable results to the established disc/contour area method and measures the actual left ventricular heart function, leaving out the papillary muscles. This results in an average 6% higher ejection fraction, which can have a significant influence on therapy decisions.

Author Contributions: Conceptualization, R.J. and K.M.; methodology, R.J. and K.M.; software, M.P.; validation, R.J.; formal analysis, K.M.; investigation, R.J. and K.M.; resources, M.U.; data curation, K.M.; writing—original draft preparation, K.M.; writing—review and editing, R.J., C.H. and S.A., M.U.; visualization, K.M. and R.J.; supervision, R.J. All authors have read and agreed to the published version of the manuscript.

Funding: This research received no external funding.

Institutional Review Board Statement: The ethics committee of the University Hospital approved the study. All procedures performed in studies with human participants complied with the ethical standards of the institutional research committee and the 1964 Helsinki Declaration and its subsequent amendments or comparable ethical standards. The need for informed consent was waived by the Ethics Committee.

Informed Consent Statement: Patient consent was waived by the ethics committee of the University Hospital Erlangen as is the routine procedure in retrospective evaluation in 2016.

Data Availability Statement: The datasets analyzed during the current study are available from the corresponding author on reasonable request.

Acknowledgments: The work reported was carried out at the University Hospital Erlangen of the Friedrich-Alexander-University Erlangen-Nürnberg (FAU), Germany. The present research was performed in fulfilment of the requirements for obtaining the degree "Dr. med." for Kerstin Michler at the Friedrich-Alexander-Universität Erlangen-Nürnberg, Germany.

Conflicts of Interest: A contributing author is the publisher of the software used by the company Chimaera.

References

1. Achenbach, S.; Barkhausen, J.; Beer, M.; Beerbaum, P.; Dill, T.; Eichhorn, J.; Fratz, S.; Gutberlet, M.; Hoffmann, M.; Huber, A.; et al. Konsensusempfehlungen der DRG/DGK/DGPK zum Einsatz der Herzbildgebung mit Computertomographie und Magnetresonanztomographie. *Der Kardiologe* **2012**, *6*, 105–125. [CrossRef]
2. Hundley, W.G.; Bluemke, D.A.; Finn, J.P.; Flamm, S.D.; Fogel, M.A.; Friedrich, M.G.; Ho, V.B.; Jerosch-Herold, M.; Kramer, C.M.; Manning, W.J.; et al. ACCF/ACR/AHA/NASCI/SCMR 2010 Expert Consensus Document on Cardiovascular Magnetic Resonance: A Report of the American College of Cardiology Foundation Task Force on Expert Consensus Documents. *J. Am. Coll. Cardiol.* **2010**, *55*, 2614–2662. [CrossRef]
3. Thiele, H.; Nagel, E.; Paetsch, I.; Schnackenburg, B.; Bornstedt, A.; Kouwenhoven, M.; Wahl, A.; Schuler, G.; Fleck, E. Functional cardiac MR imaging with steady-state free precession (SSFP) significantly improves endocardial border delineation without contrast agents. *J. Magn. Reson. Imaging* **2001**, *14*, 362–367. [CrossRef]
4. Thiele, H.; Paetsch, I.; Schnackenburg, B.; Bornstedt, A.; Grebe, O.; Wellnhofer, E.; Schuler, G.; Fleck, E.; Nagel, E. Improved accuracy of quantitative assessment of left ventricular volume and ejection fraction by geometric models with steady-state free precession. *J. Cardiovasc. Magn. Reson.* **2002**, *4*, 327–339. [CrossRef]
5. Buser, P.T.; Auffermann, W.; Holt, W.W.; Wagner, S.; Kircher, B.; Wolfe, C.; Higgins, C.B. Noninvasive evaluation of global left ventricular function with use of cine nuclear magnetic resonance. *J. Am. Coll. Cardiol.* **1989**, *13*, 1294–1300. [CrossRef]
6. Matsumura, K.; Nakase, E.; Haiyama, T.; Takeo, K.; Shimizu, K.; Yamasaki, K.; Kohno, K. Determination of cardiac ejection fraction and left ventricular volume: Contrast-enhanced ultrafast cine MR imaging vs IV digital subtraction ventriculography. *Am. J. Roentgenol.* **1993**, *160*, 979–985. [CrossRef] [PubMed]
7. Sechtem, U.; Pflugfelder, P.W.; Gould, R.G.; Cassidy, M.M.; Higgins, C.B. Measurement of right and left ventricular volumes in healthy individuals with cine MR imaging. *Radiology* **1987**, *163*, 697–702. [CrossRef]
8. Stamm, H.H. Parallele Echtzeitbildgebung zur Quantifizierung der Linksventrikulären Herzfunktion bei Freier Atmung Mittels MRT. Ph.D. Thesis, Julius-Maximilians-Universität Würzburg, Würzburg, Germany, 2010. Available online: https://opus.bibliothek.uni-wuerzburg.de/files/3965/TSENSE.pdf (accessed on 18 December 2018).
9. Mahnken, A.H.; Günther, R.W.; Krombach, G.A. Grundlagen der linksventrikulären Funktionsanalyse mittels MRT und MSCT. *Rofo* **2004**, *176*, 1365–1379. [CrossRef]
10. Vahanian, A.; Alfieri, O.; Andreotti, F.; Antunes, M.J.; Barón-Esquivias, G.; Baumgartner, H.; Borger, M.A.; Carrel, T.P.; De Bonis, M.; Evangelista, A.; et al. Guidelines on the management of valvular heart disease (version 2012). *Eur. Heart J.* **2012**, *33*, 2451–2496. [CrossRef]
11. Ehlermann, P.; Katus, H.A. Dilatative Kardiomyopathie. *Herzschrittmacherther. Elektrophysiol.* **2012**, *23*, 196–200. [CrossRef]
12. Hammon, M.; Janka, R.; Dankerl, P.; Glöckler, M.; Kammerer, F.J.; Dittrich, S.; Uder, M.; Rompel, O. Pediatric cardiac MRI: Automated left-ventricular volumes and function analysis and effects of manual adjustments. *Pediatr. Radiol.* **2014**, *45*, 651–657. [CrossRef] [PubMed]

13. Mahnken, A.H.; Mühlenbruch, G.; Koos, R.; Stanzel, S.; Busch, P.S.; Niethammer, M.; Günther, R.W.; Wildberger, J.E. Automated vs. manual assessment of left ventricular function in cardiac multidetector row computed tomography: Comparison with magnetic resonance imaging. *Eur. Radiol.* **2006**, *16*, 1416–1423. [CrossRef] [PubMed]
14. Francois, C.; Fieno, D.S.; Shors, S.M.; Finn, J.P. Left Ventricular Mass: Manual and Automatic Segmentation of True FISP and FLASH Cine MR Images in Dogs and Pigs. *Radiology* **2004**, *230*, 389–395. [CrossRef] [PubMed]
15. Bailly, A.; Lipiecki, J.; Chabrot, P.; Alfidja, A.; Garcier, J.M.; Ughetto, S.; Ponsonnaille, J.; Boyer, L. Assessment of left ventricular volumes and function by cine-MR imaging depending on the investigator's experience. *Surg. Radiol. Anat.* **2008**, *31*, 113–120. [CrossRef] [PubMed]
16. Grothues, F.; Smith, G.C.; Moon, J.C.; Bellenger, N.G.; Collins, P.; Klein, H.U.; Pennell, D.J. Comparison of interstudy reproducibility of cardiovascular magnetic resonance with two-dimensional echocardiography in normal subjects and in patients with heart failure or left ventricular hypertrophy. *Am. J. Cardiol.* **2002**, *90*, 29–34. [CrossRef]
17. Semelka, R.C.; Tomei, E.; Wagner, S.; Mayo, J.; Kondo, C.; Suzuki, J.; Caputo, G.R.; Higgins, C.B. Normal left ventricular dimensions and function: Interstudy reproducibility of measurements with cine MR imaging. *Radiology* **1990**, *174*, 763–768. [CrossRef]
18. Riffel, J.H.; Schmucker, K.; Andre, F.; Ochs, M.; Hirschberg, K.; Schaub, E.; Fritz, T.; Mueller-Hennessen, M.; Giannitsis, E.; Katus, H.A.; et al. Cardiovascular magnetic resonance of cardiac morphology and function: Impact of different strategies of contour drawing and indexing. *Clin. Res. Cardiol.* **2018**, *108*, 411–429. [CrossRef]
19. Baumgartner, H.; Falk, V.; Bax, J.J.; De Bonis, M.; Hamm, C.; Holm, P.J.; Iung, B.; Lancellotti, P.; Lansac, E.; Rodriguez Muñoz, D.; et al. 2017 ESC/EACTS Guidelines for the management of valvular heart disease. *Eur. J. Cardio-Thorac. Surg.* **2017**, *52*, 616–664, Correction: *Eur. J. Cardio-Thorac. Surg.* **2017**, *52*, 832. [CrossRef]
20. McMurray, J.J.; Adamopoulos, S.; Anker, S.D.; Auricchio, A.; Böhm, M.; Dickstein, K.; Falk, V.; Filippatos, G.; Fonseca, C.; Gomez-Sanchez, M.A.; et al. ESC Guidelines for the diagnosis and treatment of acute and chronic heart failure 2012: The Task Force for the Diagnosis and Treatment of Acute and Chronic Heart Failure 2012 of the European Society of Cardiology. Developed in collaboration with the Heart Failure Association (HFA) of the ESC. *Eur. Heart J.* **2012**, *33*, 1787–1847. [CrossRef]
21. Gage, B.F.; Waterman, A.D.; Shannon, W.; Boechler, M.; Rich, M.W.; Radford, M.J. Validation of Clinical Classification Schemes for Predicting Stroke: Results from the National Registry of Atrial Fibrillation. *JAMA* **2001**, *285*, 2864–2870. [CrossRef]
22. January, C.T.; Wann, L.S.; Alpert, J.S.; Calkins, H.; Cigarroa, J.E.; Cleveland, J.C., Jr.; Conti, J.B.; Ellinor, P.T.; Ezekowitz, M.D.; Field, M.E.; et al. 2014 AHA/ACC/HRS Guideline for the Management of Patients With Atrial Fibrillation: A report of the American College of Cardiology/American Heart Association Task Force on Practice Guidelines and the Heart Rhythm Society. *J. Am. Coll. Cardiol.* **2014**, *64*, e1–e76. [CrossRef]
23. Kawel-Boehm, N.; Maceira, A.; Valsangiacomo-Buechel, E.R.; Vogel-Claussen, J.; Turkbey, E.B.; Williams, R.; Plein, S.; Tee, M.; Eng, J.; Bluemke, D.A. Normal values for cardiovascular magnetic resonance in adults and children. *J. Cardiovasc. Magn. Reson.* **2015**, *17*, 1–33. [CrossRef] [PubMed]
24. He, B.; Kwan, A.C.; Cho, J.H.; Yuan, N.; Pollick, C.; Shiota, T.; Ebinger, J.; Bello, N.A.; Wei, J.; Josan, K.; et al. Blinded, randomized trial of sonographer versus AI cardiac function assessment. *Nature* **2023**, 1–5. [CrossRef]
25. Rominger, M.B.; Dinkel, H.-P.; Bachmann, G.F. Vergleich von schneller MR-Flussmessung in Atemanhaltetechnik in Aorta ascendens und Truncus pulmonalis mit rechts- und linksventrikulärer Cine MR-Bildgebung zur Schlagvoluminabestimmung bei Probanden. *RöFo-Fortschr. Auf Dem Geb. Der Röntgenstrahlen Und Der Bildgeb. Verfahr.* **2002**, *174*, 196–201. [CrossRef] [PubMed]
26. Kondo, G.R.C.C.; Caputo, G.R.; Semelka, R.; Foster, E.; Shimakawa, A.; Higgins, C.B.; Srichai, M.B.; Lim, R.P.; Wong, S.; Lee, V.S.; et al. Right and left ventricular stroke volume measurements with velocity-encoded cine MR imaging: In vitro and in vivo validation. *Am. J. Roentgenol.* **1991**, *157*, 9–16. [CrossRef]
27. Bertozzi, G.; Cafarelli, F.P.; Ferrara, M.; Di Fazio, N.; Guglielmi, G.; Cipolloni, L.; Manetti, F.; La Russa, R.; Fineschi, V. Sudden Cardiac Death and Ex-Situ Post-Mortem Cardiac Magnetic Resonance Imaging: A Morphological Study Based on Diagnostic Correlation Methodology. *Diagnostics* **2022**, *12*, 218. [CrossRef]

Article

Cardiovascular Magnetic Resonance Imaging Findings in Africans with Idiopathic Dilated Cardiomyopathy

Nqoba Tsabedze [1,*], Andre du Plessis [2], Dineo Mpanya [1], Anelia Vorster [2], Quinn Wells [3], Leonie Scholtz [2,†] and Pravin Manga [1,†]

1   Division of Cardiology, Department of Internal Medicine, School of Clinical Medicine, Faculty of Health Sciences, University of the Witwatersrand, Johannesburg 2193, South Africa
2   Diagnostic Radiology, Midstream Mediclinic, Centurion 1692, South Africa
3   Division of Cardiovascular Medicine, Department of Medicine, Vanderbilt University Medical Center, Nashville, TN 37232, USA
*   Correspondence: nqoba.tsabedze@wits.ac.za
†   These authors contributed equally to this work.

Abstract: In sub-Saharan Africa, idiopathic dilated cardiomyopathy (IDCM) is a common yet poorly investigated cause of heart failure. Cardiovascular magnetic resonance (CMR) imaging is the gold standard for tissue characterisation and volumetric quantification. In this paper, we present CMR findings obtained from a cohort of patients with IDCM in Southern Africa suspected of having a genetic cause of cardiomyopathy. A total of 78 IDCM study participants were referred for CMR imaging. The participants had a median left ventricular ejection fraction of 24% [interquartile range, (IQR): 18–34]. Late gadolinium enhancement (LGE) was visualised in 43 (55.1%) participants and localised in the midwall in 28 (65.0%) participants. At the time of enrolment into the study, non-survivors had a higher median left ventricular end diastolic wall mass index of 89.4 g/m^2 (IQR: 74.5–100.6) vs. 73.6 g/m^2 (IQR: 51.9–84.7), $p = 0.025$ and a higher median right ventricular end-systolic volume index of 86 mL/m^2 (IQR:74–105) vs. 41 mL/m^2 (IQR: 30–71), $p < 0.001$. After one year, 14 participants (17.9%) died. The hazard ratio for the risk of death in patients with evidence of LGE from CMR imaging was 0.435 (95% CI: 0.259–0.731; $p = 0.002$). Midwall enhancement was the most common pattern, visualised in 65% of participants. Prospective, adequately powered, and multi-centre studies across sub-Saharan Africa are required to determine the prognostic significance of CMR imaging parameters such as late gadolinium enhancement, extracellular volume fraction, and strain patterns in an African IDCM cohort.

Keywords: idiopathic dilated cardiomyopathy; magnetic resonance imaging; cardiovascular; late gadolinium enhancement; all-cause mortality

Citation: Tsabedze, N.; du Plessis, A.; Mpanya, D.; Vorster, A.; Wells, Q.; Scholtz, L.; Manga, P. Cardiovascular Magnetic Resonance Imaging Findings in Africans with Idiopathic Dilated Cardiomyopathy. *Diagnostics* **2023**, *13*, 617. https://doi.org/10.3390/diagnostics13040617

Academic Editors: Minjie Lu and Arlene Sirajuddin

Received: 4 January 2023
Revised: 31 January 2023
Accepted: 2 February 2023
Published: 8 February 2023

1. Introduction

Idiopathic dilated cardiomyopathy (IDCM) is an endemic primary myocardial disease in sub-Saharan Africa (SSA) [1,2]. It manifests clinically with ventricular dilatation and myocardial dysfunction without obstructive coronary artery disease (CAD) or abnormal loading conditions such as hypertension and valvular heart disease [3]. The clinical management of patients with IDCM includes identifying and treating reversible causes of myocardial dysfunction, improving survival, slowing disease progression, and alleviating symptoms [4]. Therefore, excluding secondary causes of dilated cardiomyopathy (DCM) is recommended, including genetic causes in all patients with unexplained ventricular dilatation and dysfunction. However, despite a comprehensive clinical workup, in regions where genetic testing is not widely available, patients without an identifiable cause for DCM are provided a working diagnosis of IDCM.

Cardiovascular magnetic resonance (CMR) imaging is considered the gold standard for anatomical and tissue characterisation and evaluating the extent of cardiac dysfunction.

Although not widely accessible in most centres across Africa, cardiac magnetic resonance imaging (MRI) plays a significant role in excluding infiltrative macrovascular and microvascular disease and defining the presence of fibrosis in patients with DCM [5]. We report CMR imaging findings in a meticulously phenotyped cohort of black Africans with a clinical working diagnosis of IDCM. The cohort described in this paper was prospectively recruited in a study designed to identify the genetic causes of myocardial dysfunction in these patients.

2. Materials and Methods

2.1. Study Design and Participants

Between July 2015 and December 2018, we screened 161 participants for idiopathic dilated cardiomyopathy. The study's participants were identified from the inpatient cardiology wards and outpatient heart failure with reduced ejection fraction (HFrEF) clinic at the Charlotte Maxeke Johannesburg Academic Hospital (CMJAH), a quaternary referral centre in Johannesburg, South Africa. The study inclusion criteria included adults 18 years and older with a left ventricular ejection fraction (LVEF) less than or equal to 40%. The exclusion criteria included organic valvular heart disease, hypertension, coronary artery disease, human immunodeficiency virus infection, myocarditis, infiltrative disease, and metabolic conditions. In addition, patients presenting with DCM in the peripartum period, post-chemotherapy, or post-radiation therapy, were also excluded.

A detailed clinical history was obtained from all participants. This included their age at the time of the index heart failure diagnosis, a family history of sudden cardiac death, comorbidities, and heart failure symptoms. In addition, a physical examination, which focused on identifying signs of heart failure, was performed on all participants. During recruitment, screening laboratory biochemical tests were performed on all study participants to exclude secondary causes of a DCM. These included haemoglobin, platelet count, C-reactive protein, a lipogram, thyroid and renal function tests, cardiac biomarkers, and micro-nutrient serum levels. A twelve-lead electrocardiogram (ECG) and a 2D transthoracic echocardiogram (General Electric Vivid 9 4D) were performed on all recruited participants. In addition, a diagnostic coronary angiogram was performed to exclude obstructive coronary artery disease.

All study participants received these investigations, including CMR imaging, performed within 72 h of recruitment. In this study, IDCM was defined as left ventricular dysfunction (LVEF $\leq$ 40%) in the absence of CAD (normal diagnostic angiography), valvular heart diseases, infiltrative disease, and metabolic abnormalities. The study complied with the Declaration of Helsinki and informed consent was obtained from the study participants. Approval to conduct the study was granted by the University of the Witwatersrand Human Research Ethics Committee (certificate number: M150467). Informed consent was obtained from all study participants.

2.2. Cardiac Magnetic Resonance Imaging Protocol and Image Analysis

A 1.5 T whole-body scanner (Philips) with an ECG triggering device was used for CMR imaging. Respiratory bellows were placed on the patient's abdomen throughout the imaging process. After the acquisition of localisation images, continuous short-axis cine images of the left ventricle were obtained using steady state free precession sequence at end-expiration.

Ventricular volumes and left- and right-ventricular ejection fractions were calculated using four-chamber and short-axis slice summation.

Images depicting late gadolinium enhancement (LGE) were acquired approximately 15 min after administering 0.1 mmol/kg of gadobenate dimeglumine (Bracco Diagnostics Inc., Princeton, NJ, USA) at an injection rate of 2 mL per second. The presence of LGE was evaluated using segmented inversion recovery prepared for true fast imaging. The images were visually analysed for the presence and extent of LGE. Two radiologists were available for image interpretation, and a single radiologist independently reviewed each set of CMR images. The visual scoring method was based on the 17-segment model. The percentage of the myocardium with LGE was calculated by counting the number of segments with LGE and dividing by 17.

2.3. Patient Follow-Up and Study Endpoints

The CMJAH is a quaternary referral specialist centre where advanced cardiac patients are preferentially referred and definitively managed. Thus, outcome data were first collected from the Electronic Health Record System, which captures all patients admitted to the cardiology wards at the CMJAH. In addition, all-cause mortality, frequency of hospitalisations (in any hospital, including CMJAH), and the occurrence of thromboembolic complications were documented after a telephonic interview with the study participants or their next-of-kin after a median follow-up duration of 12 [interquartile range (IQR): 8.8–16.8] months.

2.4. Statistical Analysis

Categorical variables are expressed as counts and percentages and were compared for the study outcome using a Chi-square test. Continuous variables with a normal and non-normal distribution are expressed as mean and standard deviation, as well as the median and IQR, respectively. The Student's *t*-test and the Wilcoxon rank sum (Mann–Whitney) test were used for comparing the mean and median, respectively. Confidence intervals were set at 95%, and $p < 0.05$ was regarded as statistically significant. A Cox proportional hazards model was created to assess for risk factors for all-cause mortality. Analyses were conducted using STATA version 16.0 (StataCorp, College Station, TX, USA).

3. Results

3.1. Baseline Clinical Characteristics

After excluding participants lost to follow-up (Figure 1), the final study cohort comprised 78 participants consisting of 53 (67.9%) males. The mean age was 47.3 ± 13.3 years (Table 1). In addition, there were 73 (93.6%) native Africans. The median left ventricular ejection fraction (LVEF) on CMR imaging was 24% (IQR: 18–34), and non-survivors had a lower median LVEF of 18% (IQR: 13–24) vs. 24% (IQR: 18–36), $p = 0.042$. Thirteen (16.7%) participants had no overt clinical symptoms of congestive cardiac failure in the study cohort. The remaining participants complained of dyspnea on minimal exertion (62.8%), paroxysmal nocturnal dyspnea (42.3%), orthopnea (58.9%), palpitations (35.9%), and syncope (7.7%), or a combination of these symptoms.

3.2. Cardiovascular Magnetic Resonance Imaging

The ventricular volumetric parameters are reported in Table 2. Non-survivors had a higher median left ventricular end diastolic wall mass index of 89.4 g/m^2 (IQR: 74.5–100.6) vs. 73.6 g/m^2 (IQR: 51.9–84.7), $p = 0.025$, a higher median right ventricular end-systolic volume index of 86 mL/m^2 (IQR: 74–105) vs. 41 mL/m^2 (IQR: 30–71), $p < 0.001$, and a higher median right ventricular end diastolic volume index 114 mL/m^2 (IQR: 101–147) vs. 70 mL/m^2 (IQR: 56–94), $p < 0.001$. The hazard ratio for the risk of death in patients presenting with evidence of LGE on CMR imaging was 0.435 (95% CI: 0.259–0.731; $p = 0.002$). However, the Cox proportional hazards model identified none of the clinical and CMR imaging parameters as statistically significant covariates for all-cause mortality. There were 43 (55.1%) participants with evidence of LGE on imaging. Late gadolinium enhancement was visualised in the midwall in 28 (65.0%) participants, and four (9.3%) had LGE in both the right ventricular insertion point and midwall (Table 3). In addition, three (7.0%) participants visualised a transmural LGE pattern, and two (4.6%) participants had areas of focal LGE. Midwall and subendocardial and epicardial LGE patterns are depicted in Figure 2. A total of nine patients had an LGE pattern commonly associated with ischaemic heart disease.

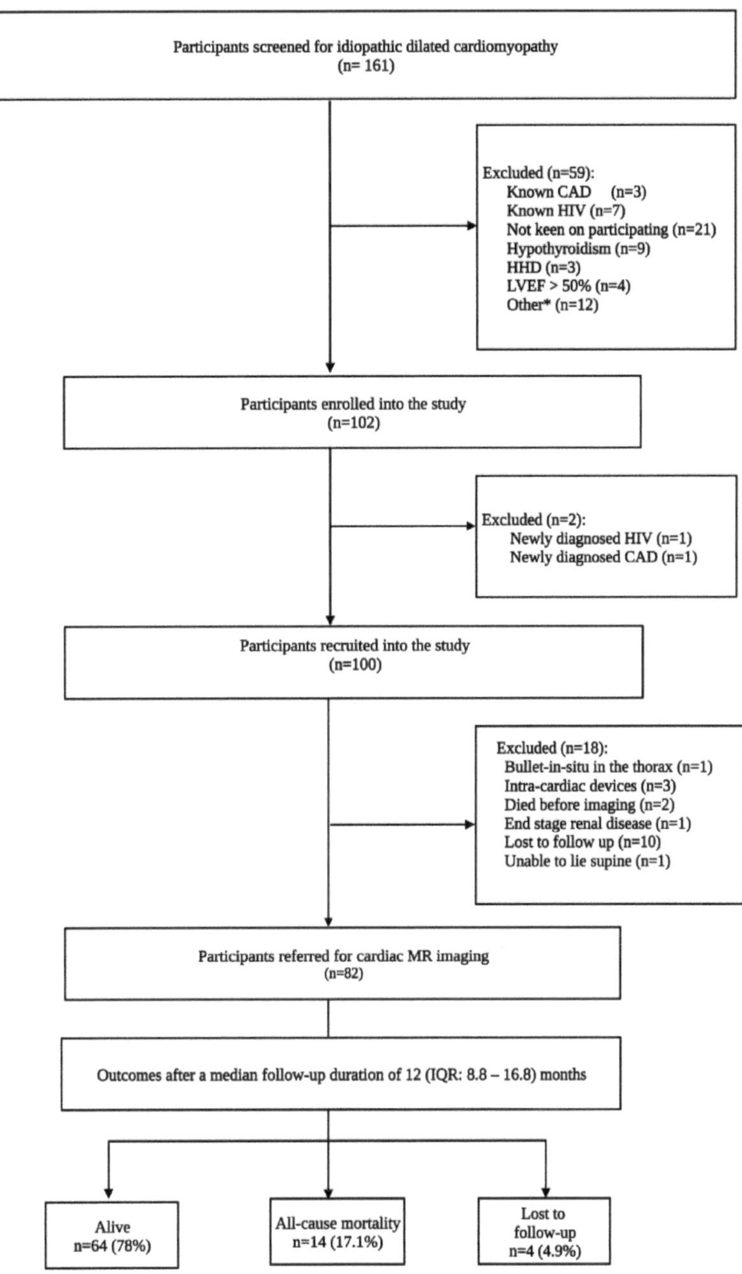

Figure 1. Flow chart detailing the identification of the study cohort. CAD = coronary artery disease, HHD = hypertensive heart disease, HIV = human immunodeficiency virus, LVEF = left ventricular ejection fraction. Other * refers to peripartum cardiomyopathy ($n = 1$), malignancy ($n = 2$), organic valvular heart diseases ($n = 2$), pacemaker in situ ($n = 1$), end-stage renal impairment ($n = 3$), recreational substance abuse ($n = 1$), unstable patient ($n = 1$), dilated cardiomyopathy precipitated by supraventricular tachycardia ($n = 1$).

Table 1. Baseline clinical characteristics of patients with idiopathic dilated cardiomyopathy stratified according to one-year outcomes.

	All Patients	All-Cause Mortality	Alive	
	(*n* = 78)	(*n* = 14)	(*n* = 64)	*p*-Value
Age, years	47.3 ± 13.3	41.6 ± 14.0	48.6 ± 13.0	0.074
Male	53 (67.9)	10 (71.4)	43 (67.2)	0.758
Smoking	15 (19.2)	6 (42.9)	9 (14.1)	0.013
BMI, kg/m^2	26.1 (23.3–30.5)	24.1 (21.4–25.3)	27.7 (23.6–31.3)	0.045
BSA, m^2	1.9 ± 0.2	1.9 ± 0.2	1.9 ± 0.2	0.328
Heart rate, bpm	80 (69–95)	89 (69–104)	80 (69–95)	0.475
Systolic BP, mmHg	119 (101–129)	115 (99–122)	119 (106–133)	0.263
Diastolic BP, mmHg	76 (67–88)	73 (60–91)	78 (67–88)	0.442
MAP, mmHg	89 (78–101)	81 (73–93)	92 (80–101)	0.124
NYHA class				0.173
1	29 (37.2)	2 (14.3)	27 (42.2)	
2	34 (43.6)	7 (50.0)	27 (42.2)	
3	12 (15.4)	4 (28.6)	8 (12.5)	
4	3 (3.8)	1 (7.1)	2 (3.1)	
Medication				
Beta-blocker	73 (93.6)	13 (92.9)	60 (93.7)	0.837
ACE inhibitors	51 (65.4)	11 (78.6)	40 (62.5)	0.706
ARB	9 (11.5)	0 (0)	9 (14.1)	0.202
MRA	55 (70.5)	10 (71.4)	45 (70.3)	0.964
Loop diuretics	71 (91.0)	13 (92.9.1)	58 (90.6)	0.795
Statin	22 (28.2)	3 (21.4)	19 (29.7)	0.842
Sodium	140 (138–143)	140 (137–141)	140 (139–143)	0.234
Potassium	4.4 ± 0.6	4.6 ± 0.8	4.4 ± 0.5	0.167
eGFR, mL/min	70.4 (49.0–92.9)	62.6 (47.0–92.9)	75.1 (49.0–94.6)	0.610
Pro BNP	1842 (526–3860)	5573 (2471–8188)	1106 (421–2826)	0.001
Troponin I	14 (8–29)	17 (13–38)	13 (8–27)	0.234
CK-MB	1.9 (1.4–3.0)	1.6 (1.3–2.6)	2.1 (1.5–3.2)	0.181
HbA1C	6.3 (6.0–6.8)	6.4 (6.0–6.8)	6.2 (6.0–6.7)	0.768
Total cholesterol	4.1 ± 1.1	3.7 ± 0.7	4.2 ± 1.2	0.137
LDL	2.6 ± 0.9	2.1 ± 0.6	2.7 ± 0.9	0.040
HDL	1.0 (0.8–1.4)	1.0 (0.8–1.3)	1.0 (0.8–1.4)	1.000
C-reactive protein	9 (9–14)	14.5 (9.5–31)	9 (9–12)	0.005

Values are expressed as mean ± standard deviation for continuous variables with a normal distribution, or as median and (interquartile range) for continuous variables with a skewed distribution and absolute value (*n*) and percentage for categorical variables. ACE, angiotensin-converting enzyme; ARB, angiotensin receptor blocker; BMI, body mass index; BNP, beta natriuretic peptide; BP, blood pressure; BSA, body surface area; CK-MB, creatine kinase myocardial band; eGFR, estimated glomerular filtration rate; HbA1c, glycated haemoglobin; HDL, high-density lipoprotein; LDL, low-density lipoprotein; MAP, mean arterial pressure; MRA, mineralocorticoid receptor antagonist; NYHA, New York Heart Association.

Late gadolinium enhancement occupied less than 50% of the left ventricle in 38 (48.7%) participants and 50–75% of the left ventricle in 5 (6.4%) participants. Late gadolinium enhancement was visualised in the basal anteroseptal, mid-inferoseptal and mid-anteroseptal segments in 85%, 70%, and 53% of participants, respectively (Figure 3). Features of left ventricular non-compaction were found in five (11.6%) participants with LGE. Furthermore, pericardial and pleural effusions were diagnosed in 23 (29.5%) and 11 (14.1%) participants in our cohort, respectively.

Table 2. Baseline cardiovascular magnetic resonance imaging parameters stratified according to all-cause mortality.

	All Patients	All-Cause Mortality	Alive	
	(*n* = 78)	(*n* = 14)	(*n* = 64)	*p*-Value
LVEF (%)	24 (18–34)	18 (13–24)	24 (18–36)	0.042
LVEDV (mL)	222 (176–297)	282 (171–313)	220 (176–297)	0.509
LVESV (mL)	175 (121–243)	230 (156–305)	168 (115–238)	0.091
LVEDVI (mL/min)	119 (95–153)	138 (96–181)	117 (95–147)	0.482
LVESVI (mL/min)	93 (63–130)	123 (89–160)	88 (63–115)	0.101
LV stroke volume (mL)	56.3 ± 17.5	50.0 ± 15.3	57.8 ± 17.7	0.132
LV stroke volume index (mL/m^2)	28.5 (22.8–35.7)	27.3 (24.1–36.6)	28.9 (22.9–35.5)	0.561
LVED wall mass (g)	131 (106–161)	156 (134–170)	123 (96–159)	0.036
LVED wall mass index (g/m^2)	74.2 (53.6–88.9)	89.4 (74.5–100.6)	73.6 (51.9–84.7)	0.025
LV total mass (g)	194.3 ± 58.7	219.2 ± 70.0	188.8 ± 55.3	0.080
LV mass index (g/m^2)	99 (85–114)	111 (90–147)	98 (82–113)	0.135
RVEDV (mL)	143 (103–201)	232 (161–316)	132 (96–178)	<0.001
RVESV (mL)	90 (58–147)	154 (127–273)	73 (55–130)	<0.001
RVEF (%)	34.2 ± 16.3	23.1 ± 12.0	36.7 ± 16.2	0.004
RVEDVI (mL/m^2)	74 (58–109)	114 (101–147)	70 (56–94)	<0.001
RVESVI (mL/m^2)	46 (32–82)	86 (74–105)	41 (30–71)	<0.001
RV stroke volume (mL)	48 (29–68)	50 (27–74)	48 (30–68)	0.889
RV ED wall mass (g)	139 ± 47.8	163.6 ± 8.7	135.2 ± 50.5	0.455
Cardiac output (L/min)	4.1 (3.2–5.3)	4.4 (3.2–5.5)	4.0 (3.1–5.3)	0.929
Cardiac index (L/min/m^2)	2.5 (2.1–3.8)	2.5 (2.4–2.7)	2.4 (2.1–4.0)	0.883
Cardiac density (g/mL)	1.0 (1.0–1.0)	1.0 (1.0–1.0)	1.0 (1.0–1.0)	0.633

Values are expressed as mean ± SD for continuous variables with a normal distribution, or median and interquartile range for continuous variables with a skewed distribution and absolute value (*n*) and percentage for categorical variables. ED = end diastolic; LV = left ventricle; LVEDV = left ventricular end diastolic volume; LVEDVI = left ventricular end diastolic volume index; LVEF = left ventricular ejection fraction; LVESV = left ventricular end systolic volume; LVESVI = left ventricular end systolic volume index; RVEDV = right ventricular end diastolic volume; RVEDVI = right ventricular end diastolic volume index; RVEF = right ventricular ejection fraction; RVESV = right ventricular end systolic volume.

Table 3. Patterns of late gadolinium enhancement in IDCM patients.

	n = 43 (%)
Midwall	28 (65)
Right ventricular insertion point (normal variant) and midwall	4 (9.3)
Right ventricular insertion point (normal variant)	3 (7)
Transmural	3 (7)
Focal	2 (4.6)
Subendocardial	2 (4.6)
Midwall and subendocardial	1 (2.3)
Midwall, subendocardial and transmural	1 (2.3)
Midwall and epicardial	1 (2.3)
Subendocardial and transmural	1 (2.3)

Values are expressed as absolute numbers and percentages.

3.3. Electrocardiogram and Echocardiogram Parameters

At the time of recruitment into the study, the median PR interval did not differ significantly between survivors and non-survivors, measuring [177 (IQR: 158–203) and 192 (IQR: 169–222), *p* = 0.156] ms, respectively. In the entire cohort, 32 (41.0%) participants had a QRS duration >110 ms. The median QRS duration was 111 (IQR: 98–148) ms in participants with LGE, and study participants who showed no evidence of LGE upon CMR imaging had a median QRS duration of 97 (IQR: 88–117) ms (*p* = 0.015).

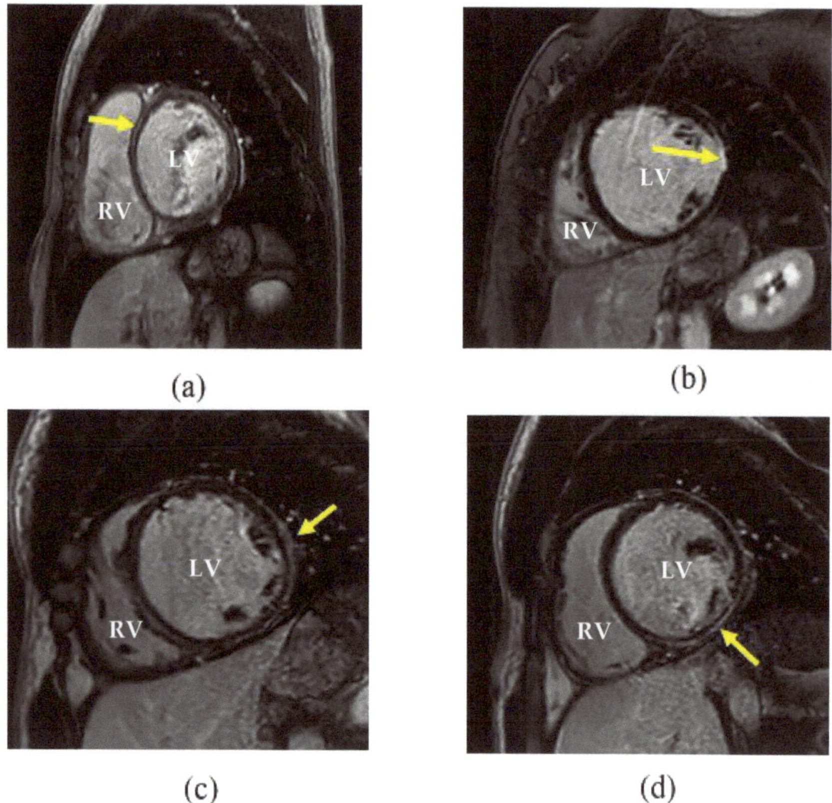

Figure 2. Late gadolinium enhancement cardiovascular magnetic resonance images showing the left and right ventricle. (**a**) Dilated cardiomyopathy with midwall enhancement of the septum (arrow); (**b**) subendocardial enhancement in the lateral ventricle free wall, typically associated with ischaemic heart disease; (**c**) shows an epicardial enhancement pattern in the lateral ventricle free wall and (**d**) inferior segment. LV = left ventricle; RV = right ventricle.

For echocardiography, the mean left atrial diameter was 43.8 ± 6.5 mm. This did not differ significantly between survivors and non-survivors (43.1 ± 6.4 vs. 46.8 ± 6.0 mm, $p = 0.05$). The mean left ventricle fractional shortening was 12.8 ± 5.7%, which differed significantly between survivors and non-survivors (13.5 ± 5.7 vs. 9.6 ± 5.0%, $p = 0.036$). The mean global longitudinal strain was −6.8 ± 4.6%, and participants that survived had a mean global longitudinal strain of −7.3 ± 4.0%. In contrast, non-survivors had a mean global longitudinal strain of −4.5 ± 6.3% ($p = 0.046$). There was no correlation between global longitudinal strain and LGE (r = 0.1173). During wall motion analysis, global hypokinesis was noted in 60 (76.9%) participants.

3.4. Follow-Up and Endpoints

After a median follow-up duration of 12 (IQR: 8.8–16.8) months, 14 (17.9%) participants died. Thromboembolic complications were uncommon and were reported in only three participants. Each participant was diagnosed with a cerebrovascular accident, lower limb deep vein thrombosis, and pulmonary embolism. Nine (11.5%) study participants were hospitalised during the follow-up period.

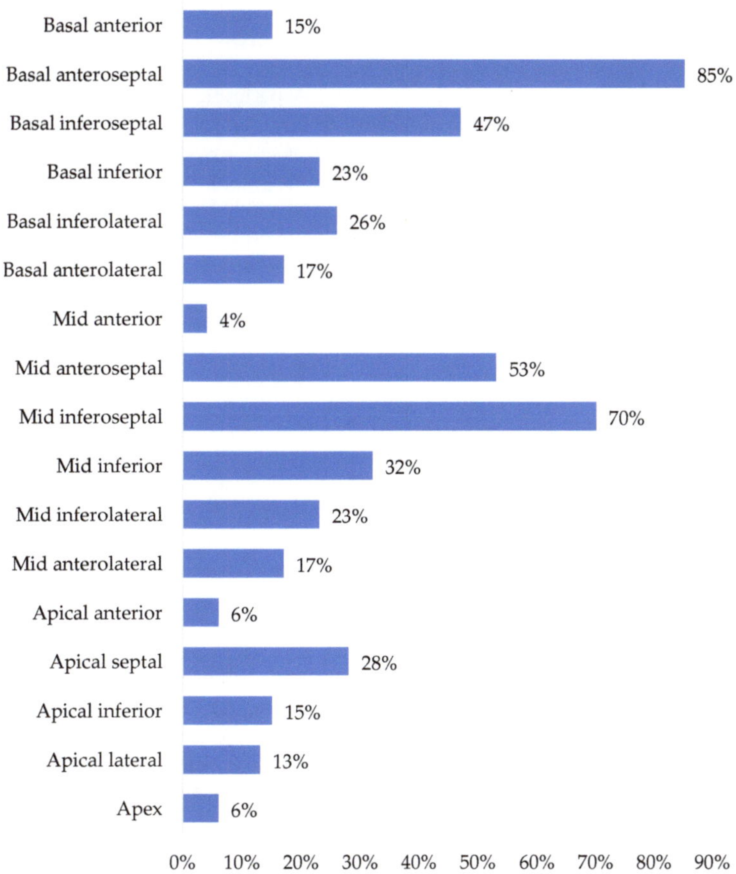

Frequency of late gadolinium enhancement in each left ventricular segment of the myocardium (%)

Figure 3. Distribution of late gadolinium enhancement in the left ventricular myocardial segments. The bars demonstrate the percentage of segments with LGE. In 85% of IDCM patients, LGE was visualised in the basal anteroseptal segment. There were 46 IDCM patients with LGE in one or more segments; therefore, the cumulative percentage will be >100%.

4. Discussion

In this study, participants with DCM were referred for CMR imaging to further exclude primary and secondary myocardial diseases. In addition, LGE was also assessed based on its prognostic value in patients with dilated cardiomyopathy. The identification of LGE in patients with non-ischaemic cardiomyopathy is associated with an increased risk of all-cause mortality, rehospitalisation, ventricular dyssynchrony, spontaneous and inducible ventricular arrhythmias, and sudden cardiac death [6–8].

More than half of the participants in our study showed evidence of LGE on CMR imaging, and 65% of the study participants presented with evidence of LGE in the midwall. Nine participants had an LGE pattern suggestive of ischaemic cardiomyopathy. However, their diagnostic angiograms showed normal epicardial coronary arteries. These findings are supported by McCrohon and colleagues, who studied 90 patients with DCM and

15 control subjects. In their study, 13% of DCM patients with unobstructed coronary arteries upon angiography presented with subendocardial and transmural LGE patterns during CMR imaging [9], implying that microvascular ischaemia may be an important cause of ischaemic LGE patterns. However, in our cohort, we did not assess the presence of microvascular disease while performing the diagnostic coronary angiograms. Furthermore, we identified isolated LGE in the right ventricular insertion point in three participants. Isolated right ventricular insertion point LGE has been reported in several studies and shown not to convey a worse prognosis [10,11].

In a meta-analysis conducted by Duan et al., evaluating the prognostic value of LGE in dilated cardiomyopathy, patients presenting with LGE upon CMR imaging had a three-fold increased risk of all-cause mortality [12]. In another systematic review, the hazard ratio for all-cause mortality in non-ischaemic cardiomyopathy patients with LGE on imaging was 2.74 (95% CI: 2.0–3.74) $p < 0.001$ [6]. Moreover, a meta-analysis by Kuruvilla and colleagues reported a higher risk of all-cause mortality in patients with non-ischaemic cardiomyopathy with LGE (odd ratio = 3.27; 95% CI: 1.94–5.51; $p < 0.00001$) [7]. In our study, survival rates did not differ significantly in participants with and without LGE, probably because of the relatively small sample size. A few studies on DCM patients with smaller sample sizes have reported a lack of association between LGE and mortality. Looi et al. studied 103 patients with DCM and found evidence of LGE in 30% of patients. After a follow-up duration of 2 years, nine deaths occurred, and there was no statistically significant difference in all-cause mortality in patients with and without LGE [13]. Moreover, the odds ratio for the occurrence of major cardiovascular events (MACEs) was 0.77 (95% CI: 0.29–1.97; $p > 0.05$), and therefore was not statistically significant. In their study, Looi and colleagues defined a MACE as the incidence of death, cardiac transplant, ventricular arrhythmias, and heart failure-related hospitalisation. They attributed their study's lack of an association between MACEs and LGE to a lower-risk patient profile. In their study, 75% of DCM patients were in NYHA class 1 and had a mean LVEF of 32 ± 12% [13]. Another study by Masci and colleagues reported a relationship between the absence of LGE and left ventricular reverse remodelling in 58 IDCM patients. Among the patients studied, eight died after two years of follow-up [14]. A higher rate of death in patients with evidence of LGE on imaging was not reported in the study by Masci et al.

Semi-quantitative and quantitative parameters generated by commercially available software for processing CMR images may be of value in improving the diagnosis and specificity of imaging findings. In our study, we used semi-quantitative indices to quantify the burden of LGE on images. This may have contributed to the lack of a linear relationship between the presence of LGE and mortality in our study. Moreover, our patients may have had diffuse fibrosis, which may be challenging to visualise in the absence of normal reference myocardium, or small microscopic fibrosis, which may be below the resolution of our scanner.

N-terminal, pro-brain natriuretic peptide (pro-BNP) levels are useful for the diagnosis of acute heart failure, the prognostication of patients with ischaemic and non-ischaemic cardiomyopathy, and for predicting the risk of cardiovascular and all-cause mortality in the general population [15,16]. In our study, pro-BNP levels were five-times higher in non-survivors, with a median of 5573 (IQR: 2471–8188) vs. 1106 (IQR: 421–2826), $p = 0.001$, suggesting that this biomarker, which is secreted in response to excessive myocyte stretching, could be a valuable marker for risk stratifying patients with IDCM.

A prolonged duration of the QRS complex, greater than 110 ms, did not independently predict mortality in our study. Unlike a study by Hombach et al., LGE in patients with idiopathic DCM who had a QRS duration >110 ms and diabetes mellitus was found to be significant predictor of cardiac death or sudden cardiac death from ventricular flutter or fibrillation [17]. Despite removing all participants with right ventricular insertion point LGE and an LGE pattern suggestive of ischaemic heart disease, none of our study's clinical and imaging variables independently predicted mortality.

Global longitudinal strain (GLS) measurements are a useful prognostic marker, since abnormalities in strain patterns are often observed, despite preserved left ventricular function. In a study involving 15 and 33 patients with ischaemic heart disease and non-ischaemic heart disease, respectively, the median GLS measurements were -15.6 (IQR: -17.9 to -11.6) and -16.0 (IQR: -19.1 to -12.7) in patients with and without LGE on imaging, respectively ($p = 0.212$) [18]. In that study, which Erley and colleagues conducted, the global circumferential strain (GSC) had a stronger association with LGE than GLS, with an area under the receiver operating characteristic curve of 0.77–078 vs. 0.67–0.72 [18]. In our study, there was no correlation between GLS generated using speckle tracking during echocardiography and the presence of LGE during CMR imaging, probably because of the limited sample size.

Cardiac tissue characterisation using CMR imaging T1 mapping and the extracellular volume fraction in DCM patients are two of the most useful techniques for detecting interstitial fibrosis [19,20]. Late gadolinium enhancement is a valuable surrogate for replacement fibrosis in the heart, whereas T1 mapping and ECV fraction can reliably demonstrate the presence of interstitial fibrosis, a common finding in patients with DCM [20]. In fibrotic hearts, both native T1 and ECV on the pre-contrast map are prolonged, and the post-contrast map shows a shortened T1 time due to the accumulation of gadolinium in the extracellular space [21,22]. The use of dedicated software for T1 mapping and the ECV fraction in our DCM cohort may have played a significant role in the risk stratification and prognostication of our patients [23].

This study had several limitations. This was a single-centre observational study with a relatively small sample size. Therefore, this study was potentially unsuitable for determining if certain clinical variables could predict all-cause mortality in our IDCM cohort. A single radiologist analysed and reported CMR images. Adding a second radiologist and comparing the CMR imaging findings would have improved the diagnostic yield and assisted in measuring inter-observer coefficients. The intensity of LGE and the proportion of the wall thickness affected by LGE were not assessed. The study follow-up period was relatively short. Furthermore, our study had less female representation, as several females with DCM detected in the peripartum period were excluded. The CMR extracellular volume fraction and strain patterns were not analysed as our reporting software did not have these capabilities.

5. Conclusions

In this study of a carefully phenotyped cohort of Africans with IDCM, midwall enhancement was the most common pattern, visualised in 65% of participants. Prospective, adequately powered, multi-centre studies across sub-Saharan Africa are required to determine the prognostic significance of LGE on CMR imaging parameters such as LGE, ECV, and strain patterns in an African IDCM cohort.

Author Contributions: Conceptualisation, P.M. and N.T.; methodology, P.M. and N.T.; software, N.T., D.M., A.d.P. and L.S.; validation, N.T. and P.M.; formal analysis, D.M. and N.T. investigation, N.T. and D.M.; data curation, N.T. writing—original draft preparation, N.T.; writing—review and editing, N.T., D.M., P.M., Q.W., L.S., A.d.P. and A.V.; visualisation, N.T. and D.M.; supervision, P.M.; project administration and funding acquisition, N.T. All authors have read and agreed to the published version of the manuscript.

Funding: The Genetics of idiopathic dilated cardiomyopathy in Johannesburg Study is funded by the Carnegie Corporation of New York Grant No. B8749.RO1, Discovery Foundation Academic Fellowship Award, South African Heart Association Research Scholarship Award, and the South African Medical Association PhD Supplementary Scholarship. None of the funders played a role in the design of the study and collection, analysis, and interpretation of data and in writing the manuscript.

Institutional Review Board Statement: The study complied with the Declaration of Helsinki and informed consent was obtained from the study participants. Approval to conduct the study was granted by the University of the Witwatersrand Human Research Ethics Committee (certificate number: M150467).

Informed Consent Statement: Written informed consent was obtained from all subjects involved in the study.

Data Availability Statement: Data are unavailable due to ethical restrictions.

Acknowledgments: The authors thank the patients and their families for participating in this study. We also acknowledge the significant role of staff members at the Charlotte Maxeke Johannesburg Academic Hospital, Division of Cardiology.

Conflicts of Interest: Nqoba Tsabedze is a cardiologist and has received consultation fees from Acino Health Care Group, AstraZeneca, Boston Scientific, Novartis Pharmaceuticals, Novo Nordisk, Pfizer, Sanofi, Phillips, Servier, Takeda, and Merck. He has also received educational and travel grants from Medtronic, Biotronik, Boston Scientific and Vertice Health Care Group. None of the other authors have any relevant financial or professional disclosures.

References

1. Agbor, V.N.; Essouma, M.; Ntusi, N.A.B.; Nyaga, U.F.; Bigna, J.J.; Noubiap, J.J. Heart failure in sub-Saharan Africa: A contemporaneous systematic review and meta-analysis. *Int. J. Cardiol.* **2018**, *257*, 207–215. [CrossRef] [PubMed]
2. Agbor, V.N.; Ntusi, N.A.B.; Noubiap, J.J. An overview of heart failure in low- and middle-income countries. *Cardiovasc. Diagn. Ther.* **2020**, *10*, 244–251. [CrossRef] [PubMed]
3. Pinto, Y.M.; Elliott, P.M.; Arbustini, E.; Adler, Y.; Anastasakis, A.; Böhm, M.; Duboc, D.; Gimeno, J.; de Groote, P.; Imazio, M.; et al. Proposal for a revised definition of dilated cardiomyopathy, hypokinetic non-dilated cardiomyopathy, and its implications for clinical practice: A position statement of the ESC working group on myocardial and pericardial diseases. *Eur. Heart J.* **2016**, *37*, 1850–1858. [CrossRef] [PubMed]
4. Hazebroek, M.; Dennert, R.; Heymans, S. Idiopathic dilated cardiomyopathy: Possible triggers and treatment strategies. *Neth. Heart J. Mon. J. Neth. Soc. Cardiol. Neth. Heart Found.* **2012**, *20*, 332–335. [CrossRef]
5. Arnold, J.R.; McCann, G.P. Cardiovascular magnetic resonance: Applications and practical considerations for the general cardiologist. *Heart* **2020**, *106*, 174–181. [CrossRef]
6. Ganesan, A.N.; Gunton, J.; Nucifora, G.; McGavigan, A.D.; Selvanayagam, J.B. Impact of Late Gadolinium Enhancement on mortality, sudden death and major adverse cardiovascular events in ischemic and nonischemic cardiomyopathy: A systematic review and meta-analysis. *Int. J Cardiol.* **2018**, *254*, 230–237. [CrossRef]
7. Kuruvilla, S.; Adenaw, N.; Katwal, A.B.; Lipinski, M.J.; Kramer, C.M.; Salerno, M. Late Gadolinium Enhancement on Cardiac Magnetic Resonance Predicts Adverse Cardiovascular Outcomes in Nonischemic Cardiomyopathy. *Circ. Cardiovasc. Imaging* **2014**, *7*, 250–258. [CrossRef]
8. Tigen, K.; Karaahmet, T.; Kirma, C.; Dundar, C.; Pala, S.; Isiklar, I.; Cevik, C.; Kilicgedik, A.; Basaran, Y. Diffuse late gadolinium enhancement by cardiovascular magnetic resonance predicts significant intraventricular systolic dyssynchrony in patients with non-ischemic dilated cardiomyopathy. *J. Am. Soc. Echocardiogr.* **2010**, *23*, 416–422. [CrossRef]
9. McCrohon, J.A.; Moon, J.C.; Prasad, S.K.; McKenna, W.J.; Lorenz, C.H.; Coats, A.J.; Pennell, D.J. Differentiation of heart failure related to dilated cardiomyopathy and coronary artery disease using gadolinium-enhanced cardiovascular magnetic resonance. *Circulation* **2003**, *108*, 54–59. [CrossRef]
10. Grigoratos, C.; Pantano, A.; Meschisi, M.; Gaeta, R.; Ait-Ali, L.; Barison, A.; Todiere, G.; Festa, P.; Sinagra, G.; Aquaro, G.D. Clinical importance of late gadolinium enhancement at right ventricular insertion points in otherwise normal hearts. *Int. J. Cardiovasc. Imaging* **2020**, *36*, 913–920. [CrossRef]
11. Yi, J.E.; Park, J.; Lee, H.J.; Shin, D.G.; Kim, Y.; Kim, M.; Kwon, K.; Pyun, W.B.; Kim, Y.J.; Joung, B. Prognostic implications of late gadolinium enhancement at the right ventricular insertion point in patients with non-ischemic dilated cardiomyopathy: A multicenter retrospective cohort study. *PLoS ONE* **2018**, *13*, e0208100. [CrossRef]
12. Duan, X.; Li, J.; Zhang, Q.; Zeng, Z.; Luo, Y.; Jiang, J.; Chen, Y. Prognostic value of late gadolinium enhancement in dilated cardiomyopathy patients: A meta-analysis. *Clin. Radiol.* **2015**, *70*, 999–1008. [CrossRef]
13. Looi, J.L.; Edwards, C.; Armstrong, G.P.; Scott, A.; Patel, H.; Hart, H.; Christiansen, J.P. Characteristics and prognostic importance of myocardial fibrosis in patients with dilated cardiomyopathy assessed by contrast-enhanced cardiac magnetic resonance imaging. *Clin. Med. Insights Cardiol.* **2010**, *4*, 129–134. [CrossRef] [PubMed]
14. Masci, P.G.; Schuurman, R.; Andrea, B.; Ripoli, A.; Coceani, M.; Chiappino, S.; Todiere, G.; Srebot, V.; Passino, C.; Aquaro, G.D.; et al. Myocardial fibrosis as a key determinant of left ventricular remodeling in idiopathic dilated cardiomyopathy: A contrast-enhanced cardiovascular magnetic study. *Circ. Cardiovasc. Imaging* **2013**, *6*, 790–799. [CrossRef] [PubMed]
15. Kim, H.; Cho, Y.K.; Jun, D.H.; Nam, C.W.; Han, S.W.; Hur, S.H.; Kim, Y.N.; Kim, K.D. Prognostic implications of the NT-ProBNP level and left atrial size in non-ischemic dilated cardiomyopathy. *Circ. J.* **2008**, *72*, 1658–1665. [CrossRef]
16. Geng, Z.; Huang, L.; Song, M.; Song, Y. N-terminal pro-brain natriuretic peptide and cardiovascular or all-cause mortality in the general population: A meta-analysis. *Sci. Rep.* **2017**, *7*, 41504. [CrossRef]

17. Hombach, V.; Merkle, N.; Torzewski, J.; Kraus, J.M.; Kunze, M.; Zimmermann, O.; Kestler, H.A.; Wohrle, J. Electrocardiographic and cardiac magnetic resonance imaging parameters as predictors of a worse outcome in patients with idiopathic dilated cardiomyopathy. *Eur. Heart J.* **2009**, *30*, 2011–2018. [CrossRef] [PubMed]
18. Erley, J.; Genovese, D.; Tapaskar, N.; Alvi, N.; Rashedi, N.; Besser, S.A.; Kawaji, K.; Goyal, N.; Kelle, S.; Lang, R.M.; et al. Echocardiography and cardiovascular magnetic resonance based evaluation of myocardial strain and relationship with late gadolinium enhancement. *J. Cardiovasc. Magn. Reson.* **2019**, *21*, 46. [CrossRef]
19. Haaf, P.; Garg, P.; Messroghli, D.R.; Broadbent, D.A.; Greenwood, J.P.; Plein, S. Cardiac T1 Mapping and Extracellular Volume (ECV) in clinical practice: A comprehensive review. *J. Cardiovasc. Magn. Reson.* **2016**, *18*, 89. [CrossRef]
20. Nakamori, S.; Dohi, K.; Ishida, M.; Goto, Y.; Imanaka-Yoshida, K.; Omori, T.; Goto, I.; Kumagai, N.; Fujimoto, N.; Ichikawa, Y.; et al. Native T1 Mapping and Extracellular Volume Mapping for the Assessment of Diffuse Myocardial Fibrosis in Dilated Cardiomyopathy. *JACC Cardiovasc. Imaging* **2018**, *11*, 48–59. [CrossRef]
21. Puntmann, V.O.; Carr-White, G.; Jabbour, A.; Yu, C.Y.; Gebker, R.; Kelle, S.; Hinojar, R.; Doltra, A.; Varma, N.; Child, N.; et al. T1-Mapping and Outcome in Nonischemic Cardiomyopathy: All-Cause Mortality and Heart Failure. *JACC Cardiovasc. Imaging* **2016**, *9*, 40–50. [CrossRef] [PubMed]
22. Dass, S.; Suttie, J.J.; Piechnik, S.K.; Ferreira, V.M.; Holloway, C.J.; Banerjee, R.; Mahmod, M.; Cochlin, L.; Karamitsos, T.D.; Robson, M.D.; et al. Myocardial tissue characterization using magnetic resonance noncontrast t1 mapping in hypertrophic and dilated cardiomyopathy. *Circ. Cardiovasc. Imaging* **2012**, *5*, 726–733. [CrossRef] [PubMed]
23. Li, S.; Zhou, D.; Sirajuddin, A.; He, J.; Xu, J.; Zhuang, B.; Huang, J.; Yin, G.; Fan, X.; Wu, W.; et al. T1 Mapping and Extracellular Volume Fraction in Dilated Cardiomyopathy: A Prognosis Study. *JACC Cardiovasc. Imaging* **2022**, *15*, 578–590. [CrossRef] [PubMed]

Article

Predictive Value of Cardiac Magnetic Resonance for Left Ventricular Remodeling of Patients with Acute Anterior Myocardial Infarction

Wenkun Ma [1], Xinni Li [1], Chengjie Gao [2], Yajie Gao [1], Yuting Liu [1], Sang Kang [1] and Jingwei Pan [1,*]

[1] Department of Cardiovasology, Shanghai Sixth People's Hospital Affiliated to Shanghai Jiao Tong University School of Medicine, Shanghai 200233, China
[2] Department of Geriatrics, Shanghai Sixth People's Hospital Affiliated to Shanghai Jiao Tong University School of Medicine, Shanghai 200233, China
* Correspondence: jwpan@sjtu.edu.cn; Tel.: +86-021-64369181-8411

Citation: Ma, W.; Li, X.; Gao, C.; Gao, Y.; Liu, Y.; Kang, S.; Pan, J. Predictive Value of Cardiac Magnetic Resonance for Left Ventricular Remodeling of Patients with Acute Anterior Myocardial Infarction. *Diagnostics* **2022**, *12*, 2780. https://doi.org/10.3390/diagnostics12112780

Academic Editors: Minjie Lu and Arlene Sirajuddin

Received: 27 October 2022
Accepted: 9 November 2022
Published: 14 November 2022

Publisher's Note: MDPI stays neutral with regard to jurisdictional claims in published maps and institutional affiliations.

Abstract: Background: Heart failure is a serious complication resulting from left ventricular remodeling (LVR), especially in patients experiencing acute anterior myocardial infarction (AAMI). It is crucial to explore the predictive parameters for LVR following primary percutaneous coronary intervention (PPCI) in patients with AAMI. Methods: A total of 128 AAMI patients who were reperfused successfully by PPCI were enrolled sequentially from June 2018 to December 2019. Cardiovascular magnetic resonance (CMR) was performed at the early stage (<7 days) and after the 6-month follow-up. The patients were divided into LVR and non-LVR groups according to the increase of left ventricular end diastolic volume (LVEDV) measured by the second cardiac magnetic resonance examination ≥20% from baseline. (3) Results: The left ventricular ejection fraction (LVEF), the global longitudinal strain (GLS), the peak circumferential strain in infarcted segments, and the infarct size (IS) remained significantly different in the multivariate logistic regression analysis (all $p < 0.05$). The area under the receiver operating characteristic curve of Model 1, wherein the GLS was added to the LVEF, was 0.832 (95% CI 0.758–0.907, $p < 0.001$). The C-statistics for Model 2, which included the infarct-related regional parameters (IS and the peak circumferential strain in infarcted segments)was 0.917 (95% CI 0.870–0.965, $p < 0.001$). Model 2 was statistically superior to Model 1 in predicting LVR (IDI: 0.190, $p = 0.002$). (4) Conclusions: Both the global and regional CMR parameters were valuable in predicting LVR in patients with AAMI following the PPCI. The local parameters of the infarct zones were superior to those of the global ones.

Keywords: acute myocardial infarction; cardiac magnetic resonance; left ventricular remodeling; strain

1. Introduction

Acute myocardial infarction (AMI) can lead to irreversible myocardial necrosis. Although the mortality at the acute stage has been notably reduced by primary percutaneous coronary intervention (PPCI), the instances of long-term adverse events, especially chronic heart dysfunction, remain high [1]. Necrotic myocardium impairs cardiac contractile function with a decreased ejection fraction and compensatory cardiac dilatation, which attributes to refractory heart failure, especially acute anterior myocardial infarction (AAMI). AAMI is caused by the acute block of the left anterior descending coronary artery (LAD), which perfuses the anterior septum and anterior wall of the left ventricle (LV). Under the similar myocardial salvage, the patients with AAMI have a larger infarcted size than the non-AAMI patients do [2], which more likely leads to adverse left ventricular remodeling (LVR). Predicting LVR after AAMI is crucial because LVR is closely related to an adverse prognosis.

The previous risk stratification of AMI, such as Global Registry of Acute Coronary Events (GRACE) and Thrombolysis in Myocardial Infarction (TIMI) scores [3,4], which are established before a routine PPCI, provide limited indicative power in patients with AMI that is treated with PPCI. The parameters that are included in the previous risk stratification cannot precisely quantify the damage of the LV. Cardiovascular Magnetic Resonance (CMR) has been conclusively demonstrated as an accurate and effective method for identifying patients who are suffering from AMI with poor clinical outcomes [5,6]. CMR can provide a comprehensive assessment of the functional and structural damage that is caused by AMI [7]. Standard parameters such as the left ventricular ejection fraction (LVEF) and the infarct size based on gadolinium-enhanced sequences in the early stages of AMI are valuable prognostic factors [8]. While the myocardial strain is another informative parameter that can reveal cardiac deformation both globally and within the MI zone [9,10]. Early studies have explored the CMR multiparameter risk score as a prognostic indicator [11,12]. Moreover, it is still unclear whether adverse LVR is associated with the loss of viable myocardium in the MI zone or excessive compensation from the non-MI region. Which factors are more effective in predicting the outcomes in patients with AAMI: the global parameters or the regional ones? The global parameters consider the integrated function of the left ventricle, while the regional parameters are focused on the influence of the infarct zones [13]. Hence, we conducted this study to explore the delineated question.

2. Patients and Methods

2.1. Study Population and Data Collection

This prospective study was performed between June 2018 and December 2019 at the cardiology center of the Shanghai Sixth People's Hospital. The enrolled patients were diagnosed with AAMI and treated with PPCI. The study inclusion criteria were as follows: (1) AMI which was diagnosed according to the AMI diagnostic criteria [14] (i.e., clinical symptoms of myocardial ischemia lasting <12 h, ST segment elevation >0.2 mV in ≥2 adjacent precordial leads, and/or an acute myocardial biomarker serum troponin level >99% of the upper limit of the normal value); (2) PPCI which was performed within 12 h after the onset of chest pain (the PPCI window was extended to 36 h if the patients had endured hemodynamic instability, such as in cardiogenic shock); (3) the culprit coronary artery that caused myocardial infarction was the LAD. Study exclusion criteria were as follows: (1) non-ST segment elevation myocardial infarction or previous myocardial infarction; (2) active myocarditis or cardiomyopathy (i.e., dilated cardiomyopathy, hypertrophic cardiomyopathy, or restrictive cardiomyopathy); (3) valvular disease of moderate or greater severity; (4) cardiac pacemaker implantation; (5) malignant arrhythmia; (6) chronic obstructive pulmonary disease, severe anemia, malnutrition, severe kidney dysfunction, and/or an estimated glomerular filtration rate ≤30 mL/min/1.73 m^2; (7) claustrophobia or magnetic resonance imaging contraindications (such as an allergy to the contrast agent).

The patients underwent CMR examinations during both the early phase following AAMI (i.e., within 7 days of the onset) and after the six-month follow-up visit. Demographic data and medical history were collected from the patients in the study. The collected data included information on laboratory examinations, coronary vascular disease risk factors, concomitant diseases, and medicines. Ethical approval for this study was obtained from Shanghai Jiao Tong University which is affiliated to the Shanghai Sixth People's Hospital (2017-KY-003 (K)). All of the patients provided written informed consent.

2.2. Definition of LVR

The definition of LVR was an increase in left ventricular end-diastolic volume (LVEDV) at $\geq20\%$ over approximately six months of follow-ups (ΔLVEDV% $\geq$ 20%) [13].

2.3. CMR Technique

The ECG-gated CMR examinations were performed using a 3-T system (Philips Healthcare, Best, The Netherlands) within seven days of the onset and at the six-month follow-up visit. The CMR protocol included a localizer, T2-weighted, cine, first-pass perfusion, and delay-enhancement sequences. Each long-axis slice was subjected to balanced, steady-state free precision (bSSFP) cines (with a 60° separation around the long axis of the LV (two, three, and four chambers)). A total of 10–12 continuous slices of short-axis cines were obtained after the injection of the contrast agent (gadolinium, 0.2 mmol/kg), covering the entire LV region from the ring of the mitral valve to the apex (parallel slices 8 mm wide without gap; TE = 1.6 ms, TR = 3.2 ms, flip angle 45, voxel size $2.0 \times 1.6 \times 8$ mm^3, field of view 350×350 mm; 30 phases in each cardiac cycle). The gadolinium enhancement sequences were acquired 10 min after the contrast injection (Magnevist, Bayer Healthcare, Berlin, Germany).

2.4. CMR Analysis

The LVEF, LV mass (LVM), LV end-diastolic volume (LVEDV), and LV end-systolic volume (LVESV) were analyzed using standardized protocols. The LVM was derived by subtracting the papillary muscles at the end diastole [15]. CVI42 (Circle Cardiovascular Imaging, Calgary, AB, Canada) was used for CMR feature tracking analysis of the bSSFP cine images. At the end diastole, the borders of the LV endocardium and epicardium were automatically tracked and manually corrected in the short-axis series with three long-axis slices. The long-axis cines were used to extract the global longitudinal strain (GLS) of the LV, whereas the short-axis cines were used to calculate the global circumferential strain (GCS) and global radial strain (GRS) [16,17]. Using the 16-segment American Heart Association model, the segments containing positive late gadolinium enhancement (LGE) were defined as infarct segments. The remote non-infarcted segments were identified as unenhanced segments which were separated from the infarct segments by a single unenhanced border segment. The regional peak strains were defined as the average of each segmental peak value [18] (Figure 1). The quantification of the infarct size (IS) was performed using LGE images which were obtained from the short axis. The IS was delineated using a semiautomated computer-aided threshold detection protocol (>5 standard deviations [SDs] of remote myocardium) and was calculated by dividing the sum of infarct size from all of the sections by the mass of LV areas (including those without infarct scar) and multiplying by 100.

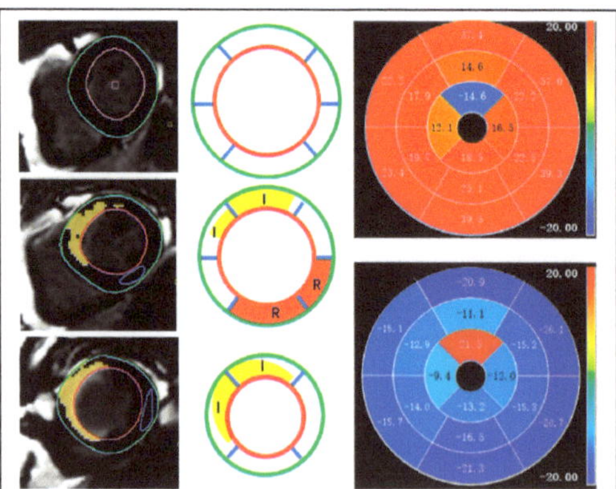

Figure 1. According to late gadolinium enhancement, the infarcted segments and remote non-infarcted segments were defined. Red and green were used to outline the endocardial and epicardial contours of the left ventricle, respectively. Myocardium was divided into infarcted (I, yellow) and remote non-infarcted segments (R, red) based on the American Heart Association 16-segment mode after the healthy myocardium was delineated by a blue circle. Remote non-infarcted segments were defined as unenhanced zone with one unenhanced border segments between them and infarcted segments.

2.5. Angiographic Assessments

The angiographic analysis was performed following standard protocols, including those that have been delineated for TIMI [19]. All of the images were reviewed by a qualified interventional cardiologist (Prof. PAN).

2.6. Reproducibility Analysis

The data from 10 LVR patients and 10 non-LVR subjects were sampled to evaluate the inter- and intra-observer variability. The CMR analyses were performed independently by two cardiologists (Dr. MA and Dr. GAO) who were blinded to each other's recordings. The inter-observer variability was tested using the data from different acquisitions. The observers reanalyzed their own recordings to check for intra-observer variability; these evaluations were conducted at 2 weeks apart.

2.7. Statistical Analysis

For the continuous variables, the means $\pm$ SD or medians and interquartile ranges (i.e., 25th to 75th percentiles) are reported, and the variable distributions were evaluated using the Kolmogorov–Smirnov test. Wilcoxon rank-sum tests was used for inter-group comparisons of the non-normally distributed variables, whereas the *t*-tests were used for inter-group comparisons of the normally distributed continuous variables. Fisher's exact tests or chi-square tests were used to compare the categorical variables. The results are displayed as numbers and percentages. Moreover, univariate and multivariate logistic regression analyses were conducted to identify the prognostic predictors of LVR. A receiver operating characteristic (ROC) curve analysis was used to identify the parameters that were best suited for checking the LVR model. The optimal cut-off values were determined according to the maximum Youden index. The Delong test was used to compare the areas under the ROC curves (AUCs). We also calculated integrated discrimination improvement (IDI) by comparing global and regional parameters in order to find the strongest predictors of LVR following AAMI. The model calibration was assessed using the Hosmer–Lemeshow goodness-of-fit test, and the inter- and intra-observer reproducibility were determined using intraclass correlation coefficients. Standard statistical software (i.e., SPSS Statistics Version 26.0 (IBM Corp. Armonk, NY, USA), R version 4.2.0 (R Foundation for Statistical Computing, Vienna, Austria), and GraphPad Prism 9 (GraphPad Software, San Diego, CA, USA)) were used to perform all of the calculations. A two-sided *p*-value of <0.05 was considered to be the threshold for statistical significance.

3. Results

3.1. Patient Characteristics

Of the 128 included study participants, five of them were excluded due to them having a poor CMR image quality, and seven patients did not complete the follow-up. Consequently, 116 patients with AAMI were included in the final analysis; these patients were divided into an LVR group (n = 39) and a non-LVR group (n = 77). The average age was 58.22 $\pm$ 12.30 years, and 86.20% of the included patients were male. The patients in the LVR group had a higher serum level of peak hypersensitive cardiac troponin I (hs-cTnI). There were no statistically significant differences in the sex, age, heart rate, body mass index, pain to balloon time, coronary heart disease risk factors, Killip classification at the point of admission, peak pro-brain natriuretic peptide (pro-BNP) levels, TIMI flow pre-PPCI, TIMI flow after PPCI, and medication use between the study groups (all p > 0.05). The demographic and clinical characteristics of the included patients with AAMI are shown in Table 1.

Table 1. Demographic and clinical characteristics of the included patients with AAMI.

	All (n = 116)	Non-LVR (n = 77)	LVR (n = 39)	*p*-Value
Age, year	58.22 ± 12.30	56.96 ± 12.95	60.72 ± 10.62	0.121
Male, n (%)	100 (86.20)	67 (87.00)	33(84.60)	0.724
BMI, kg/m^2	23.52 ± 4.24	23.07 ± 4.87	24.40 ± 2.44	0.110
Heat rate, bpm	72.75 ± 8.57	72.26 ± 8.25	73.72 ± 9.20	0.389
Pain to balloon time, h	11.12 (7.13,15.20)	10.67 (7.51,14.56)	12.02 (5.75,16.20)	0.916
CHD risk factors, n (%)				
Smoking	65 (56.00)	43 (54.80)	22 (56.40)	0.954
Hypertension	59 (50.90)	39 (50.60)	20 (51.30)	0.949
Hyperlipidemia	44 (37.9)	29 (37.70)	15 (38.50)	0.933
Diabetes	53 (45.70)	34 (44.20)	19 (48.70)	0.641
Killip classification on admission, n (%)				0.122
1	95 (81.90)	67 (87.00)	28 (71.80)	
2	16 (13.80)	8 (10.40)	8 (20.50)	
3–4	5 (4.30)	2 (2.60)	3 (7.70)	
Peak hs-cTnI, ug/L	48.15 ± 28.08	38.40 ± 26.30	67.40 ± 20.74	**<0.001**
Peak pro-BNP, ng/L	799.30 (476.00,1796.50)	724.30 (427.35,1628.00)	1066.00 (561.10,2400.50)	0.168
TIMI flow pre-PPCI, n (%)				0.943
0	71 (61.20)	46 (59.70)	25 (64.10)	
1–2	28 (24.20)	19 (24.70)	9 (23.10)	
3	17 (14.70)	12 (15.60)	5 (12.80)	
TIMI flow post-PPCI, n (%)				0.123
0–2	11 (9.50)	5 (6.50)	6 (15.40)	
3	105 (90.5)	72 (93.50)	33 (84.60)	
Medication, n (%)				
Statin	101 (87.10)	66 (85.70)	35 (89.70)	0.541
β-blocker	110 (94.80)	73 (94.80)	37 (94.90)	0.988
ACEI/ARB	54 (46.60)	35 (45.50)	19 (48.70)	0.739
Diuretic	54 (46.60)	34 (44.20)	20 (51.30)	0.467
ARNI	31 (26.70)	18 (23.40)	13 (33.30)	0.252
Nitrates	36 (31.00)	24 (31.20)	12 (30.80)	0.965

AAMI, acute anterior myocardial infarction; LVR, left ventricular remodeling; BMI, body mass index; CHD, coronary heart disease; hs-cTnI, hypersensitive serum cardiac troponin I; pro-BNP, pro-brain natriuretic peptide; TIMI, thrombolysis in myocardial infarction; PPCI, primary percutaneous coronary intervention; ACEI, angiotensin converting enzyme inhibitor; ARB, angiotensin II receptor blocker; ARNI, angiotensin receptor-neprilysin inhibitor; *p*-values of factors with bold values are less than 0.05.

3.2. CMR Parameter Analysis

In this study, the initial CMR examinations were performed at 3 days (interquartile range, IQR: 2–5 days) after the AAMI onset, and the follow-up CMR tests were conducted at an average of 10 months (IQR: 8–12 months) after the AAMI onset. Compared with the non-LVR group, the patients who developed LVR presented with a lower LVEF and a more extensive IS. The GLS and the GCS of LV were statistically significantly decreased in the LVR group, while there was no difference in LVEDV, LVESV, and GRS. Using a segment strain analysis, the peak radial strain in infarcted segments (RS$_{infarct}$) and the peak circumferential strain in infarcted segments (CS$_{infarct}$) were statistically significantly lower in the LVR group ($p < 0.001$). There were no differences in the peak circumferential strain in remote non-infarcted segments (CS$_{remote}$) or the peak radial strain in remote non-infarcted segments (RS$_{remote}$) between the study groups. The CMR characteristics are listed in Table 2.

Table 2. Intergroup comparison of CMR indexes in patients with AAMI.

	All (n = 116)	Non-LVR (n = 77)	LVR (n = 39)	*p*-Value
LVEDV, mL	155.46 ± 32.95	156.15 ± 36.17	154.12 ± 25.79	0.756
LVESV, mL	83.84 ± 21.36	81.15 ± 22.22	89.15 ± 18.69	0.056
LVEF, %	46.32 ± 6.35	48.34 ± 5.87	42.31 ± 5.32	**<0.001**
GLS, %	−10.15 ± 3.13	−11.26 ± 2.58	−7.96 ± 2.98	**<0.001**
GCS, %	−14.09 ± 3.58	−15.10 ± 3.28	−12.08 ± 3.35	**<0.001**
GRS, %	20.36 ± 5.42	21.03 ± 5.80	19.05 ± 4.38	0.063
Peak $CS_{infarct}$,%	−7.34 ± 3.88	−8.98 ± 3.42	−4.09 ± 2.43	**<0.001**
Peak CS_{remote},%	−17.10 ± 3.39	−17.00 ± 3.20	−17.32 ± 3.78	0.634
Peak $RS_{infarct}$,%	11.61 ± 6.52	14.17 ± 6.14	6.57 ± 3.73	**<0.001**
Peak RS_{remote},%	27.60 ± 7.01	27.78 ± 6.67	27.26 ± 7.73	0.710
IS, %LVMM	20.24 ± 8.91	17.30 ± 8.75	26.05 ± 5.91	**<0.001**

CMR, cardiovascular magnetic resonance; AAMI, acute anterior myocardial infarction; LVR, left ventricular remodeling; LVEDV, left ventricular end-diastolic volume; LVESV, left ventricular end-systolic volume; LVEF, left ventricular ejection fraction; GLS, global longitudinal strain; GCS, global circumferential strain; GRS, global radial strain; $CS_{infarct}$, circumferential strain in infarcted segments; CS_{remote}, circumferential strain in remote non-infarcted segments; $RS_{infarct}$, radial strain in infarcted segments; RS_{remote}, radial strain in remote non-infarcted segments; IS, infarct size; LVMM, left ventricular myocardial mass. Significant *p*-values ($p < 0.05$) marked in bold.

3.3. Cardiovascular Outcomes

The results of univariate and multivariate logistic regression analyses (i.e., in regard to the predictors of LVR) are presented in Table 3. In the univariate analysis, the predictors associated with LVR occurrence were as follows: peak hs-cTnI, LVEF, GLS, GCS, peak $CS_{infarct}$, peak $RS_{infarct}$, and IS (all $p < 0.001$). The peak $CS_{infarct}$ was highly correlated with the GCS and peak $RS_{infarct}$, (r = 0.72, $p = 0.006$; r = 0.805, $p < 0.001$). Therefore, the multivariable models included the following parameters: LVEF, GLS, peak $CS_{infarct}$, and IS. After testing, these four parameters (LVEF, GLS, $CS_{infarct}$, and IS) independently predicted the occurrence of LVR in patients with AAMI ($p < 0.05$).

Table 3. Logistic regression analysis of predictors for LVR.

Parameters	Univariate Analysis OR (95%CI)	*p*-Value	Multivariate Analysis OR (95%CI)	*p*-Value
Sex, n (%)				
Male	Reference			
Female	1.218 (0.386, 3.575)	0.724		
Age, year	1.026 (0.994, 1.062)	0.122		
BMI, kg/m^2	1.103 (0.989, 1.260)	0.116		
HR, bpm	1.020 (0.975, 1.069)	0.386		
Pain to balloon time, h	1.006 (0.949, 1.065)	0.839		
CHD risk factors, n (%)				
Hypertension	1.026 (0.474,2.226)	0.949		
Hyperlipidemia	1.034 (0.463,2.277)	0.933		
Smoker	1.023 (0.471,2.242)	0.954		
Diabetes	1.201 (0.553,2.610)	0.641		
Killip classification on admission, n (%)				
1	Reference			
2	2.393 (0.806, 7.132)	0.112		
3–4	3.569 (0.566, 28.381)	0.174		
TIMI flow post-PPCI				
0–2	2.618 (0.739,9.678)	0.133		
3	Reference			
Peak pro-BNP, ng/L	1.000 (0.999,1.000)	0.523		
Peak hs-cTnI, ug/L	1.045 (1.027,1.067)	**<0.001**		

Table 3. *Cont.*

Parameters	Univariate Analysis		Multivariate Analysis	
	OR (95%CI)	*p*-Value	OR (95%CI)	*p*-Value
CMR parameters				
LVEF, mL	0.839 (0.771,0.903)	**<0.001**	0.882 (0.777, 0.990)	**0.038**
GLS, %	1.634 (1.354,2.043)	**<0.001**	1.347 (1.031,1.818)	**0.039**
GCS, %	1.339 (1.168,1.572)	**<0.001**		
GRS, %	0.931 (0.861,1.003)	0.062		
Peak $CS_{infarct}$, %	1.873 (1.499,2.419)	**<0.001**	1.726 (1.365, 2.354)	**<0.001**
Peak CS_{remote}, %	0.972 (0.865,1.090)	0.631		
Peak $RS_{infarct}$, %	0.694 (0.582,0.794)	**<0.001**		
Peak RS_{remote}, %	0.989 (0.935,1.046)	0.707		
IS, %LVMM	1.156 (1.090,1.241)	**<0.001**	1.098 (1.016,1.202)	**0.027**

LVR, left ventricular remodeling; BMI, body mass index; HR, heart rate; CHD, coronary heart disease; TIMI, thrombolysis In myocardial infarction; PPCI, primary percutaneous coronary intervention; pro-BNP, pro-brain natriuretic peptide; hs-cTnI, hypersensitive serum cardiac troponin I; CMR, cardiovascular magnetic resonance; LVEF, left ventricular ejection fraction; GLS, global longitudinal strain; GCS, global circumferential strain; GRS, global radial strain; $CS_{infarct}$, circumferential strain in infarcted segments; CS_{remote}, circumferential strain in remote non-infarcted segments; $RS_{infarct}$, radial strain in infarcted segments; RS_{remote}, radial strain in remote non-infarcted segments; IS, infarct size; LVMM, left ventricular myocardial mass. Significant *p*-values ($p < 0.05$) marked in bold.

3.4. ROC Curve Analysis of the Risk Score Model

To predict LVR in the early stages of AAMI, four independent influencing factors of CMR were allocated between the models in order to construct two predictive models. Model 1, which included LVEF and GLS, evaluated the global change in functionality following AAMI, while Model 2, consisting of peak $CS_{infarct}$ and IS, emphasized the impairment of the infarct segments. In Model 1, the GLS predicted LVR with an AUC of 0.812 (95% CI 0.731–0.893). The addition of the GLS to the LVEF resulted in a better prognostic value and an improved risk evaluation, with an increase in the AUC from the sole LVEF evaluation (AUC, 0.789, 95% CI 0.708–0.870) to the LVEF + GLS (AUC, 0.832, 95% CI 0.758–0.907) (Figure 2a). In Model 2, the AUC for the peak $CS_{infarct}$ and IS in regard to LVR in the patients with AAMI were 0.861 (95% CI 0.793–0.930) and 0.778 (95% CI 0.696–0.860), respectively. After combining the two parameters ($CS_{infarct}$+IS), the AUC increased to 0.917 (95% CI 0.870–0.965) (Figure 2b). Each model passed the Hosmer–Lemeshow goodness-of-fit test (all $p > 0.05$).

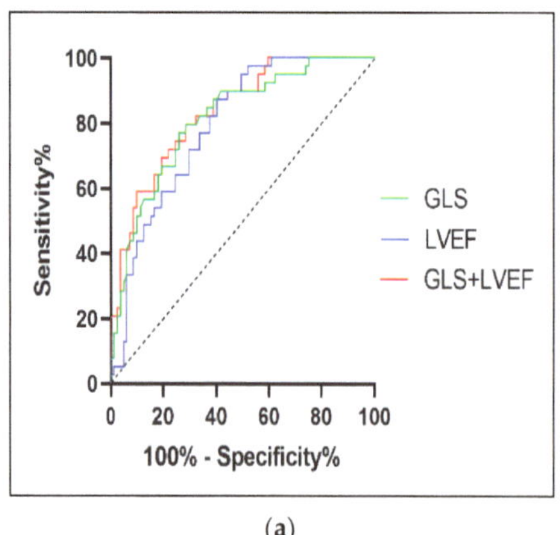

(a)

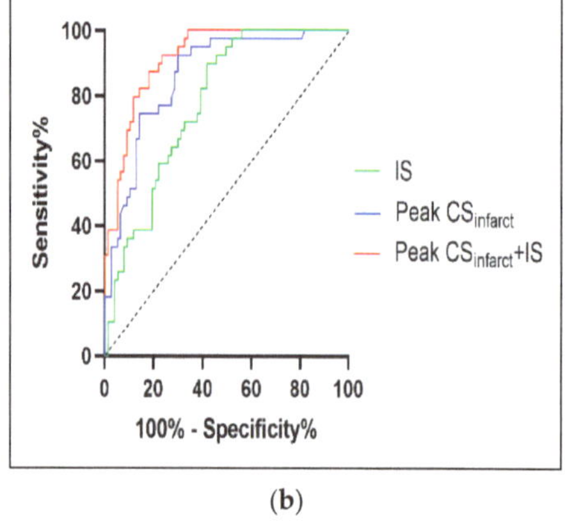

(b)

Figure 2. The area under ROC curves of the global parameters and regional parameters as a marker to predict LVR. (**a**) The AUCs of the LVEF, GLS, GLS + LVEF for predicting the occurrence of were

0.789 (95% CI 0.708–0.870), 0.812 (95% CI 0.731–0.893), and 0.832 (95% CI 0.758–0.907), respectively, all $p < 0.001$. (**b**) The AUCs of $CS_{infarct}$, IS, IS+$CS_{infarct}$ for predicting the occurrence of were 0.861 (95% CI 0.793–0.930), 0.778 (95% CI 0.696–0.860), and 0.917 (95% CI 0.870–0.965), respectively, all $p < 0.001$. ROC, receiver operating characteristic; LVR, left ventricular remodeling; AUCs, the area under ROC curves; LVEF, left ventricular ejection fraction; GLS, global longitudinal strain; $CS_{infarct}$, circumferential strain in infarcted segments; IS, infarct size.

3.5. Incremental Effects of Global and Regional Parameters

Figure 3 suggests that when it was compared with Model 1 (LVEF + GLS), Model 2 ($CS_{infarct}$+IS) showed a better prognostic value for LVR based on the Delong test ($p < 0.05$). Moreover, Model 2 demonstrated statistically improved discrimination and reclassification, with an IDI of 0.190 ($p = 0.002$).

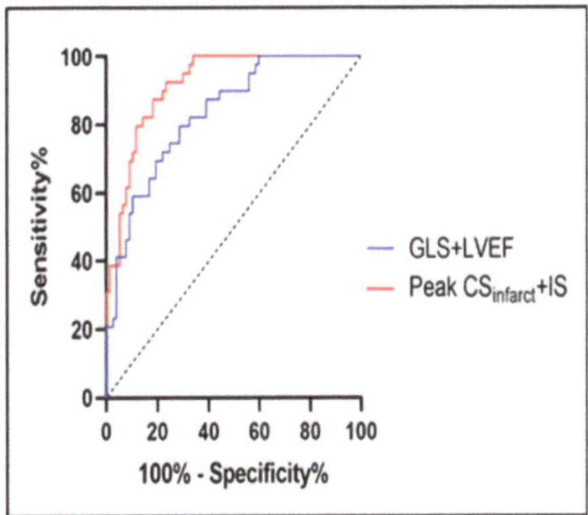

Figure 3. Discriminative prognostic power of GLS + LVEF and peak $CS_{infarct}$+IS. Area under ROC curves of GLS + LVEF in comparison to peak $CS_{infarct}$+IS for the prediction of LVR. Peak $CS_{infarct}$+IS revealed a significantly higher AUC than GLS + LVEF did (0.917, 95% CI: 0.870–0.965 $p < 0.001$ vs. 0.832, 95% CI 0.758–0.907, $p < 0.001$. AUC difference: 0.085, $p < 0.05$). GLS, global longitudinal strain; LVEF, left ventricular ejection fraction; $CS_{infarct}$, circumferential strain in infarcted segments; IS, infarct size; ROC, receiver operating characteristic; LVR, left ventricular remodeling; AUC, the area under ROC curve.

3.6. Reproducibility

We found good intra- and inter-observer variabilities for the evaluated CMR measurements. Supplement Table S1 and Supplement Figure S1 show the relevant detailed Bland–Altman graphics and the correlation coefficient data.

4. Discussion

This study focused on the prediction of adverse LVR with CMR parameters. LVR is a major cause of heart failure following AMI. Strain imaging can detect early or subtle changes in LV function and can effectively predict the AMI prognosis. This study revealed that the analysis of both the regional and global parameters in regard to myocardial strain is valuable for predicting LVR in patients with AAMI who are undergoing PPCI. By investigating two predictive models further, which were composed of the CMR parameters, we found that Model 2, which combined peak $CS_{infarct}$ and IS, manifested more predictive power than Model 1 did, which included a combination of global parameters, including LVEF and GLS. The parameters of the MI zone were more efficient and specific for LVR prediction following AAMI in the present study.

To evaluate LVR following AMI, we investigated AAMI that was caused by LAD occlusion in order to avoid confusing different myocardial injuries that are perfused by different coronary arteries. LVR resulting from AAMI impairs the heart function more severely than the remodeling of non-AAMI, and even the level of myocardial necrosis biomarkers serum troponin I are the same [2]. All of the patients in this study underwent PPCI treatment. Revascularization over time can save more myocardium which is helpful to reserve the heart contraction, on the other hand, oxygen free radicals releasing from successful reperfusion may aggravate myocardial edema or injury. Many factors, such as the severity of the local ischemia, local deformation, local hormone, stress, and autophagy, affect the myocyte salvage and collagen synthesis. Following AMI, the contraction becomes imbalanced and unsynchronized [20]. LVEF is reduced because the loss of vital myocyte is replaced with the abnormal distribution of collagen, meanwhile the reserved myocardium compensates the need for circulation by accelerating the segmental motion in non-infarcted segments, which also changes the mechanical properties of remote non-infarcted myocardium. This progressively contributes to negative remodeling [20,21]. LVR following AMI represents a compensatory balance between the infarcted and non-infarcted regions, and the global parameters may efficiently integrate the changes in the infarcted and non-infarcted regions. LVEF is a well-known factor contributing to systolic function. Additionally, it has been associated with a poor prognosis in patients with AMI [22]. However, LVEF had a limited value in predicting the patient mortality when it was above 45% [23]. Several studies have verified that the GLS is an important indicator of the prognosis of AMI [10,24]. More importantly, this study revealed that the addition of GLS to LVEF in a prediction model improved its predictive power in regard to LVR following AAMI. It seems that GLS detects more potential damage as well as subclinical damage in regard to LV function when it is compared with LVEF, which has been reported by Altiok and Ben Driss [25,26].

This study assessed both the global and segmental strain using the feature tracking module of CVI42, and we have identified GLS as a more accurate indicator of LVR than LVEF is. However, the factors describing the circumferential motion of the infarct zones seemed to predict LVR more effectively than GLS did in the present study. GCS is well recognized as a determinant of AMI prognosis based on the findings of two other studies. More specifically, Mangion's team found that the circumferential strain was an independent prognostic factor in patients with ST-segment elevation myocardial infarction [27], and Buss et al. revealed that GCS could predict LVEF alterations at six months following AMI [28]. There are three layers of myocardium in the left ventricle: the oblique inner and outer layers and the circular middle layer. The circumferential myofibers are less predisposed to ischemia and a subsequent infarction owing to its anatomic location. The inner layer of the myocardium moves longitudinally during systole, whereas the middle and outer layers move circumferentially. Therefore, the middle and outer layers have a greater influence on the LV circumferential motion, which may help to maintain the shape and size of the LV, and it may serve as a functional biomarker of myocardial salvage and the propensity to reserve the LV pump function in the long term [27]. Circumferential motion can cause LVR due to the degree of damage. We found that peak $CS_{infarct}$ and IS could predict LVR better than LVEF and GLS could (IDI, 0.190 (0.0712–0.3089) $p = 0.002$). The evidence demonstrated that the myocardial biomechanical changes that follow MI were initially limited to the MI region, but they gradually extended to the neighboring border zones and the remote myocardium over the 28 days post-MI [29]. This might account for why the regional parameters are more sensitive than the global one in predicting the LVR in patients with AAMI. In addition, we found that the peak $CS_{infarct}$+ LVEF showed a similar predictive power as the peak $CS_{infarct}$+IS did (Supplement Figure S2); the former combination will be available for patients with impaired renal function who undergo an abbreviated CMR protocol without contrast.

It is vital to explore the predictors of LVR in regard to the clinical adverse outcomes in patients with AMI. This information would be extremely helpful for patients in the high-risk group, who could receive further novel therapies in addition to routine post-AMI medications, as well as more aggressive up-titration of prognostic medicines (e.g., beta blockers, ACEI, and mineralocorticoid receptor antagonists) and more in-depth follow-ups in order to identify an adverse LVR. For example, this information could aid the prescription of primary prevention implantable cardioverter-defibrillators which aim to reduce the major clinical events that are associated with AMI.

5. Limitations

We acknowledge several limitations of the present research. Firstly, this was a single-center study, and therefore, the generalizability of our findings is unclear. Moreover, only 128 patients were diagnosed with AAMI and treated with PPCI. Hence, the enrolled sample size was small, which might have affected the accuracy of the cutoff values that were obtained in the multivariate analysis. The sample size should be expanded to obtain more reliable clinical conclusions, and the follow-up time should be extended. New parameter such as the sphericity index of LV will be explored in future studies.

6. Conclusions

The current study demonstrated that both the global (GLS, LVEF) and regional parameters (peak $CS_{infarct}$, IS) were strong independent predictors of LVR in patients with AAMI. Moreover, the peak $CS_{infarct}$ and IS might play a stronger role in predicting LVR in patients with AAMI than the global parameters do.

Supplementary Materials: The following supporting information can be downloaded at: https://www.mdpi.com/article/10.3390/diagnostics12112780/s1, Table S1: Intraclass correlation coefficient of CMR strain analysis; Figure S1: Bland–Altman analysis of LV strain indexes; Figure S2: Receiver operating characteristic analysis for predication of LVR after PPCI.

Author Contributions: Conceptualization, W.M. and J.P.; methodology, Y.G.; formal analysis, C.G. and W.M.; investigation, X.L. and S.K.; writing—review and editing, Y.L. and W.M.; supervision, J.P. All authors approved the final manuscript. All authors have read and agreed to the published version of the manuscript.

Funding: This research was supported by the Shanghai Medical Innovation Research Special Project of Science and Technology Innovation Action Plan in 2021 (grant number: 21Y11909400) funded by Shanghai Science and Technology Committee and Hospital-level exploratory clinical research projects in 2021 (grant number: ynts202107) funded by Shanghai Sixth People's Hospital.

Institutional Review Board Statement: Not applicable.

Informed Consent Statement: Informed consent was obtained from all subjects involved in the study.

Data Availability Statement: The data presented in this study are available on request from the corresponding author on reasonable request.

Conflicts of Interest: The authors declare no conflict of interest.

References

1. Szummer, K.; Wallentin, L.; Lindhagen, L.; Alfredsson, J.; Erlinge, D.; Held, C.; James, S.; Kellerth, T.; Lindahl, B.; Ravn-Fischer, A.; et al. Improved outcomes in patients with ST-elevation myocardial infarction during the last 20 years are related to implementation of evidence-based treatments: Experiences from the SWEDEHEART registry 1995–2014. *Eur. Heart J.* **2017**, *38*, 3056–3065. [CrossRef]
2. Masci, P.G.; Ganame, J.; Francone, M.; Desmet, W.; Lorenzoni, V.; Iacucci, I.; Barison, A.; Carbone, I.; Lombardi, M.; Agati, L.; et al. Relationship between location and size of myocardial infarction and their reciprocal influences on post-infarction left ventricular remodelling. *Eur. Heart J.* **2011**, *32*, 1640–1648. [CrossRef]
3. Granger, C.B.; Goldberg, R.J.; Dabbous, O.; Pieper, K.S.; Eagle, K.A.; Cannon, C.P.; Van de Werf, F.; Avezum, A.; Goodman, S.G.; Flather, M.D.; et al. Predictors of Hospital Mortality in the Global Registry of Acute Coronary Events. *Arch. Intern. Med.* **2003**, *163*, 2345–2353. [CrossRef]
4. Morrow, D.A.; Antman, E.M.; Charlesworth, A.; Cairns, R.; Murphy, S.A.; de Lemos, J.A.; Giugliano, R.P.; McCabe, C.H.; Braunwald, E. TIMI Risk Score for ST-Elevation Myocardial Infarction: A Convenient, Bedside, Clinical Score for Risk Assessment at Presentation. *Circulation* **2000**, *102*, 2031–2037. [CrossRef]
5. Pontone, G.; Guaricci, A.I.; Andreini, D.; Ferro, G.; Guglielmo, M.; Baggiano, A.; Fusini, L.; Muscogiuri, G.; Lorenzoni, V.; Mushtaq, S.; et al. Prognostic Stratification of Patients With ST-Segment–Elevation Myocardial Infarction (PROSPECT). *Circ. Cardiovasc. Imaging* **2017**, *10*, e006428. [CrossRef]

6.   Ibanez, B.; Aletras, A.H.; Arai, A.E.; Arheden, H.; Bax, J.; Berry, C.; Bucciarelli-Ducci, C.; Croisille, P.; Dall'Armellina, E.; Dharmakumar, R.; et al. Cardiac MRI Endpoints in Myocardial Infarction Experimental and Clinical Trials: JACC Scientific Expert Panel. *J. Am. Coll. Cardiol.* **2019**, *74*, 238–256. [CrossRef]

7.   Reindl, M.; Eitel, I.; Reinstadler, S.J. Role of Cardiac Magnetic Resonance to Improve Risk Prediction following Acute ST-elevation Myocardial Infarction. *J. Clin. Med.* **2020**, *9*, 1041. [CrossRef]

8.   Solomon, S.D.; Skali, H.; Anavekar, N.S.; Bourgoun, M.; Barvik, S.; Ghali, J.K.; Warnica, J.W.; Khrakovskaya, M.; Arnold, J.M.O.; Schwartz, Y.; et al. Changes in Ventricular Size and Function in Patients Treated With Valsartan, Captopril, or Both After Myocardial Infarction. *Circulation* **2005**, *111*, 3411–3419. [CrossRef]

9.   Ibanez, B.; James, S.; Agewall, S.; Antunes, M.J.; Bucciarelli-Ducci, C.; Bueno, H.; Caforio, A.L.P.; Crea, F.; Goudevenos, J.A.; Halvorsen, S.; et al. 2017 ESC Guidelines for the management of acute myocardial infarction in patients presenting with ST-segment elevation. *Eur. Heart J.* **2018**, *39*, 119–177. [CrossRef]

10.   Holzknecht, M.; Reindl, M.; Tiller, C.; Reinstadler, S.J.; Lechner, I.; Pamminger, M.; Schwaiger, J.P.; Klug, G.; Bauer, A.; Metzler, B.; et al. Global longitudinal strain improves risk assessment after ST-segment elevation myocardial infarction: A comparative prognostic evaluation of left ventricular functional parameters. *Clin. Res. Cardiol.* **2021**, *110*, 1599–1611. [CrossRef]

11.   Bulluck, H.; Carberry, J.; Carrick, D.; McCartney, P.J.; Maznyczka, A.M.; Greenwood, J.P.; Maredia, N.; Chowdhary, S.; Gershlick, A.H.; Appleby, C.; et al. A Noncontrast CMR Risk Score for Long-Term Risk Stratification in Reperfused ST-Segment Elevation Myocardial Infarction. *JACC Cardiovasc. Imaging* **2022**, *15*, 431–440. [CrossRef]

12.   Stiermaier, T.; Jobs, A.; de Waha, S.; Fuernau, G.; Pöss, J.; Desch, S.; Thiele, H.; Eitel, I. Optimized Prognosis Assessment in ST-Segment–Elevation Myocardial Infarction Using a Cardiac Magnetic Resonance Imaging Risk Score. *Circ. Cardiovasc. Imaging* **2017**, *10*, e006774. [CrossRef]

13.   Legallois, D.; Hodzic, A.; Alexandre, J.; Dolladille, C.; Saloux, E.; Manrique, A.; Roule, V.; Labombarda, F.; Milliez, P.; Beygui, F. Definition of left ventricular remodelling following ST-elevation myocardial infarction: A systematic review of cardiac magnetic resonance studies in the past decade. *Heart Fail. Rev.* **2022**, *27*, 37–48. [CrossRef]

14.   Thygesen, K.; Alpert, J.S.; Jaffe, A.S.; Simoons, M.L.; Chaitman, B.R.; White, H.D.; Thygesen, K.; Alpert, J.S.; White, H.D.; Jaffe, A.S.; et al. Third universal definition of myocardial infarction. *Eur. Heart J.* **2012**, *33*, 2551–2567. [CrossRef]

15.   Gao, C.; Tao, Y.; Pan, J.; Shen, C.; Zhang, J.; Xia, Z.; Wan, Q.; Wu, H.; Gao, Y.; Shen, H.; et al. Evaluation of elevated left ventricular end diastolic pressure in patients with preserved ejection fraction using cardiac magnetic resonance. *Eur. Radiol.* **2019**, *29*, 2360–2368. [CrossRef]

16.   Bondarenko, O.; Beek, A.M.; Hofman, M.B.M.; Kühl, H.P.; Twisk, J.W.R.; Van Dockum, W.G.; Visser, C.A.; Van Rossum, A.C. Standardizing the Definition of Hyperenhancement in the Quantitative Assessment of Infarct Size and Myocardial Viability Using Delayed Contrast-Enhanced CMR. *J. Cardiovasc. Magn. Reson.* **2005**, *7*, 481–485. [CrossRef]

17.   Gao, C.; Gao, Y.; Hang, J.; Wei, M.; Li, J.; Wan, Q.; Tao, Y.; Wu, H.; Xia, Z.; Shen, C.; et al. Strain parameters for predicting the prognosis of non-ischemic dilated cardiomyopathy using cardiovascular magnetic resonance tissue feature tracking. *J. Cardiovasc. Magn. Reson.* **2021**, *23*, 21. [CrossRef]

18.   Lange, T.; Stiermaier, T.; Backhaus, S.J.; Boom, P.C.; Kowallick, J.T.; de Waha-Thiele, S.; Lotz, J.; Kutty, S.; Bigalke, B.; Gutberlet, M.; et al. Functional and prognostic implications of cardiac magnetic resonance feature tracking-derived remote myocardial strain analyses in patients following acute myocardial infarction. *Clin. Res. Cardiol.* **2021**, *110*, 270–280. [CrossRef]

19.   Sheehan, F.H.; Braunwald, E.; Canner, P.; Dodge, H.T.; Gore, J.; Van Natta, P.; Passamani, E.R.; Williams, D.O.; Zaret, B. The effect of intravenous thrombolytic therapy on left ventricular function: A report on tissue-type plasminogen activator and streptokinase from the Thrombolysis in Myocardial Infarction (TIMI Phase I) trial. *Circulation* **1987**, *75*, 817–829. [CrossRef]

20.   Prabhu, S.D.; Frangogiannis, N.G. The Biological Basis for Cardiac Repair After Myocardial Infarction. *Circ. Res.* **2016**, *119*, 91–112. [CrossRef]

21.   Liehn, E.A.; Postea, O.; Curaj, A.; Marx, N. Repair after Myocardial Infarction, between Fantasy and Reality: The Role of Chemokines. *J. Am. Coll. Cardiol.* **2011**, *58*, 2357–2362. [CrossRef]

22.   The Multicenter Postinfarction Research Group Risk Stratification and Survival after Myocardial Infarction. *N. Engl. J. Med.* **1983**, *309*, 331–336. [CrossRef]

23.   Curtis, J.P.; Sokol, S.I.; Wang, Y.; Rathore, S.S.; Ko, D.T.; Jadbabaie, F.; Portnay, E.L.; Marshalko, S.J.; Radford, M.J.; Krumholz, H.M. The association of left ventricular ejection fraction, mortality, and cause of death in stable outpatients with heart failure. *J. Am. Coll. Cardiol.* **2003**, *42*, 736–742. [CrossRef]

24.   Reindl, M.; Tiller, C.; Holzknecht, M.; Lechner, I.; Eisner, D.; Riepl, L.; Pamminger, M.; Henninger, B.; Mayr, A.; Schwaiger, J.P.; et al. Global longitudinal strain by feature tracking for optimized prediction of adverse remodeling after ST-elevation myocardial infarction. *Clin. Res. Cardiol.* **2021**, *110*, 61–71. [CrossRef]

25.   Altiok, E.; Tiemann, S.; Becker, M.; Koos, R.; Zwicker, C.; Schroeder, J.; Kraemer, N.; Schoth, F.; Adam, D.; Friedman, Z.; et al. Myocardial Deformation Imaging by Two-Dimensional Speckle-Tracking Echocardiography for Prediction of Global and Segmental Functional Changes after Acute Myocardial Infarction: A Comparison with Late Gadolinium Enhancement Cardiac Magnetic Resonance. *J. Am. Soc. Echocardiogr.* **2014**, *27*, 249–257. [CrossRef]

26.   Ben Driss, A.; Lepage, C.B.D.; Sfaxi, A.; Hakim, M.; Elhadad, S.; Tabet, J.Y.; Salhi, A.; Carreira, V.B.; Hattab, M.; Meurin, P.; et al. Strain predicts left ventricular functional recovery after acute myocardial infarction with systolic dysfunction. *Int. J. Cardiol.* **2020**, *307*, 1–7. [CrossRef]

27. Mangion, K.; Carrick, D.; Carberry, J.; Mahrous, A.; McComb, C.; Oldroyd, K.G.; Eteiba, H.; Lindsay, M.; McEntegart, M.; Hood, S.; et al. Circumferential Strain Predicts Major Adverse Cardiovascular Events Following an Acute ST-Segment–Elevation Myocardial Infarction. *Radiology* **2019**, *290*, 329–337. [CrossRef]
28. Buss, S.J.; Krautz, B.; Hofmann, N.; Sander, Y.; Rust, L.; Giusca, S.; Galuschky, C.; Seitz, S.; Giannitsis, E.; Pleger, S.; et al. Prediction of functional recovery by cardiac magnetic resonance feature tracking imaging in first time ST-elevation myocardial infarction. Comparison to infarct size and transmurality by late gadolinium enhancement. *Int. J. Cardiol.* **2015**, *183*, 162–170. [CrossRef]
29. Torres, W.M.; Jacobs, J.; Doviak, H.; Barlow, S.C.; Zile, M.; Shazly, T.; Spinale, F.G. Regional and temporal changes in left ventricular strain and stiffness in a porcine model of myocardial infarction. *Am. J. Physiol. Circ. Physiol.* **2018**, *315*, H958–H967. [CrossRef]

Article

Prognostic Value of Late Gadolinium Enhancement in Left Ventricular Noncompaction: A Multicenter Study

Wei Huang [1,†], Ran Sun [1,†], Wenbin Liu [2], Rong Xu [1], Ziqi Zhou [1], Wei Bai [1], Ruilai Hou [1], Huayan Xu [1], Yingkun Guo [1], Li Yu [3,*,†] and Lu Ye [4,*,†]

[1] Department of Radiology, Key Laboratory of Birth Defects and Related Diseases of Women and Children of Ministry of Education, West China Second University Hospital, Sichuan University, Chengdu 610017, China
[2] Department of Radiology, Hospital of Chengdu University of Traditional Chinese Medicine, Chengdu 610075, China
[3] Department of Pediatric Cardiology, West China Second University Hospital, Sichuan University, Chengdu 610017, China
[4] Department of Ultrasound, West China Second University Hospital, Sichuan University, Chengdu 610017, China
* Correspondence: yulischuaxi@scu.edu.cn (L.Y.); yelu2001@scu.edu.cn (L.Y.)
† These authors contributed equally to this work.

Abstract: Current diagnostic criteria for left ventricular noncompaction (LVNC) may be poorly related to adverse prognosis. Late gadolinium enhancement (LGE) is a predictor of major adverse cardiovascular events (MACE), but risk stratification of LGE in patients with LVNC remains unclear. We retrospectively analyzed the clinical and cardiovascular magnetic resonance (CMR) data of 75 patients from three institutes and examined the correlation between different LGE types and MACE based on the extent, pattern (including a specific ring-like pattern), and locations of LGE in LVNC. A total of 51 patients (68%) presented LGE. A specific ring-like pattern was observed in 9 (12%). MACE occurred in 29 (38.7%) at 4.3 years of follow-up (interquartile range: 2.1–5.7 years). The adjusted hazard ratio (HR) for patients with ring-like LGE were 6.10 (95% CI, 1.39–26.75, $p < 0.05$). Free-wall or mid-wall LGE was associated with an increased risk of MACE after adjustment (HR 2.85, 95% CI, 1.31–6.21; HR 4.35, 95% CI, 1.23–15.37, respectively, $p < 0.05$). The risk of MACE in LVNC significantly increased when the LGE extent was greater than 7.5% and ring-like, multiple segments, and free-wall LGE were associated with MACE. These results suggest the value of LGE risk stratification in patients with LVNC.

Keywords: cardiovascular magnetic resonance; left ventricular noncompaction; hypertrabeculation; diagnostic criteria; risk stratification; late gadolinium enhancement

Citation: Huang, W.; Sun, R.; Liu, W.; Xu, R.; Zhou, Z.; Bai, W.; Hou, R.; Xu, H.; Guo, Y.; Yu, L.; et al. Prognostic Value of Late Gadolinium Enhancement in Left Ventricular Noncompaction: A Multicenter Study. *Diagnostics* **2022**, *12*, 2457. https://doi.org/10.3390/diagnostics12102457

Academic Editors: Minjie Lu and Arlene Sirajuddin

Received: 31 July 2022
Accepted: 27 September 2022
Published: 11 October 2022

Publisher's Note: MDPI stays neutral with regard to jurisdictional claims in published maps and institutional affiliations.

1. Introduction

Left ventricular noncompaction (LVNC), an uncommon cardiomyopathy characterized by a thickened endocardial layer with prominent trabeculae and a thinned, compacted epicardial layer [1], can occur as an isolated anomaly or associated with left ventricular dilation or hypertrophy, or various forms of congenital heart disease [2–4], and may lead to serious outcomes such as heart failure, thromboembolism, implantable cardioverter-defibrillator (ICD) therapy, heart transplantation, or sudden cardiac death [5,6]. Although several definitions have been proposed, currently diagnosis is mainly based on morphologic findings, and have the risk of overdiagnosis [1,7–9]. Several studies show a poor correlation between diagnostic criteria and adverse clinical events [10–12]. Furthermore, risk stratification in LVNC is particularly challenging and not available. A recent JACC paper proposes a risk prediction model of LVNC, in which late gadolinium enhancement (LGE) being one of the main prognostic factors [13,14]. LGE has negative prognostic implications in heart diseases, including hypertrophic cardiomyopathy, dilated cardiomyopathy, and LVNC [12,15–20]; however, previous studies have focused primarily on the presence of LGE but do not provide a detailed risk stratification analysis. The relationship between dose-response, LGE location and pattern, and specific clinical outcomes are poorly understood [14]. Therefore, we conducted a multicenter study to evaluate the association between the extent, location, or pattern of LGE and the impact on prognosis in patients with LVNC and hypertrabeculation patients. It is of great significance for clinicians and radiologists to judge the risk stratification of LVNC and hypertrabeculation patients intuitively and concisely. The guidelines of LVNC were lacking currently, and this evaluation may contribute to the establishment of hypertrabeculated LVNC guidelines and direct clinical management strategies (Figure 1).

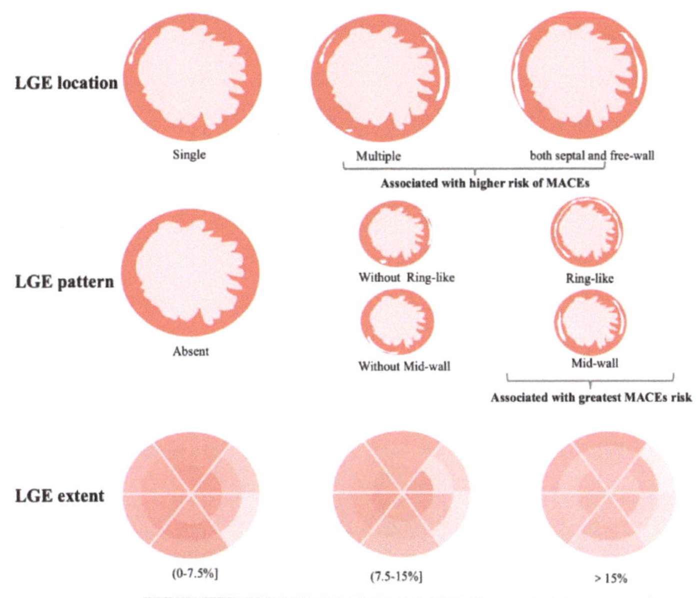

Figure 1. The graphic abstract. CMR, cardiovascular magnetic resonance; NC/C noncompaction/compacted; LGE, late gadolinium enhancement.

2. Materials and Methods

2.1. Study Population

We searched the clinical and cardiovascular magnetic resonance (CMR) databases at three institutions for patients' diagnosis described as LVNC or hypertrabeculation between January 2013 and December 2018, and the total number is 85. The CMR images of these patients were measured again according to Petersen's criteria by two radiologists with at least two years of experience. The inclusion criteria followed Petersen et al.'s CMR criteria: (1) CMR images with a distinct two-layered appearance of trabeculated and compacted myocardium; (2) There are prominent myocardial trabeculations in the noncompacted myocardium and deep intertrabecular recesses communicating with the LV; and (3) Subjects with increased LV trabeculation as measured by a noncompaction/compacted (NC/C) ratio $\geq$ 1.0 anywhere in the myocardial segments on the CMR images. Current diagnosis is mainly based on this criterion [8,14]. Both LVNC and hypertrabeculation patients were included in the cohort. The exclusion criteria were as follows: (1) presence of other known coexisting cardiac abnormalities, including congenital heart disease, coronary heart disease, valvular heart disease, hypertrophic cardiomyopathy, dilated cardiomyopathy or other types of cardiomyopathy, and myocarditis; (2) non-enhanced or poor image quality; and (3) incomplete clinical records. A total of 75 patients were eventually included in the cohort. The Ethics Committee of Clinical Trials and Biomedicine at the West China Hospital of Sichuan University and the ethics committees of other authors' hospitals approved this study. This study was performed in accordance with the Declaration of Helsinki (2000). Informed consent was obtained from all participants before study participation.

2.2. CMR Protocol

Standard gadolinium-enhanced CMR scanning was performed by a 3.0-T whole-body scanner (Skyra; Siemens Medical Solutions, Erlangen, Germany) with a two-element cardiac phased-array coil and ECG-triggering device. After acquiring localization images, 8–12 continuous short-axis cine images that covered the whole left ventricle were obtained by steady-state free-precession sequence. All cine images were acquired at end-expiration. Gadobenate dimeglumine (MultiHance 0.5 mmol/mL; Bracco, Milan, Italy) was injected intravenously at a dose of 0.1–0.2 mL/kg body weight and a flow rate of 2.5–3.0 mL/s after a 20–25 mL saline flush at a rate of 3.0 mL/s. LGE images were obtained using the inversion recovery TrueFISP sequence 10–15 min after contrast injection.

2.3. Image Analysis

All CMR data were analyzed using the commercially available software cvi42 (Circle Cardiovascular Imaging, Inc., Calgary, AB, Canada). Image analysis was performed to evaluate conventional cardiac function. For the cine imaging analysis, left ventricular structure and function parameters were measured on the short and long axis at end-diastole and end-systole, respectively. The left ventricular geometric parameters included the ratio of non-compacted to compacted myocardium, the end-diastolic volume (EDV), and end-systolic volume (ESV) mass. Cardiac function measurements, including the LV EDV, LV ESV, LV ejection fraction (EF), LV stroke volume, and LV mass, were analyzed by manually tracing the endocardial and epicardial contours. When delineated, the papillary muscles were excluded from the compacted myocardium. The end-diastolic and end-systolic phases were defined as those with maximum and minimum visual areas, respectively. The endocardial and epicardial boundaries of all images were manually delineated by a radiologist with at least two years of experience who was blinded to the clinical information. In addition, the ratio of NC/C was calculated in end-diastole. In each of three diastolic long-axis views (i.e., horizontal and vertical long-axis and LV outflow tract), the segment with the most pronounced trabeculations was chosen for measurement of the thickness of the non-compacted and the compacted myocardium perpendicular to the compacted myocardium, and only the maximal ratio was then used for analysis. The presence and extent of LGE were assessed and quantified on short-axis images, and LGE was deemed present if myocardial enhancement was confirmed on short-axis areas by using a signal intensity threshold of 5 standard deviations (SD) [21,22] above the mean signal of the remote normal myocardium, expressed as a percentage of scar mass/total LV mass. The presence of LGE was determined by two independent radiologists, with a third providing adjudication if necessary. An experienced radiologist categorized the LGE location and pattern. Visual assessment for the presence and distribution of LGE areas for each left ventricular (LV) myocardial segment was done using a standard 17-segment cardiac model. The definition of ring-like was that LGE present in at least three contiguous segments in the same short-axis slice [23].

2.4. Follow-Up

All patients were followed up on the telephone by using the standard questionnaire interview and the clinical medical records, which was performed by experienced physicians blinded to the clinical and CMR data. The clinical endpoint of this study was MACEs, defined as HF hospitalizations, thromboembolic events, appropriate ICD therapy, heart transplant, and sudden cardiac death. The duration of follow-up was calculated from the date of first CMR examination to the first occurrence of an endpoint. The median follow-up was 4.3 years (interquartile range: 2.1–5.7 years).

2.5. Statistical Analysis

Statistical analysis was performed with SPSS software (version 26.0, IBM Corp., Armonk, NY, USA). Continuous variables are expressed as means $\pm$ SD and categorical data as percentages. Baseline characteristics were compared using the Kruskal–Wallis rank test for continuous data and the Fisher exact test for categorical data. The Kaplan–Meier method was used to evaluate survival, and multivariate Cox regression analyses were performed to compute the hazard ratio (HR) and 95% confidence interval (CI). A p-value < 0.05 was significant. The proportional hazard models were adjusted for LVEF, sex, and age, which may confound the association between LGE and results.

3. Results

3.1. Study Population

The final cohort comprised 75 patients; a total of 52 (69.3%) were men, the median LVEF was 29.1% (IQ range: 17.5–37.7%), and LGE was present in 51 (68%). Patients with and without LGE had similar baseline ages. Patients with LGE had higher diastolic blood pressures (p = 0.023), lower LVEF (p = 0.005), greater LVEDV (p = 0.001) and LVESV (p < 0.001), larger left ventricular systolic mass (p = 0.001), and diastolic mass (p = 0.001). NYHA class $\geq$ III (n = 38, 50.7%) were common in all participants, and patients with LGE tended to have a poor NYHA functional (p = 0.01). Baseline characteristics are presented in Table 1. Two experienced radiologists determined LGE, and there was no significant difference between the two diagnoses.

Of the patients with LGE, 9 (12%) patients displayed a ring-like pattern and 42 (56%) a non-ring-like scar pattern. LGE was present only in the septum in 27 (36%) patients, only in the LV free wall in 16 (21.3%), and in both locations in 8 (10.7%). LGE was present only in a single segment in 37 (49.3%) patients, and in multiple segments in 14 (18.7%). LGE was categorized as mid-wall in 43 (57.3%) patients, and non-mid-wall in 8 (10.7%). Additionally, the LGE extent was categorized as three groups (>0 and $\leq$7.5%, >7.5% and $\leq$15%, and >15%).

Table 1. Baseline Characteristic.

		LGE			
	NO LGE (n = 24)	**0–7.5** (n = 23)	**7.5–15** (n = 10)	**>15** (n = 18)	p **Value** *
Age	42.9 $\pm$ 20.8	49.0 $\pm$ 14.0	50.8 $\pm$ 17.7	42.5 $\pm$ 14.2	0.382
Male	14 (58.3%)	19 (82.6%)	7 (70%)	12 (66.7%)	0.159
Height	161.9 $\pm$ 8.8	165.7 $\pm$ 6.1	167.3 $\pm$ 6.8	165.8 $\pm$ 7.8	0.035
Weight	60.4 $\pm$ 12.5	67.7 $\pm$ 11.6	67.5 $\pm$ 12.0	62.4 $\pm$ 9.0	0.108
Heart rate	88.8 $\pm$ 21.7	81.8 $\pm$ 20.2	77.0 $\pm$ 8.2	87.6 $\pm$ 31.7	0.185
Systolic blood pressure	112.3 $\pm$ 15.5	121.56 $\pm$ 16.3	119.7 $\pm$ 17.9	113.5 $\pm$ 18.1	0.178
Diastolic blood pressure	71.4 $\pm$ 13.0	78.6 $\pm$ 11.4	77.7 $\pm$ 8.8	75.2 $\pm$ 12.8	0.023
Hypertension	2 (8.3%)	3 (13.0%)	0	3 (16.7%)	0.656
Alcohol	5 (20.8%)	9 (39.1%)	2 (20%)	7 (38.9%)	0.208
Smoke	5 (20.8%)	11 (47.8%)	1 (10%)	7 (38.9%)	0.158
Medications					
ACE inhibitor	13 (54.2%)	15 (65.2%)	4 (40%)	7 (38.9%)	0.798
Beta-blocker	14 (58.3%)	22 (95.7%)	8 (80%)	15 (83.3%)	0.003
ARB	4 (16.7%)	5 (21.7%)	4 (40%)	6 (33.3%)	0.240
NYHA class $\geq$ 3	7 (29.2%)	14 (60.9%)	5 (50%)	12 (66.7%)	0.011

Table 1. *Cont.*

	NO LGE (n = 24)	0–7.5 (n = 23)	7.5–15 (n = 10)	>15 (n = 18)	p Value *
		LGE			
		CMR measurements			
LVEF (%)	35.2 ± 13.4	25.3 ± 12.6	28.0 ± 14.5	26.2 ± 15.7	0.005
LVEDV (mL)	209.4 ± 65.1	297.4 ± 100.5	289.1 ± 89.0	258.3 ± 85.9	0.001
LVEDVi (mL/m^2)	120.7 ± 32.3	161.5 ± 52.2	155.4 ± 39.0	145.2 ± 45.5	0.002
LVESV (mL)	140.5 ± 66.8	227.6 ± 89.6	216.7 ± 94.3	200.4 ± 87.3	0.0005
LVESVi (mL/m^2)	80.8 ± 35.4	123.3 ± 46.2	116.1 ± 45.3	112.5 ± 47.0	0.0009
LVSV (mL)	68.9 ± 25.9	69.8 ± 33.3	72.4 ± 22.4	61.9 ± 21.2	0.883
LVSVi (mL/m^2)	39.9 ± 13.9	38.2 ± 18.2	39.3 ± 11.8	35.2 ± 12.9	0.388
LV Mass, ED (g)	105.8 ± 33.8	161.3 ± 44.5	127.4 ± 32.5	128.3 ± 43.8	0.001
LV Mass index, ED (g/m^2)	60.8 ± 16.7	87.3 ± 21.9	68.5 ± 14.1	71.9 ± 22.8	0.001
LV Mass, ES (g)	116.6 ± 37.2	174.9 ± 44.6	142.6 ± 34.2	145.6 ± 54.5	0.001
LV Mass index, ES (g/m^2)	67.2 ± 19.4	94.8 ± 22.2	76.7 ± 13.9	81.4 ± 28.2	0.001
NC/C	3.0 ± 1.4	2.6 ± 1.6	2.5 ± 1.0	3.0 ± 2.0	0.116

Values are mean ± standard deviation or number (percentage). * Kruskal-Wallis Rank Test for continuous variables; Fisher Exact Test for categorical variables. ACE, angiotensin-converting enzyme; ARB, angiotensin II receptor blocker; NYHA, New York Heart Association; LVEF, left ventricular ejection fraction; LVEDV, left ventricular end diastolic volume; LVESV, left ventricular end systolic volume; LVSV, left ventricular stroke volume; NC/C, non-compacted/compacted ratio.

3.2. Outcome of Follow-Up

Over a median follow-up period of 4.3 years (IQ range: 2.1–5.7 years), two patients were lost. Of the 73 patients who completed follow-up, MACEs occurred in 29 (38.7%) patients. A total of 9 (31.0%) had HF hospitalizations, 8 (27.6%) underwent primary prevention ICD implantation, 2 (6.9%) had cardiac transplantation, and 10 (34.5%) had sudden cardiac death. MACEs occurred in 25 (50%) patients with LGE and in 4 (17.4%) without LGE (HR: 5.39; 95% CI: 1.59–18.31; $p = 0.007$). After adjustment of LVEF, age, and sex, LGE was associated with a higher risk of MACEs (HR: 3.84; 95% CI: 1.10–13.40; $p = 0.035$, Table 2 and Figure 2A).

Extent of LGE. Estimated adjusted HRs for patients with LGE extents of 0–7.5%, 7.5–15%, and >15% were 2.01 (95% CI: 0.50–8.03; $p = 0.323$), 7.42 (95% CI: 1.76–31.3; $p = 0.006$), and 9.02 (95% CI: 2.11–38.52; $p = 0.003$), respectively, compared to the patients without LGE (Table 2 and Figure 2B).

Table 2. Individual proportional hazard models investigating the association between major adverse cardiovascular events and late gadolinium enhancement (Presence and Extent).

		n	Endpoint	HR (95% CI)	Individual p Value	Overall p Value
				Adjusted for LVEF, Sex, and Age		
			Presence and extent			
LGE	No	23	4 (17.4%)	1.00	-	0.002
	Any	50	25 (50.0%)	3.84 (1.10–13.40)	0.035	
LGE	No	23	4 (17.4%)	1.00	-	<0.001
	(0–7.5%)	23	8 (34.8%)	2.01 (0.50–8.03)	0.323	
	(7.5–15%)	10	7 (70%)	7.42 (1.76–31.3)	0.006	
	>15%	17	10 (58.8%)	9.02 (2.11–38.52)	0.003	

Values are *n* or *n* (%) unless otherwise indicated. *p* values are quoted for each model overall and for the individual components. LVEF, left ventricular ejection fraction; HR, hazard ratio; CI, confidence interval.

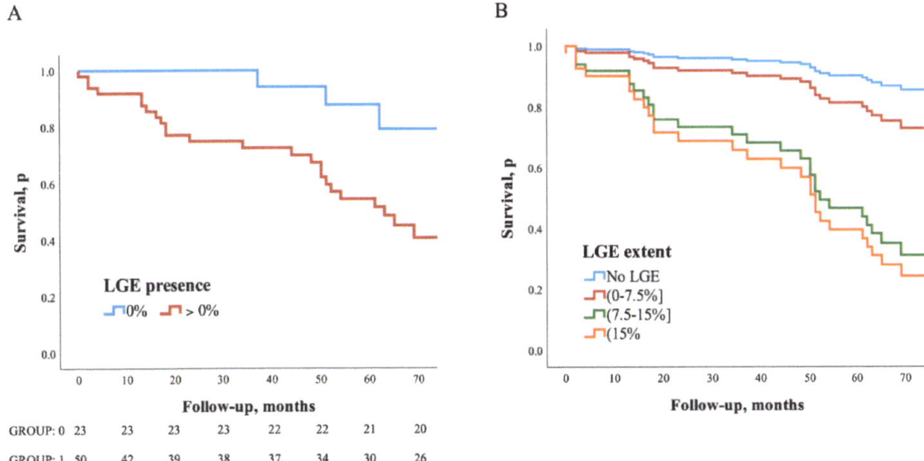

Figure 2. Impact of LGE presence and extent of left ventricular on long–term outcomes. Kaplan–Meier and COX regression analysis survival curves depicting time to MACE. Kaplan–Meier curve for LGE presence (**A**) and Cox regression analysis for LGE extent (**B**); MACE, major adverse cardiovascular events.

Pattern of LGE. Estimated adjusted HRs for patients with ring-like and non-ring-like scar were 6.10 (95% CI: 1.39–26.75; $p = 0.016$) and 3.59 (95% CI: 0.99–12.39; $p = 0.053$) compared to those patients without LGE. Patients with LGE only in single segment and in multiple segments had adjusted HRs for the MACEs of 2.96 (95% CI: 0.82–10.69; $p = 0.098$) and 8.35 (95% CI: 2.10–33.17; $p = 0.003$) compared to those patients without LGE (Table 3 and Figure 3A).

Table 3. Individual proportional hazard models investigating the association between major ad–verse cardiovascular events and late gadolinium enhancement (Location and Pattern).

				Adjusted for LVEF, Sex, and Age		
		n	Endpoint	HR (95% CI)	Individual *p* Value	Overall *p* Value
Location and pattern						
LGE (ring-like)	Absent	23	4 (17.4%)	1.00	-	0.004
	No ring-like	41	20 (48.8%)	3.59 (0.99–12.39)	0.053	
	Ring-like	9	5 (55.6%)	6.10 (1.39–26.75)	0.016	
LGE (segment)	Absent	23	4 (17.4%)	1.00	-	0.001
	Single	36	16 (44.4%)	2.96 (0.82–10.69)	0.098	
	Multiple	14	9 (64.3%)	8.35 (2.10–33.17)	0.003	
LGE (Free-wall)	No	51	17 (33.3%)	1.00		0.002
	Yes	22	12 (54.5%)	2.85 (1.31–6.21)	0.008	
LGE (Free-wall)	Absent	23	4 (17.4%)	1.00	-	0.001
	Septal only	27	12 (44.4%)	2.57 (0.69–9.60)	0.160	
	Free-wall only	15	7 (46.7%)	4.92 (1.18–20.58)	0.029	
	Both	8	6 (75%)	10.29 (2.42–43.75)	0.002	
LGE (Mid-wall)	Absent	23	4 (17.4%)	1.00	-	
	Without Mid-wall	8	2 (25%)	1.86 (0.30–11.61)	0.507	
	Mid-wall	42	23 (54.8%)	4.35 (1.23–15.37)	0.023	

Values are *n* or *n* (%) unless otherwise indicated. *p* values are quoted for each model overall and for the individual components. LVEF, left ventricular ejection fraction; HR, hazard ratio; CI, confidence interval.

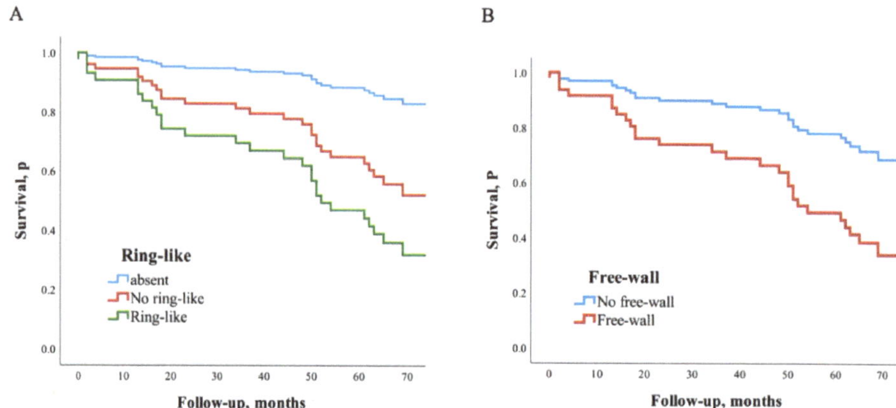

Figure 3. Impact of ring–like and free–wall LGE of left ventricular on long–term outcomes. Cox re–gression analysis for ring–like LGE (**A**) and free–wall LGE (**B**).

Location of LGE. Estimated adjusted HRs for patients with LGE only in the septum, only in the free-wall, and in both locations were 2.57 (95% CI: 0.69–9.60; $p = 0.16$), 4.92 (95% CI: 1.18–20.58; $p = 0.029$), and 10.29 (95% CI: 2.42–43.75; $p = 0.002$), respectively, compared to the patients without LGE. Patients with free-wall LGE had an estimated adjusted HR of 2.85 (95% CI: 1.31–6.21; $p = 0.008$) compared to those without free-wall LGE. Patients with mid-wall LGE and other myocardial locations had adjusted HRs for the MACEs of 4.35 (95% CI: 1.23–15.37; $p = 0.023$) and 1.86 (95% CI: 0.30–11.61; $p = 0.507$), respectively, compared to those without LGE (Table 3 and Figure 3B).

NC/C ratio. The NC/C ratio was categorized as hypertrabeculation (NC/C ratio $\geq$1 and <2.3) and non-compaction (NC/C ratio $\geq$ 2.3); the risk between the two groups did not reach statistical significance (HR: 1.03 95% CI, 0.49–2.17; $p = 0.93$) (Table 4).

Table 4. Individual proportional hazard models investigating the association between major adverse cardiovascular events and NC/C ratio.

NC/C	Adjusted for LVEF, Sex, and Age				
	n	Endpoint	HR (95% CI)	Individual p Value	Overall p Value
Myocardial hypertra-beculation(NC/C ration $\geq$ 1 and <2.3)	29	13 (44.8%)	1.00	-	0.93
Noncompaction(NC/C ratio $\geq$ 2.3)	44	16 (36.4%)	1.03 (0.49–2.17)	0.93	

Values are n or n (%) unless otherwise indicated. p values are quoted for each model overall and for the individual components. NC/C, noncompaction/compacted; HR, hazard ratio; CI, confidence interval.

4. Discussion

The major clinical messages arising from our study were as follows:

(1) The >7.5% LGE extent may be associated with a significantly poor long-term prognosis in hypertrabeculation and LVNC patients.
(2) Ring-like LGE and multiple segments LGE were associated with a particularly high risk of MACEs, which deserves extra clinical attention.
(3) The NC/C ratio poorly correlates with clinical outcomes, LGE should be considered in diagnoses as a risk predictor, and our study provided useful risk stratification.

Myocardial fibrosis can significantly affect patients' prognoses. LGE is of great significance in identifying high-risk patients. This finding has been widely confirmed in heart diseases [24,25]. However, the detail risk stratification about LGE and specific LGE pattern in the prognosis of patients with myocardial hypertrabeculation and LVNC are not yet clear [20]. Seventy-five patients from three study centers with CMR-confirmed hypertrabeculation and LVNC were enrolled. During the median follow-up of 4.3 years (IQ range: 2.1–5.7 years), major cardiovascular events (cardiac death, heart failure, thromboembolism, appropriate ICD therapy, and cardiac transplantation) were endpoints. The risk stratification of LGE extent, pattern, and location was determined. The results showed that LGE had a significant impact on the prognosis of patients with hypertrabeculation or LVNC. Risk of MACEs increased significantly with a greater extent of LGE. Our data also showed that patients with different LGE locations or patterns had a different MACE risk, which facilitates the use of CMR for prognostic risk stratification in hypertrabeculation or LVNC patients with LGE.

Patients with LGE were placed in three groups according to the extent of enhancement. After being adjusted for LVEF, sex, and age, the absence of LGE was still associated with a lower risk of MACE. The risks increased significantly when the LGE extent was greater than 7.5%; HRs were 7.42 and 9.02 in the 7.5–15% and >15% groups, respectively. In a prior study, a ≥15% LGE extent was considered a potentially clinically relevant threshold in HCM [26]; however, our data indicate that the prognosis relevant threshold may be lower in patients with LVNC.

A recent study points out that a specific ring-like LGE pattern is associated with a particularly high risk of malignant arrhythmic events, which are significant and independent of the total LGE burden and the presence of other additional risk factors; the HR of a ring-like pattern in this study was 68.98 (95% CI, 14.67–324.39; $p < 0.01$) compared to the absence of LGE [27]. In our study, the LGE pattern was classified as a ring-like and non-ring-like scar. The definition of ring-like was that LGE was present in at least three contiguous segments in the same short-axis slice (Figure 4) [23]. The result of our study is in line with previous reports. The risk of MACE was significantly higher in ring-like LGE patients than patients with non-ring-like LGE and without LGE by Kaplan–Meier analysis. After multivariate adjustment, the presence of ring-like LGE remained associated with an increased risk of the endpoint (HR: 6.10 95% CI, 1.39–26.75; $p = 0.016$). This finding indicates that ring-like LGE is also a predictor of adverse events in patients with LVNC or hypertrabeculation. In particular, ring-like LGE also presented even the extent of LGE was low (<7.5%). The worse prognosis may relate to insults of the conduction system. In this regard, qualitative indicators are more convenient for radiologists to diagnose than quantitative analyses by CVi.

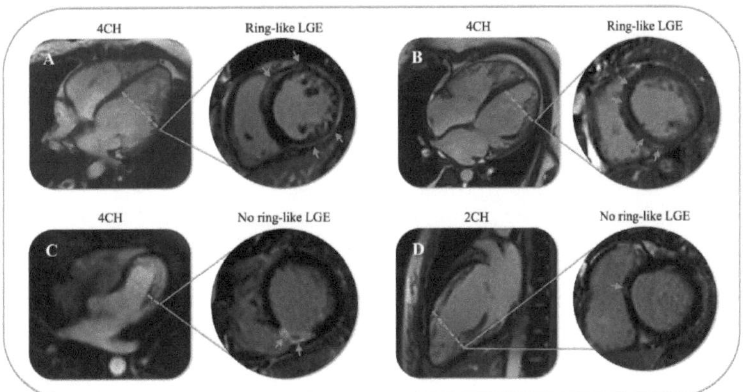

Figure 4. Examples of 4 patients with ring-like or no ring-like LGE. Ring-like LGE (**A,B**) and no ring-like LGE (**C,D**); 4CH: the four-cavity heart; 2CH: the two-cavity heart.

Analogously, it is more intuitive to determine whether LGE is present in multiple segments of the left ventricle. We showed that patients with multiple LGE segments were at higher risk of MACEs. In contrast, those with single-segment LGE were at similar risk to those without LGE. In patients with multiple LGE segments, LGE is not only present in consecutive segments at the same short-axis level but also scattered in a total of 17 segments.

Mahrholdt et al. showed that in a setting of HHV6 and combined PVB19/HHV6 myocarditis, LGE is predominantly located in the anteroseptal region [28]. Aquaro et al. found that LGE present in the anteroseptal wall in patients with acute myocarditis was associated with a worse prognosis [17]. This finding is similar in DCM in those patients with septal LGE had a higher risk of all-cause mortality [15]. However, we showed that free-wall LGE was associated with increased MACEs in LVNC. A greater risk was seen in concomitant free-wall and septum LGE, which indicates that the prognosis of patients with LVNC or hypertrabeculation may have greater relevance with free-wall LGE. In agreement with previous studies, we observed that the most common distribution of LGE in LVNC or hypertrabeculation patients was mid-myocardial and associated with a poor prognosis, similar to other cardiac diseases [29].

Current criteria for the diagnosis of LVNC lead to high disease prevalence in patients referred for CMR, and the NC/C ratio > 2.3 is common in a large population-based cohort, indicating that the NC/C ratio alone for LVNC or hypertrabeculation diagnosis may have low specificity [10,30]. Some studies suggest that the NC/C ratio and the extent of trabeculation do not correlate with adverse outcomes [31], which is in line with our data. These results suggest that a more comprehensive criteria model including LGE should be used. Our study provides several risk stratification models of LGE that are significantly associated with the prognosis in patients with LVNC.

Study limitation. This study has some limitations. Although this is a multicenter study, the low number of patients limited statistical power. Secondly, all the three centers are large referral hospitals. Some patients who came to our hospitals were referred by multiple primary hospitals with more severe symptoms, therefore, they had a lower LVEF overall, and there may be some selection bias. However, this study also provides more prognostic information for patients with relatively severe symptoms.

5. Conclusions

In LVNC or hypertrabeculation, the risk of MACE increases significantly when the LGE extent is greater than 7.5% or presence as ring-like LGE. Moreover, multiple segments and free-wall LGE are associated with MACE. The detailed study and risk stratification of LGE in LVNC or hypertrabeculation patients will help improve the diagnostic criteria and make this criterion more closely to clinical prognosis. This study provided useful models based on the extent, pattern, and location of LGE, which provide a much-needed approach to quantify risk.

Author Contributions: Conceptualization, Y.G.; Data curation, W.H., R.S., Z.Z., W.B., W.L., R.H. and L.Y. (Li Yu); Methodology, L.Y. (Lu Ye); Project administration, Y.G.; Writing—Original draft, W.H.; Writing—Review and editing, R.X. and H.X. All authors have read and agreed to the published version of the manuscript.

Funding: This work was supported by National Natural Science Foundation of China (82120108015, 82102020, 82071874, 81971586) and Sichuan Science and Technology Program (2020YJ0029, 2017TD0005); Clinical Research Grant of Chinese Society of Cardiovascular Disease (CSC) of 2019 (No. HFCSC2019B01); and the Fundamental Research Funds for the Central Universities (SCU2020D4132).

Institutional Review Board Statement: The study was conducted according to the guidelines of the Decla-ration of Helsinki, and approved by the Institutional Ethics Review Board of Sichuan University (K2019059), West China Hospital; Sichuan University (756/2019; 27-09-2019); and Peking Union Medical College Hospital and was carried out in accordance with the tenets of the Declaration of Helsinki. The written informed consent was obtained from all participants before they were included in the study.

Informed Consent Statement: Informed consent was obtained from all subjects involved in the study.

Data Availability Statement: Not applicable.

Conflicts of Interest: The authors declare no conflict of interest.

References

1. Chin, T.K.; Perloff, J.K.; Williams, R.G.; Jue, K.; Mohrmann, R. Isolated noncompaction of left ventricular myocardium. A study of eight cases. *Circulation* **1990**, *82*, 507–513. [CrossRef]
2. Oechslin, E.; Attenhofer Jost, C.; Rojas, J.; Kaufmann, P.; Jenni, R. Long-term follow-up of 34 adults with isolated left ventricular noncompaction: A distinct cardiomyopathy with poor prognosis. *J. Am. Coll. Cardiol.* **2000**, *36*, 493–500. [CrossRef]
3. Stöllberger, C.; Finsterer, J.; Blazek, G. Left ventricular hypertrabeculation/noncompaction and association with additional cardiac abnormalities and neuromuscular disorders. *Am. J. Cardiol.* **2002**, *90*, 899–902. [CrossRef]
4. Towbin, J.; Lorts, A.; Jefferies, J. Left ventricular non-compaction cardiomyopathy. *Lancet* **2015**, *386*, 813–825. [CrossRef]
5. Brescia, S.; Rossano, J.; Pignatelli, R.; Jefferies, J.; Price, J.; Decker, J.; Denfield, S.; Dreyer, W.; Smith, O.; Towbin, J.; et al. Mortality and sudden death in pediatric left ventricular noncompaction in a tertiary referral center. *Circulation* **2013**, *127*, 2202–2208. [CrossRef] [PubMed]
6. Stanton, C.; Bruce, C.; Connolly, H.; Brady, P.; Syed, I.; Hodge, D.; Asirvatham, S.; Friedman, P. Isolated left ventricular noncompaction syndrome. *Am. J. Cardiol.* **2009**, *104*, 1135–1138. [CrossRef] [PubMed]
7. Jenni, R.; Oechslin, E.; Schneider, J.; Attenhofer Jost, C.; Kaufmann, P. Echocardiographic and pathoanatomical characteristics of isolated left ventricular non-compaction: A step towards classification as a distinct cardiomyopathy. *Heart* **2001**, *86*, 666–671. [CrossRef] [PubMed]
8. Petersen, S.; Selvanayagam, J.; Wiesmann, F.; Robson, M.; Francis, J.; Anderson, R.; Watkins, H.; Neubauer, S. Left ventricular non-compaction: Insights from cardiovascular magnetic resonance imaging. *J. Am. Coll. Cardiol.* **2005**, *46*, 101–105. [CrossRef]
9. Jacquier, A.; Thuny, F.; Jop, B.; Giorgi, R.; Cohen, F.; Gaubert, J.; Vidal, V.; Bartoli, J.; Habib, G.; Moulin, G. Measurement of trabeculated left ventricular mass using cardiac magnetic resonance imaging in the diagnosis of left ventricular non-compaction. *Eur. Heart J.* **2010**, *31*, 1098–1104. [CrossRef] [PubMed]
10. Ivanov, A.; Dabiesingh, D.; Bhumireddy, G.; Mohamed, A.; Asfour, A.; Briggs, W.; Ho, J.; Khan, S.; Grossman, A.; Klem, I.; et al. Prevalence and Prognostic Significance of Left Ventricular Noncompaction in Patients Referred for Cardiac Magnetic Resonance Imaging. *Circ. Cardiovasc. Imaging* **2017**, *10*, e006174. [CrossRef]
11. Ross, S.; Jones, K.; Blanch, B.; Puranik, R.; McGeechan, K.; Barratt, A.; Semsarian, C. A systematic review and meta-analysis of the prevalence of left ventricular non-compaction in adults. *Eur. Heart J.* **2020**, *41*, 1428–1436. [CrossRef] [PubMed]
12. Andreini, D.; Pontone, G.; Bogaert, J.; Roghi, A.; Barison, A.; Schwitter, J.; Mushtaq, S.; Vovas, G.; Sormani, P.; Aquaro, G.D.; et al. Long-Term Prognostic Value of Cardiac Magnetic Resonance in Left Ventricle Noncompaction: A Prospective Multicenter Study. *J. Am. Coll. Cardiol.* **2016**, *68*, 2166–2181. [CrossRef] [PubMed]
13. Jefferies, J. Risk Prediction in a Debated Diagnosis: Is it Time for LVNC Guidelines? *J. Am. Coll. Cardiol.* **2021**, *78*, 663–665. [CrossRef] [PubMed]
14. Casas, G.; Limeres, J.; Oristrell, G.; Gutierrez-Garcia, L.; Andreini, D.; Borregan, M.; Larrañaga-Moreira, J.; Lopez-Sainz, A.; Codina-Solà, M.; Teixido-Tura, G.; et al. Clinical Risk Prediction in Patients with Left Ventricular Myocardial Noncompaction. *J. Am. Coll. Cardiol.* **2021**, *78*, 643–662. [CrossRef]
15. Halliday, B.; Baksi, A.; Gulati, A.; Ali, A.; Newsome, S.; Izgi, C.; Arzanauskaite, M.; Lota, A.; Tayal, U.; Vassiliou, V.; et al. Outcome in Dilated Cardiomyopathy Related to the Extent, Location, and Pattern of Late Gadolinium Enhancement. *JACC Cardiovasc. Imaging* **2019**, *12*, 1645–1655. [CrossRef]
16. Mentias, A.; Raeisi-Giglou, P.; Smedira, N.; Feng, K.; Sato, K.; Wazni, O.; Kanj, M.; Flamm, S.; Thamilarasan, M.; Popovic, Z.; et al. Late Gadolinium Enhancement in Patients with Hypertrophic Cardiomyopathy and Preserved Systolic Function. *J. Am. Coll. Cardiol.* **2018**, *72*, 857–870. [CrossRef]
17. Aquaro, G.; Perfetti, M.; Camastra, G.; Monti, L.; Dellegrottaglie, S.; Moro, C.; Pepe, A.; Todiere, G.; Lanzillo, C.; Scatteia, A.; et al. Cardiac MR with Late Gadolinium Enhancement in Acute Myocarditis with Preserved Systolic Function: ITAMY Study. *J. Am. Coll. Cardiol.* **2017**, *70*, 1977–1987. [CrossRef]
18. Di Marco, A.; Anguera, I.; Schmitt, M.; Klem, I.; Neilan, T.; White, J.; Sramko, M.; Masci, P.; Barison, A.; Mckenna, P.; et al. Late Gadolinium Enhancement and the Risk for Ventricular Arrhythmias or Sudden Death in Dilated Cardiomyopathy: Systematic Review and Meta-Analysis. *JACC Heart Fail.* **2017**, *5*, 28–38. [CrossRef]
19. Halliday, B.; Gulati, A.; Ali, A.; Guha, K.; Newsome, S.; Arzanauskaite, M.; Vassiliou, V.; Lota, A.; Izgi, C.; Tayal, U.; et al. Association Between Midwall Late Gadolinium Enhancement and Sudden Cardiac Death in Patients with Dilated Cardiomyopathy and Mild and Moderate Left Ventricular Systolic Dysfunction. *Circulation* **2017**, *135*, 2106–2115. [CrossRef]
20. Grigoratos, C.; Barison, A.; Ivanov, A.; Andreini, D.; Amzulescu, M.; Mazurkiewicz, L.; De Luca, A.; Grzybowski, J.; Masci, P.; Marczak, M.; et al. Meta-Analysis of the Prognostic Role of Late Gadolinium Enhancement and Global Systolic Impairment in Left Ventricular Noncompaction. *JACC Cardiovasc. Imaging* **2019**, *12*, 2141–2151. [CrossRef]
21. Bondarenko, O.; Beek, A.; Hofman, M.; Kühl, H.; Twisk, J.; van Dockum, W.; Visser, C.; van Rossum, A. Standardizing the definition of hyperenhancement in the quantitative assessment of infarct size and myocardial viability using delayed contrast-enhanced CMR. *J. Cardiovasc. Magn. Reson.* **2005**, *7*, 481–485. [CrossRef] [PubMed]
22. Schulz-Menger, J.; Bluemke, D.; Bremerich, J.; Flamm, S.; Fogel, M.; Friedrich, M.; Kim, R.; von Knobelsdorff-Brenkenhoff, F.; Kramer, C.; Pennell, D.; et al. Standardized image interpretation and post processing in cardiovascular magnetic resonance:

Society for Cardiovascular Magnetic Resonance (SCMR) board of trustees task force on standardized post processing. *J. Cardiovasc. Magn. Reson.* **2013**, *15*, 35. [CrossRef] [PubMed]

23. Augusto, J.; Eiros, R.; Nakou, E.; Moura-Ferreira, S.; Treibel, T.; Captur, G.; Akhtar, M.; Protonotarios, A.; Gossios, T.; Savvatis, K.; et al. Dilated cardiomyopathy and arrhythmogenic left ventricular cardiomyopathy: A comprehensive genotype-imaging phenotype study. *Eur. Heart J. Cardiovasc. Imaging* **2020**, *21*, 326–336. [CrossRef]

24. Alba, A.C.; Gaztanaga, J.; Foroutan, F.; Thavendiranathan, P.; Merlo, M.; Alonso-Rodriguez, D.; Vallejo-Garcia, V.; Vidal-Perez, R.; Corros-Vicente, C.; Barreiro-Perez, M.; et al. Prognostic Value of Late Gadolinium Enhancement for the Prediction of Cardiovascular Outcomes in Dilated Cardiomyopathy: An International, Multi-Institutional Study of the MINICOR Group. *Circ. Cardiovasc. Imaging* **2020**, *13*, e010105. [CrossRef] [PubMed]

25. Disertori, M.; Rigoni, M.; Pace, N.; Casolo, G.; Masè, M.; Gonzini, L.; Lucci, D.; Nollo, G.; Ravelli, F. Myocardial Fibrosis Assessment by LGE Is a Powerful Predictor of Ventricular Tachyarrhythmias in Ischemic and Nonischemic LV Dysfunction: A Meta-Analysis. *JACC Cardiovasc. Imaging* **2016**, *9*, 1046–1055. [CrossRef] [PubMed]

26. Habib, M.; Adler, A.; Fardfini, K.; Hoss, S.; Hanneman, K.; Rowin, E.; Maron, M.; Maron, B.; Rakowski, H.; Chan, R. Progression of Myocardial Fibrosis in Hypertrophic Cardiomyopathy: A Cardiac Magnetic Resonance Study. *JACC Cardiovasc. Imaging* **2021**, *14*, 947–958. [CrossRef]

27. Muser, D.; Nucifora, G.; Pieroni, M.; Castro, S.; Casado Arroyo, R.; Maeda, S.; Benhayon, D.; Liuba, I.; Sadek, M.; Magnani, S.; et al. Prognostic Value of Non-Ischemic Ring-Like Left Ventricular Scar in Patients with Apparently Idiopathic Non-Sustained Ventricular Arrhythmias. *Circulation* **2021**, *143*, 1359–1373. [CrossRef]

28. Mahrholdt, H.; Wagner, A.; Deluigi, C.; Kispert, E.; Hager, S.; Meinhardt, G.; Vogelsberg, H.; Fritz, P.; Dippon, J.; Bock, C.; et al. Presentation, patterns of myocardial damage, and clinical course of viral myocarditis. *Circulation* **2006**, *114*, 1581–1590. [CrossRef]

29. Wan, J.; Zhao, S.; Cheng, H.; Lu, M.; Jiang, S.; Yin, G.; Gao, X.; Yang, Y. Varied distributions of late gadolinium enhancement found among patients meeting cardiovascular magnetic resonance criteria for isolated left ventricular non-compaction. *J. Cardiovasc. Magn. Reson.* **2013**, *15*, 20. [CrossRef]

30. Kawel, N.; Nacif, M.; Arai, A.E.; Gomes, A.S.; Hundley, W.G.; Johnson, W.C.; Prince, M.R.; Stacey, R.B.; Lima, J.A.; Bluemke, D.A. Trabeculated (noncompacted) and compact myocardium in adults: The multi-ethnic study of atherosclerosis. *Circ. Cardiovasc. Imaging* **2012**, *5*, 357–366. [CrossRef]

31. Aung, N.; Doimo, S.; Ricci, F.; Sanghvi, M.; Pedrosa, C.; Woodbridge, S.; Al-Balah, A.; Zemrak, F.; Khanji, M.; Munroe, P.; et al. Prognostic Significance of Left Ventricular Noncompaction: Systematic Review and Meta-Analysis of Observational Studies. *Circ. Cardiovasc. Imaging* **2020**, *13*, e009712. [CrossRef] [PubMed]

Article

Clinical Application of Cardiac Magnetic Resonance in ART-Treated AIDS Males with Short Disease Duration

Keke Hou [1,†], Hang Fu [2,†], Wei Xiong [1], Yueqin Gao [1], Liqiu Xie [1], Jianglin He [1], Xianbiao Feng [1], Tao Zeng [3], Lin Cai [4], Lei Xiong [1], Nan Jiang [1], Min Jiang [1], Bin Kang [1], Haiyan Zheng [1], Na Zhang [1,*,‡] and Yingkun Guo [2,*,‡]

[1] Department of Radiology, Public Health Clinical Center of Chengdu, Chengdu 610061, China
[2] Key Laboratory of Obstetric & Gynecologic and Pediatric Diseases and Birth Defects of Ministry of Education, Department of Radiology, West China Second University Hospital, Sichuan University, 20# South Renmin Road, Chengdu 610017, China
[3] Department of Ultrasound, Public Health Clinical Center of Chengdu, Chengdu 610061, China
[4] Department of Infectious Disease, Public Health Clinical Center of Chengdu, Chengdu 610061, China
* Correspondence: 13281190036@163.com (N.Z.); gykpanda@163.com (Y.G.)
† These authors contributed equally to this work.
‡ These authors contributed equally to this work.

Abstract: Cardiac complications are common in antiretroviral therapy-treated (ART-treated) acquired immune deficiency syndrome (AIDS) patients, and the incidence increases with age. Myocardial injury in ART-treated AIDS patients with a relatively longer disease duration has been evaluated. However, there is no relevant study on whether patients with a short AIDS duration have cardiac dysfunction. Thirty-seven ART-treated males with AIDS and eighteen healthy controls (HCs) were prospectively included for CMR scanning. Clinical data and laboratory examination results were collected. The ART-treated males with AIDS did not have significantly reduced biventricular ejection fraction, myocardial edema, or late gadolinium enhancement. Compared with the HCs, the biventricular volume parameters and left ventricle myocardial strain indices in ART-treated males with AIDS were not significantly reduced (all $p > 0.05$). ART-treated males with AIDS were divided into subgroups according to their CD4+ T-cell counts (<350 cells/µL and $\geq$350 cells/µL) and duration of disease (1–12 months, 13–24 months, and 25–36 months). There was no significant decrease in left or right ventricular volume parameters or myocardial strain indices among the subgroups (all $p > 0.05$). In Pearson correlation analysis, CD4+ T-cell counts were not significantly correlated with biventricular volume parameters or left ventricular myocardial strain indices. In conclusion, ART-treated males with AIDS receiving ART therapy with a short disease duration (less than 3 years) might not develop obvious cardiac dysfunction as evaluated by routine CMR, so it is reasonable to appropriately extend the interval between cardiovascular follow-ups to more than 3 years.

Keywords: acquired immune deficiency syndrome; CMR; cardiovascular complications

Citation: Hou, K.; Fu, H.; Xiong, W.; Gao, Y.; Xie, L.; He, J.; Feng, X.; Zeng, T.; Cai, L.; Xiong, L.; et al. Clinical Application of Cardiac Magnetic Resonance in ART-Treated AIDS Males with Short Disease Duration. *Diagnostics* **2022**, *12*, 2417. https://doi.org/10.3390/diagnostics12102417

Academic Editors: Minjie Lu and Arlene Sirajuddin

Received: 28 July 2022
Accepted: 4 October 2022
Published: 6 October 2022

1. Introduction

Acquired immune deficiency syndrome (AIDS) is a serious global public health concern. In AIDS patients, mortality rates have decreased and survival duration has increased with increased use of antiretroviral treatment (ART). Therefore, the management of HIV/AIDS has become a long-term process. During this long process, multiple factors, including myocardial viral infections, immune activation, inflammation, metabolic abnormalities, nutritional deficiency, and so on, have been proposed to have negative impacts on the heart and to cause cardiac complications [1]. Except for a strong association between HIV and atherosclerosis, myocardial diseases, especially dilated cardiomyopathy and myocarditis, have been reported frequently [2]. Previous studies reported that the incidence of cardiac complications in AIDS patients is approximately 25–75%, and the incidence increases as AIDS patients age [3–6]. In addition, a review concluded that AIDS patients affected by cardiac complications always had a poor prognosis [7]. AIDS-related symptomatic heart failure will become one of the leading causes of HIV/AIDS deaths worldwide in the future [8]. Therefore, timely evaluation of cardiac complications is of paramount significance in AIDS patients.

Echocardiography is the main modality for evaluating cardiac complications in AIDS patients in long-term follow-up with a relatively lower cost. However, due to the intrinsic limitations of echocardiography, such as a lower accuracy of cardiac function evaluation, poor repeatability, and an inability to characterize the myocardium, cardiac magnetic resonance (CMR) is often recommended for the assessment of myocardial involvement and accurate function evaluation in AIDS patients, especially in those in whom cardiac complications are suspected and those having clinical symptoms. A CMR study indicated that HIV infection treated with ART is more likely to result in changes in myocardial structure and function and a higher prevalence of subclinical myocardial edema, myocardial fibrosis, and frequent pericardial effusions [9]. Holloway and colleagues found that cardiac steatosis and fibrosis are highly associated with cardiac dysfunction, cardiovascular morbidity, and mortality in HIV patients [10]. Early evaluation of cardiac dysfunction could provide objective evidence for timely intervention. However, the patients included in the above studies have different and heterogeneous disease courses, with relatively longer disease courses in general. There is no unified conclusion on the time point of cardiac complication onset in patients receiving ART treatment. In addition, the number of AIDS cases diagnosed in males each year is significantly higher than that in females, and males have a more favorable response to ART treatment and have a higher cardiovascular risk than females [11–14]. Therefore, in this study, we aimed to perform an early evaluation of cardiac complications by CMR in male AIDS patients treated with ART with a short disease duration.

2. Materials and Methods

2.1. Study Population

After approval by the Ethics Committee of Chengdu Public Health Clinical Medical Center, 39 male human immunodeficiency virus (HIV)-positive patients diagnosed with AIDS were prospectively enrolled and underwent enhanced CMR examination. Inclusion criteria were as follows: (1) 18 years < age < 70 years; (2) disease duration $\leq$ 36 months; (3) ART therapy initiation after diagnosis; (4) sinus heart rate and no cardiovascular history, including congenital heart disease, heart valve disease, cardiomyopathies, coronary heart disease, and so on; and (5) no malignant tumors or non-HIV-related respiratory diseases. Exclusion criteria were as follows: (1) CMR contraindications; (2) acute kidney injury or severe chronic kidney disease (GFR < 30 mL/min/1.73 m^2); and (3) poor image quality impeding postprocessing. Finally, 2 patients who could not hold their breath, which resulted in poor image quality, were excluded, and 37 patients were included. Eighteen age-matched healthy males were enrolled as healthy controls (HCs). All participants signed informed consent forms. Basic information and laboratory data were collected.

2.2. CMR Scanning Protocol and Imaging Postprocessing

All participants underwent a CMR scan using a 1.5-T whole-body scanner (Signa HDxt; GE Medical Systems, United States of America) in the supine position. Image acquisition was performed during the breath-holding period at the end of inspiration. A series of 8–12 continuous short-axis views of the left ventricle (LV) from the mitral valve to the level of the LV apex were obtained using steady-state free-precession sequences with the following parameters: repetition time, 4 ms; echo time, 2 ms; slice thickness, 6.0–8.0 mm; flip angle, 39°; field of view, 360 × 360 mm^2; and matrix size, 256 × 256. Two-chamber and four-chamber long-axis cine images were also acquired. T2WI black blood images were obtained with a triple-inversion recovery sequence including a left ventricle four-chamber long-axis section and left ventricle short-axis basal, middle, and apical sections. Ten to fifteen minutes after intravenous injection of Meglumine Gadopentetate (Beilu, Beijing,

China) (0.2 mL/kg body weight, flow rate of 3–5 mL/s), late gadolinium enhancement (LGE) images were obtained using the inversion recovery TrueFISP sequence (inversion time was based on TI scout) for the short-axis (slice thickness 8 mm) and 2-chamber and 4-chamber long-axis sections.

2.3. Data Analyses

All CMR data were analyzed using the commercially available software cvi42 (Circle Cardiovascular Imaging, Inc., Calgary, AB, Canada). The series of short-axis cine images and 2-chamber and 4-chamber short-axis cine images were loaded into the short 3D module and issue feature tracking module. Epicardial and endocardial borders were traced manually to compute cardiac function parameters and myocardial deformation parameters. Biventricular function parameters included left ventricular ejection fraction (LVEF), left ventricular end-diastolic volume (LVEDV), left ventricular end-systolic volume (LVESV), left ventricular stroke volume (LVSV), left ventricular myocardial mass (LVM), right ventricular ejection fraction (RVEF), right ventricular end-diastolic volume (RVEDV), right ventricular end-systolic volume (RVESV), and right ventricular stroke volume (RVSV). LV remodeling was defined as the ratio of LVEDV to LV mass (LVRI). Myocardial strain indices included global radial strain (GRS), global circumferential strain (GCS), global longitudinal strain (GLS), global diastolic strain rate radial (GDSR), global diastolic strain rate circumferential (GDSC), and global diastolic strain rate longitudinal (GDSL). Segmental myocardial strain indices for the basal, middle, and apical parts of the LV were also acquired.

2.4. Statistical Analysis

Statistical analyses were performed with SPSS (version 21.0 for Windows; SPSS, Inc., Chicago, IL, USA). Continuous variables are expressed as the mean $\pm$ standard deviation or the median and interquartile range. Categorical variables are presented as frequencies (percentages) and were compared using the chi-square test. Normal distribution was tested with the Kolmogorov–Smirnov test. Continuous variables were compared using ANOVA and independent t tests (normal distribution) or Kruskal–Wallis and Mann–Whitney U tests. Pearson's test was performed to evaluate the relationship between CD4+ T-cell counts and parameters of cardiac function and myocardial strain. p-value < 0.05 was considered statistically significant.

3. Results

3.1. Baseline Characteristics

Thirty-seven ART-treated males with AIDS and eighteen HCs were included in this study. The baseline clinical characteristics are presented in Table 1. There was no significant difference in age, blood pressure, or body mass index between ART-treated males with AIDS and HCs (all $p > 0.05$). The mean age of ART-treated males with AIDS was 37.62 ± 11.10 years old. All patients received ART therapy, and the duration was 20.58 ± 3.72 months. Blood CD4+ T-cell counts were 358.21 ± 57.41 cells/μL. Regarding HIV-related comorbidities, we found that syphilis was present in 5 (13.51%) patients, granulocytopenia was present in 3 (8.11%) patients, HIV-related pneumonia was present in 9 (24.32%) patients, metabolic syndrome was present in 10 (27.03%) patients, and oral fungal infection was present in 2 (5.41%) patients. No patients had cardiovascular symptoms, and electrocardiography and echocardiography findings were negative in all patients. Regarding myocardial injury biomarkers, no significant abnormality was found in the myocardial enzyme spectrum or high-sensitivity c-TnT in 37 patients.

3.2. Comparison between the AIDS Group and the Healthy Control Group

No ART-treated males with AIDS had biventricular ejection fraction reduction, myocardial edema, or LGE. Biventricular function and LV deformation indices were compared between ART-treated males with AIDS and HCs (as shown in Table 2). Compared with the LVM of healthy controls, the LVM of ART-treated males with AIDS tended to be lower (78.14 ± 15.41 g vs. 85.26 ± 15.13 g, $p = 0.053$), while LVMi (43.37 ± 8.98 g/m^2 vs. 45.01 ± 6.34 g/m^2, $p = 0.236$) was similar between the groups after standardization by body surface area.

Table 1. Baseline characteristics.

Characteristics	AIDS Males ($n = 37$)	Healthy Controls ($n = 18$)	p-Value
	General Characteristics		
Age, y	37.62 ± 11.10	39.56 ± 11.24	0.549
Body mass index, kg/m^2	21.88 ± 3.99	23.57 ± 2.53	0.106
Systolic BP, mmHg	126.41 ± 11.37	127.44 ± 10.03	0.743
Diastolic BP, mmHg	81.19 ± 6.79	83.11 ± 4.390	0.280
Electrocardiogram	negative		
	Risk for HIV infection		
Heterosexual, %	19 (51.35)	MSM, %	14 (37.84)
IVDU, %	2 (5.41)	Other, %	2 (5.41)
	Complication for AIDS patients		
Diabetes, %	0 (0.00)	Cardiovascular disease, %	0 (0.00)
Granulocytopenia, %	3 (8.11)	Metabolic syndrome, %	10 (27.03)
Erythra, %	1 (2.70)	Oral fungal infections, %	2 (5.41)
Pneumonia, %	9 (24.32)	HBV co-infection, %	3 (8.11)
HCV co-infection, %	2 (5.41)	Liver dysfunction, %	22 (59.46)
Renal dysfunction, %	1 (2.70)	Tuberculosis, %	6 (16.22)
Syphilis, %	5 (13.51)	malignant tumor, %	0 (0.00)
	Plasma metabolites in AIDS patients		
TSHD, months	20.58 ± 3.72	CD4+ T-cell counts, cells/μL	358.21 ± 57.41
Total duration of ART, %	37 (100.00%)	CHOL, mmol/L	4.57 ± 0.68
Hs-cTnT < 3.00, ng/mL, %	37 (100.00%)	CK-MB, ng/mL	0.52 ± 0.14
Myo < 21.00, ng/mL, %	37 (100.00%)	TG, mmol/L	1.58 ± 0.11
CK-MB, μL	21.45 ± 2.79	LDH, μL	187.06 ± 19.56
HBDH, μL	159.72 ± 21.34	CK, μL	80.19 ± 5.83

Data are summarized by mean $\pm$ SD if they are normal distribution, or median (first and third quartiles) if they are abnormal distribution and n (%) for categorical variables. p-values are obtained using the Student t-test, or Mann–Whitney U test (for non-normal data), X^2 test, or Fisher exact test. MSM, male who has sex with males; IVDU, intravenous drug user; TSHD, time since HIV diagnosis; ART, antiretroviral therapy; Hs-cTnT, high-sensitivity troponin T; Myo, myoglobin; CK-MB, creatine kinase isoenzyme; HBDH, hydroxybutyrate dehydrogenase; CHOL, cholesterol; TG, triglycerides; LDH, lactate dehydrogenase; CK, creatine kinase.

Table 2. CMR characteristics of ART-treated AIDS males and healthy controls.

Cardiac Function	ART-Treated AIDS Males ($n = 37$)	Healthy Controls ($n = 18$)	p-Value
LVEDV (mL)	122.79 ± 21.95	124.39 ± 26.18	0.706
LVESV (mL)	47.35 ± 9.30	49.73 ± 13.29	0.603
LVSV (mL)	75.43 ± 15.72	74.67 ± 15.26	0.816
LVEF (%)	61.75 ± 4.90	60.33 ± 4.66	0.307
LVM (g)	78.14 ± 15.41	85.26 ± 15.13	0.053
LVRI (mL/g)	1.59 ± 0.24	1.48 ± 0.27	0.200
RVEDV (mL)	127.78 ± 26.31	127.09 ± 24.34	0.802
RVESV (mL)	62.22 ± 18.77	62.01 ± 11.72	0.957
RVSV (mL)	65.56 ± 14.06	65.08 ± 17.38	0.513
RVEF (%)	51.94 ± 8.75	50.89 ± 6.08	0.872
LVEDVi (mL/m^2)	68.44 ± 14.84	65.83 ± 12.55	0.720
LVESVi (mL/m^2)	26.39 ± 6.13	26.31 ± 6.64	0.914
LVSVi (mL/m^2)	42.05 ± 10.13	39.53 ± 7.29	0.404
LVMi (g/m^2)	43.37 ± 8.98	45.01 ± 6.34	0.236
RVEDVi (mL/m^2)	71.29 ± 17.19	67.41 ± 12.19	0.554

Table 2. *Cont.*

Cardiac Function	ART-Treated AIDS Males (*n* = 37)	Healthy Controls (*n* = 18)	*p*-Value
RVESVi (mL/m^2)	34.74 ± 11.53	32.95 ± 6.29	0.788
RVSVi (mL/m^2)	36.55 ± 8.97	34.46 ± 8.57	0.441
GRS (%)	24.89 ± 5.60	22.67 ± 4.00	0.216
GCS (%)	−17.39 ± 1.90	−17.08 ± 1.83	0.518
GLS (%)	−17.97 ± 1.99	−11.75 ± 1.35	0.375
BRS (%)	31.23 ± 6.99	30.75 ± 7.26	0.693
BCS (%)	−14.69 ± 2.20	−14.58 ± 1.78	0.673
BLS (%)	−8.73 ± 2.94	−8.86 ± 2.24	0.851
MRS (%)	24.89 ± 5.67	20.12 ± 5.08	0.360
MCS (%)	−17.30 ± 2.09	−16.44 ± 2.40	0.159
MLS (%)	−11.90 ± 2.82	−11.33 ± 2.29	0.216
ARS (%)	25.00 ± 10.02	20.94 ± 9.86	0.441
ACS (%)	−20.30 ± 2.95	−20.48 ± 2.70	0.693
ALS (%)	−14.65 ± 1.46	−14.62 ± 1.60	0.425
BDSR (1/s)	−2.43 ± 0.77	−2.05 ± 0.48	0.060
BDSC (1/s)	0.91 ± 0.19	0.92 ± 0.22	0.828
BDSL (1/s)	0.49 ± 0.43	0.62 ± 0.22	0.206
MDSR (1/s)	−1.57 ± 0.42	−1.40 ± 0.53	0.215
MDSC (1/s)	1.58 ± 0.24	1.02 ± 0.29	0.066
MDSL (1/s)	0.74 ± 0.17	0.76 ± 1.67	0.663
ADSR (1/s)	−2.06 ± 1.10	−1.84 ± 1.14	0.504
ADSC (1/s)	1.44 ± 0.40	1.37 ± 0.37	0.500
ADSL (1/s)	0.88 ± 0.17	0.85 ± 0.15	0.532
GDSR (1/s)	−1.67 ± 0.67	−1.51 ± 0.55	0.368
GDSC (1/s)	1.07 ± 0.22	1.01 ± 0.21	0.360
GDSL (1/s)	0.69 ± 0.16	0.72 ± 0.16	0.459

Data are summarized by mean ± SD if they are normal distribution or median (first and third quartiles) if they are abnormal distribution and n (%) for categorical variables. *p*-values are obtained using the Student *t*-test, or Mann–Whitney U test (for non-normal data), X^2 test, or Fisher exact test. LVEDV, left ventricular end-diastolic volume; LVESV, left ventricular end-systolic volume; LVSV, left ventricular systolic volume; LVEF, left ventricular ejection fraction; LVM, left ventricular mass; LVRI, left ventricular remodeling index; RVEDV, right ventricular end-diastolic volume; RVESV, right ventricular end-systolic volume; RVSV, right ventricular systolic volume; RVEF, right ventricular ejection fraction; LVEDVi, left ventricular end-diastolic volume index; LVESVi, left ventricular end-systolic volume; LVMi, Left Ventricular Mass Index; RVEDVi, right ventricular end-diastolic volume index; RVESVi, right ventricular end-systolic volume; RVSVi, right ventricular systolic volume index; GRS, global radial strain; GCS, global circumferential strain; GLS, global longitudinal strain; BRS, basal radial strain; BCS, basal circumferential strain; BLS, global longitudinal strain; MRS, middle radial strain; MCS, middle circumferential strain; MLS, middle longitudinal strain; ARS, apical radial strain; ACS, apical circumferential strain; ALS, apical longitudinal strain; BDSR, basal peak diastolic strain rate radial; BDSC, basal peak diastolic strain rate circumferential; BDSL, basal peak diastolic strain rate longitudinal; MDSR, middle peak diastolic strain rate radial; MDSC, middle peak diastolic strain rate circumferential; MDSL, middle peak diastolic strain rate longitudinal; ADSR, apical peak diastolic strain rate radial; ADSC, apical peak diastolic strain rate circumferential; ADSL, apical peak diastolic strain rate longitudinal; GDSR, global peak diastolic strain rate radial; GDSC, global peak diastolic strain rate circumferential; GDSL, global peak diastolic strain rate longitudinal.

In the group of ART-treated males with AIDS, LVEF (61.75 ± 4.90% vs. 60.33 ± 4.66%, *p* = 0.307), RVEF (51.94 ± 8.75% vs. 50.89 ± 6.08%, *p* = 0.872), and LVRI (1.59 ± 0.24 mL/g vs. 1.48 ± 0.27 mL/g, *p* = 0.200) were not significantly decreased compared to those of HCs, nor were biventricular volume parameters significantly reduced (all *p* > 0.05). Figure 1 shows that the myocardial deformation parameters GRS, GCS, and GLS were not significantly different between the AIDS group and the HC group (all *p* > 0.05, see also Figure 2). Moreover, the diastolic function indices of males with AIDS, including GDSR, GDSC, and GDSL, were not significantly reduced. Segmental analysis in the basal, middle, and apical segments also did not show significant decreases.

3.3. Subgroup Analysis According to CD4+ T-Cell Counts

According to the CD4+ T-cell counts, ART-treated males with AIDS were divided into two groups: those with CD4+ T-cell counts < 350 cells/μL (moderate or severe reduction of immune function) and those with CD4+ T-cell counts $\geq$ 350 cells/μL (normal or slightly decreased immune function). The results of a comparison of cardiac function and myocardial strain are shown in Table 3. Compared with HCs, no significant decrease in biventricular ejection fraction or volume was found in ART-treated males with AIDS with different levels of CD4+ T-cell counts (all $p > 0.05$). LVRI was not significantly different in the three groups (1.54 $\pm$ 0.19 mL/g vs. 1.64 $\pm$ 0.27 mL/g vs. 1.48 $\pm$ 0.27 mL/g, $p = 0.151$). Figure 3 reports the results of the subgroup comparison: ART-treated males with AIDS with lower CD4+ T-cell counts did not have lower peak strain in any of the three directions compared to ART-treated males with AIDS with CD4+ T-cell counts $\geq$ 350 cells/μL and HCs (all $p > 0.05$). A diastolic function comparison showed that GDSR (-1.68 ± 0.44 1/s vs. -1.67 ± 0.84 1/s vs. -1.51 ± 0.55 1/s, $p = 0.669$), GDSC (1.02 $\pm$ 0.2 1/s vs. 1.11 $\pm$ 0.23 1/s vs. 1.01 $\pm$ 0.21 1/s, $p = 0.294$), and GDSL (0.67 $\pm$ 0.11 1/s vs. 0.7 $\pm$ 0.19 1/s vs. 0.72 $\pm$ 0.16 1/s, $p = 0.677$) were similar among the ART-treated AIDS male subgroups and HCs. Segmental analysis in basal, middle, and apical segments did not show significant decreases.

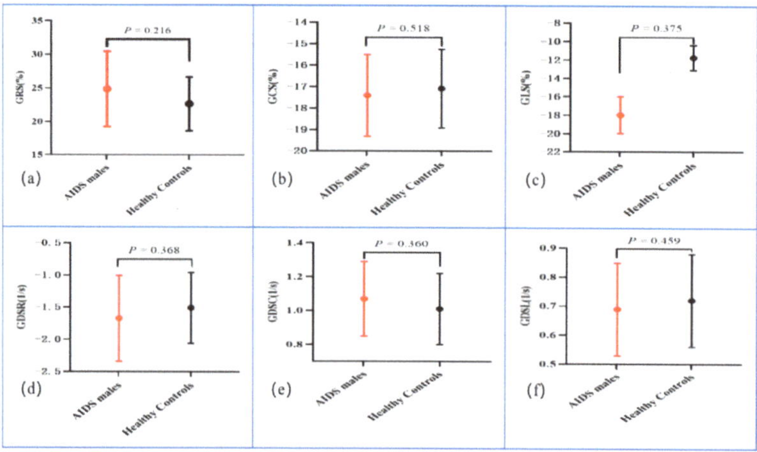

Figure 1. Plots for comparison of left ventricular global radial strain (**a**), global circumferential strain (**b**), global longitudinal strain (**c**), global diastolic strain rate radial(**d**), global diastolic strain rate circumferential (**e**), and global diastolic strain rate longitudinal (**f**).

3.4. Subgroup Analysis According to Disease Duration

According to disease duration, the ART-treated males with AIDS were divided into three subgroups: those with a disease duration of 1–12 months, 13–24 months, and 25–36 months. There was no significant difference in the remaining left and right ventricular function and structure in these three subgroups compared to the healthy control group (shown in Table 4, all $p > 0.05$). Figure 4 shows the results of the subgroup analysis based on disease duration. None of the myocardial strain parameters were decreased in ART-treated AIDS patients with different disease durations compared to HCs (all $p > 0.05$).

3.5. Correlation Analysis

The results of the correlation analysis between CD4+ T-cell counts and CMR parameters are presented in Table 5. In Pearson correlation analysis, CD4+ T-cell counts were not significantly correlated with biventricular volume parameters or left ventricular myocardial strain indices (all $p > 0.05$).

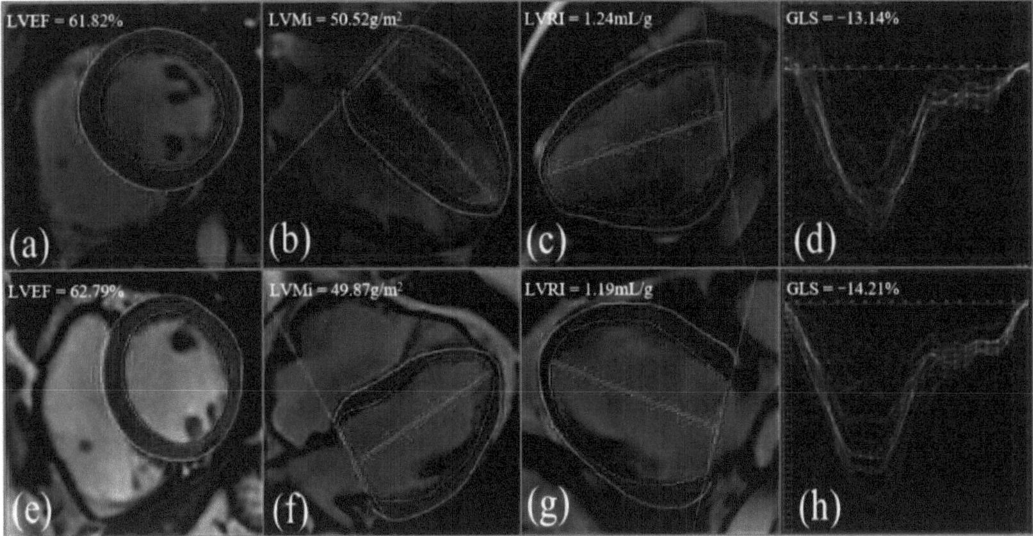

Figure 2. Left ventricular global longitudinal strain between ART-treated AIDS males and HCs. (**a–d**) AIDS, male, 27 years old. (**e–h**) HC, male, 30 years old. (**a,e**) Short axis; (**b,f**) four chamber; (**c,g**) two chamber; (**d,h**) GLS curve.

Table 3. Subgroup analysis according to the CD4+ T-cell counts.

Cardiac Function	CD4 < 350 cells/μL (*n* = 17)	CD4 ≥ 350 cells/μL (*n* = 20)	Healthy Controls (*n* = 18)	*p*-Value
LVEDV (mL)	117.04 ± 17.23	127.67 ± 24.66	124.39 ± 26.18	0.376
LVESV (mL)	46.14 ± 5.54	48.38 ± 11.65	49.73 ± 13.29	0.616
LVSV (mL)	70.89 ± 16.38	79.29 ± 14.42	74.67 ± 15.26	0.256
LVEF (%)	61.10 ± 6.18	62.30 ± 3.55	60.33 ± 4.66	0.456
LVM (g)	76.60 ± 12.60	79.44 ± 17.67	85.26 ± 15.13	0.245
LVRI (mL/g)	1.54 ± 0.19	1.64 ± 0.27	1.48 ± 0.27	0.151
RVEDV (mL)	125.18 ± 23.00	130.00 ± 29.24	127.09 ± 24.34	0.849
RVESV (mL)	61.76 ± 15.40	62.62 ± 21.63	62.01 ± 11.72	0.987
RVSV (mL)	63.42 ± 13.00	67.38 ± 14.99	65.08 ± 17.38	0.730
RVEF (%)	51.07 ± 8.63	52.67 ± 9.01	50.89 ± 6.08	0.754
LVEDVi (mL/m^2)	64.81 ± 10.31	71.52 ± 17.49	65.83 ± 12.55	0.290
LVESVi (mL/m^2)	25.48 ± 2.80	27.16 ± 7.96	26.31 ± 6.64	0.722
LVSVi (mL/m^2)	39.33 ± 9.68	44.36 ± 10.16	39.53 ± 7.29	0.169
LVMi (g/m^2)	42.25 ± 6.00	44.32 ± 10.97	45.01 ± 6.34	0.593
RVEDVi (mL/m^2)	69.40 ± 13.86	72.90 ± 19.81	67.41 ± 12.19	0.560
RVESVi (mL/m^2)	34.19 ± 8.69	35.21 ± 13.71	32.95 ± 6.29	0.794
RVSVi (mL/m^2)	35.21 ± 7.98	37.69 ± 9.79	34.46 ± 8.57	0.504
GRS (%)	23.57 ± 3.96	26.01 ± 6.57	22.67 ± 4.00	0.120
GCS (%)	−17.32 ± 1.70	−17.45 ± 2.10	−17.08 ± 1.83	0.832
GLS (%)	−11.73 ± 2.21	−12.17 ± 1.82	−11.75 ± 1.35	0.705
BRS (%)	29.27 ± 6.11	32.89 ± 7.40	30.75 ± 7.26	0.292
BCS (%)	−14.74 ± 2.29	−14.64 ± 2.19	−14.58 ± 1.78	0.973
BLS (%)	−8.77 ± 2.92	−8.70 ± 3.03	−8.86 ± 2.24	0.984
MRS (%)	21.13 ± 4.55	22.53 ± 6.53	20.12 ± 5.08	0.404
MCS (%)	−16.93 ± 1.82	−17.62 ± 2.30	−16.44 ± 2.40	0.262
MLS (%)	−11.43 ± 3.45	−12.31 ± 2.17	−11.33 ± 2.29	0.467
ARS (%)	23.67 ± 6.41	26.13 ± 12.37	20.94 ± 9.86	0.288

Table 3. *Cont.*

Cardiac Function	CD4 < 350 cells/μL (*n* = 17)	CD4 ≥ 350 cells/μL (*n* = 20)	Healthy Controls (*n* = 18)	*p*-Value
ACS (%)	−20.54 ± 2.64	−20.09 ± 3.25	−20.48 ± 2.70	0.877
ALS (%)	−14.31 ± 1.45	−14.95 ± 1.43	−14.62 ± 1.60	0.441
BDSR (1/s)	−2.28 ± 0.60	−2.55 ± 0.88	−2.05 ± 0.48	0.088
BDSC (1/s)	0.94 ± 0.20	0.88 ± 0.19	0.92 ± 0.22	0.686
BDSL (1/s)	0.60 (0.49, 0.68)	0.58 (−0.06, 0.64)	0.59 (0.46, 0.72)	0.067
MDSR (1/s)	−1.52 ± 0.39	−1.61 ± 0.45	−1.40 ± 0.53	0.402
MDSC (1/s)	1.09 ± 0.19	1.21 ± 0.26	1.02 ± 0.29	0.063
MDSL (1/s)	0.70 ± 0.14	0.77 ± 0.19	0.76 ± 1.67	0.406
ADSR (1/s)	−1.98 ± 0.68	−2.12 ± 1.37	−1.84 ± 1.14	0.745
ADSC (1/s)	1.39 ± 0.36	1.49 ± 0.44	1.37 ± 0.37	0.581
ADSL (1/s)	0.85 ± 0.16	0.91 ± 0.18	0.85 ± 0.15	0.426
GDSR (1/s)	−1.68 ± 0.44	−1.67 ± 0.84	−1.51 ± 0.55	0.669
GDSC (1/s)	1.02 ± 0.20	1.11 ± 0.23	1.01 ± 0.21	0.294
GDSL (1/s)	0.67 ± 0.11	0.70 ± 0.19	0.72 ± 0.16	0.677

abbreviations as in Table 2. The distribution of some groups of data is not satisfied but close to the normal distribution, so the median is used to represent its concentration.

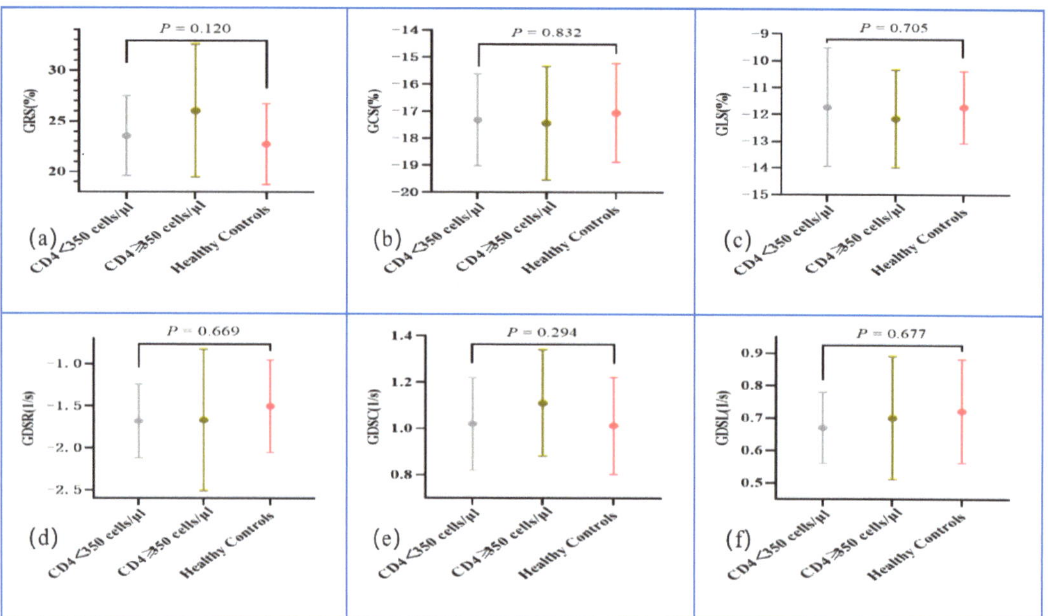

Figure 3. Plots for comparison of left ventricular global radial strain (**a**), global circumferential strain (**b**), global longitudinal strain (**c**), global diastolic strain rate radial (**d**), global diastolic strain rate circumferential (**e**), and global diastolic strain rate longitudinal (**f**).

Table 4. Subgroup analysis based on disease duration.

Cardiac Function	1–12 Months (*n* = 10)	13–24 Months (*n* = 12)	25–36 Months (*n* = 15)	Healthy Controls (*n* = 18)	*p*-Value
LVEDV (mL)	116.12 ± 15.77	131.62 ± 25.00	120.16 ± 21.96	124.39 ± 26.18	0.496
LVESV (mL)	48.37 ± 5.76	48.67 ± 11.35	45.62 ± 9.72	49.73 ± 13.29	0.671

Table 4. *Cont.*

Cardiac Function	1–12 Months (*n* = 10)	13–24 Months (*n* = 12)	25–36 Months (*n* = 15)	Healthy Controls (*n* = 18)	*p*-Value
LVSV (mL)	67.75 ± 13.07	82.95 ± 16.12	74.55 ± 15.16	74.67 ± 15.26	0.176
LVEF (%)	58.05 ± 4.75	62.75 ± 3.38	63.41 ± 4.97	60.33 ± 4.66	0.050
LVM (g)	79.72 ± 14.86	79.69 ± 19.35	75.85 ± 12.84	85.26 ± 15.13	0.232
LVRI (mL/g)	1.48 ± 0.21	1.68 ± 0.23	1.60 ± 0.25	1.48 ± 0.27	0.171
RVEDV (mL)	121.65 ± 26.58	139.47 ± 28.33	122.52 ± 22.84	127.09 ± 24.34	0.398
RVESV (mL)	61.33 ± 18.59	67.36 ± 16.62	58.71 ± 20.76	62.01 ± 11.72	0.524
RVSV (mL)	60.32 ± 12.83	72.12 ± 17.40	63.81 ± 10.25	65.08 ± 17.38	0.334
RVEF (%)	50.24 ± 8.77	51.72 ± 6.78	52.23 ± 10.37	50.89 ± 6.08	0.709
LVEDVi (mL/m^2)	65.05 ± 11.69	73.09 ± 16.82	66.97 ± 15.06	65.83 ± 12.55	0.661
LVESVi (mL/m^2)	27.03 ± 4.11	27.01 ± 7.20	25.47 ± 6.61	26.31 ± 6.64	0.807
LVSVi (mL/m^2)	38.02 ± 8.80	46.09 ± 10.82	41.50 ± 9.81	39.53 ± 7.29	0.254
LVMi (g/m^2)	44.30 ± 7.46	44.15 ± 11.63	42.12 ± 7.89	45.01 ± 6.34	0.571
RVEDVi (mL/m^2)	68.02 ± 16.36	77.56 ± 18.61	68.47 ± 16.31	67.41 ± 12.19	0.482
RVESVi (mL/m^2)	34.13 ± 10.50	37.58 ± 10.92	32.88 ± 12.91	32.95 ± 6.29	0.500
RVSVi (mL/m^2)	33.88 ± 8.65	39.68 ± 10.49	35.58 ± 7.49	34.46 ± 8.57	0.453
GRS (%)	24.91 ± 3.22	24.64 ± 4.94	25.07 ± 7.39	22.67 ± 4.00	0.509
GCS (%)	−17.56 ± 1.47	−16.90 ± 1.75	−17.67 ± 2.28	−17.08 ± 1.83	0.583
GLS (%)	−11.34 ± 2.07	−12.05 ± 2.22	−12.32 ± 1.77	−11.75 ± 1.35	0.600
BRS (%)	28.39 ± 6.31	32.69 ± 8.56	31.95 ± 5.84	30.75 ± 7.26	0.527
BCS (%)	−14.54 ± 1.76	−13.54 ± 2.60	−15.71 ± 1.71	−14.58 ± 1.78	0.156
BLS (%)	−8.03 ± 2.23	−8.63 ± 3.41	−9.28 ± 3.03	−8.86 ± 2.24	0.809
MRS (%)	21.94 ± 3.74	22.51 ± 1.78	21.35 ± 7.07	20.12 ± 5.08	0.598
MCS (%)	−17.39 ± 1.79	−16.97 ± 1.78	−17.51 ± 2.57	−16.44 ± 2.40	0.543
MLS (%)	−11.18 ± 3.23	−11.83 ± 2.47	−12.45 ± 2.89	−11.33 ± 2.29	0.458
ARS (%)	26.87 ± 4.26	21.69 ± 7.91	26.40 ± 13.55	20.94 ± 9.86	0.405
ACS (%)	−20.85 ± 2.67	−20.00 ± 2.57	−20.17 ± 3.57	−20.48 ± 2.70	0.744
ALS (%)	−14.26 ± 2.35	−15.02 ± 1.66	−14.62 ± 0.97	−14.62 ± 1.60	0.814
BDSR (1/s)	−2.22 ± 0.75	−2.52 ± 0.85	−2.49 ± 0.74	−2.05 ± 0.48	0.198
BDSC (1/s)	0.96 ± 0.16	0.80 ± 0.17	0.96 ± 0.20	0.92 ± 0.219	0.122
BDSL (1/s)	0.60 (0.47, 0.69)	0.52 (−0.06, 0.67)	0.60 (0.51, 0.67)	0.59 (0.46, 0.72)	0.237
MDSR (1/s)	−1.66 ± 0.34	−1.56 ± 0.30	−1.51 ± 0.55	−1.40 ± 0.53	0.549
MDSC (1/s)	1.24 ± 0.22	1.10 ± 0.16	1.15 ± 0.29	1.02 ± 0.29	0.183
MDSL (1/s)	0.76 ± 0.16	0.72 ± 0.16	0.73 ± 0.19	0.76 ± 1.67	0.914
ADSR (1/s)	−2.31 ± 0.63	−1.75 ± 0.84	−2.13 ± 1.47	−1.84 ± 1.14	0.593
ADSC (1/s)	1.52 ± 0.48	1.41 ± 0.22	1.42 ± 0.46	1.37 ± 0.37	0.812
ADSL (1/s)	0.87 ± 0.15	0.88 ± 0.13	0.89 ± 0.22	0.85 ± 0.15	0.932
GDSR (1/s)	−1.81 ± 0.15	−1.72 ± 0.38	−1.56 ± 1.05	−1.51 ± 0.55	0.622
GDSC (1/s)	1.13 ± 0.21	1.03 ± 0.14	1.05 ± 0.28	1.01 ± 0.21	0.581
GDSL (1/s)	0.67 ± 0.13	0.39 ± 0.14	0.70 ± 0.19	0.72 ± 0.16	0.867

abbreviations as in Table 2. The distribution of some groups of data is not satisfied but close to the normal distribution, so the median is used to represent its concentration.

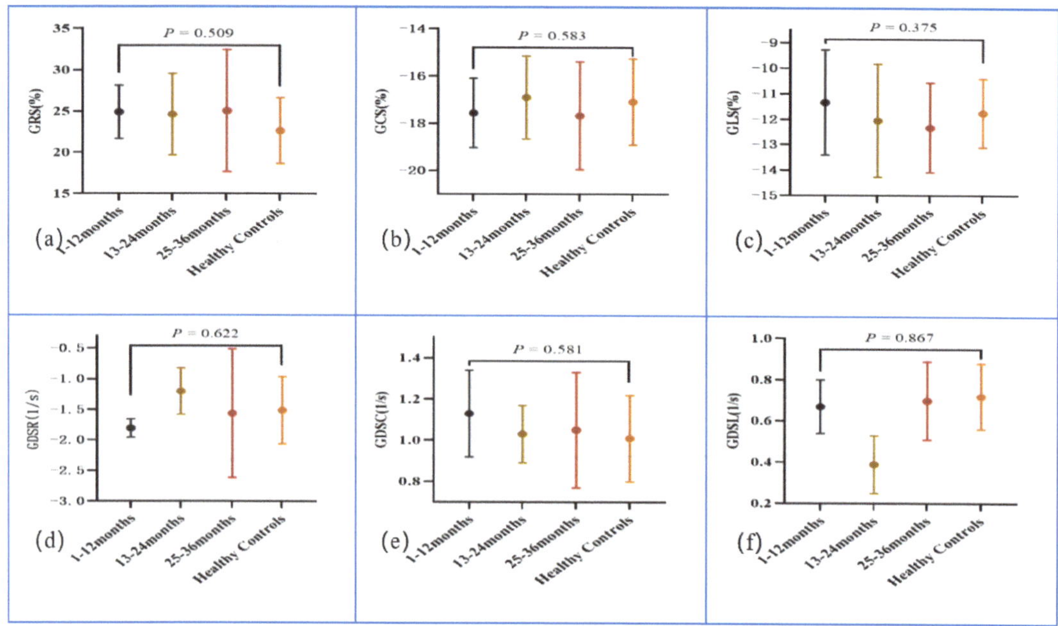

Figure 4. Plots for comparison of left ventricular global radial strain (**a**), global circumferential strain (**b**), global longitudinal strain (**c**), global diastolic strain rate radial (**d**), global diastolic strain rate circumferential (**e**), and global diastolic strain rate longitudinal (**f**).

Table 5. Correlation between CD4+ T-cell counts and CMR indices.

Cardiac Function	CD4+ T-Cell Counts (Cells/ μL)		Cardiac Function	CD4+ T-Cell Counts (Cells/ μL)	
	r	*p*-Value		r	*p*-Value
LVEDV (mL)	0.157	0.355	BCS (%)	−0.156	0.358
LVESV (mL)	0.075	0.659	BLS (%)	−0.086	0.614
LVSV (mL)	0.174	0.302	MRS (%)	−0.059	0.727
LVEF (%)	0.102	0.546	MCS (%)	0.009	0.960
LVM (g)	0.110	0.518	MLS (%)	−0.023	0.892
LVRI (mL/g)	0.025	0.883	ARS (%)	0.080	0.639
RVEDV (mL)	0.034	0.841	ACS (%)	−0.078	0.645
RVESV (mL)	0.022	0.899	ALS (%)	0.014	0.937
RVSV (mL)	0.035	0.836	BDSR (1/s)	0.014	0.936
RVEF (%)	0.009	0.958	BDSC (1/s)	−0.144	0.395
LVEDVi (mL/m^2)	0.118	0.488	BDSL (1/s)	−0.085	0.618
LVESVi (mL/m^2)	0.062	0.714	MDSR (1/s)	−0.069	0.684
LVSVi (mL/m^2)	0.135	0.427	MDSC (1/s)	0.026	0.877
LVMi (g/m^2)	0.143	0.397	MDSL (1/s)	0.013	0.941
RVEDVi (mL/m^2)	0.027	0.875	ADSR (1/s)	0.091	0.591
RVESVi (mL/m^2)	0.029	0.867	ADSC (1/s)	0.000	0.999
RVSVi (mL/m^2)	0.015	0.931	ADSL (1/s)	0.041	0.810
GRS (%)	0.046	0.787	GDSR (1/s)	−0.009	0.959
GCS (%)	−0.072	0.670	GDSC (1/s)	−0.011	0.948
GLS (%)	−0.094	0.581	GDSL (1/s)	−0.088	0.603
BRS (%)	0.085	0.618			

abbreviations as in Table 2.

4. Discussion

This was the first CMR study that focused on the assessment of cardiac complications in ART-treated males with AIDS with a short disease duration (within 3 years), and no obvious cardiac dysfunction was found. ART is relatively safe for the heart over a short course of treatment. We recommend that the interval between cardiac follow-ups after the diagnosis of AIDS be appropriately extended for these patients.

Previous studies that enrolled ART-treated AIDS patients with a mean disease duration of 90 months determined that AIDS patients can develop heart enlargement, ventricular septal thickening, and increased volume and myocardial mass, which further progress to diastolic and systolic dysfunction and even heart failure [15–17]. The results of a study on cardiac function in 156 AIDS patients with a median disease duration of 10.4 years [18] also suggested that left ventricular remodeling, revealed by an increase in LVMi (65 g/m^2 [$49\text{–}77 \text{ g/m}^2$] vs. 57 g/m^2 [$49\text{–}64 \text{ g/m}^2$]), is closely related to cardiovascular adverse events, even though LVEF was still within the normal range. Nevertheless, neither LVRI nor biventricular functional parameters in ART-treated males with AIDS were different from those in HCs in this study. Subgroup analysis according to different disease severities based on CD4+ cell counts and disease duration did not show significant differences. Myocardial strain reflects the change in the length of myocardial fibers in the process of contraction and relaxation, and it is an early sensitive index to evaluate subclinical cardiac dysfunction [19,20]. Case reports and cohort studies both found that ART-treated AIDS patients with a mean treatment course of 9.3 years developed significantly decreased global longitudinal strain and circumferential strain with normal LVEF [21]. Even in children and young adults with a treatment course of 6.8 years, significantly reduced global longitudinal strain was revealed [22]. The disease duration of AIDS patients in these studies was relatively longer (more than 5 years) than that of the patients in our study (within 3 years), which might be the foundation of the lack of myocardial strain observed in our study. Cardiac dysfunction is the result of a long-term cumulative effect of many factors [23–30], and the exact duration over which enough negative effects would accumulate and lead to cardiac dysfunction in AIDS needs further study.

In addition to cardiac function and myocardial strain, myocardial characterization in ART-treated males with AIDS was also performed. No evident myocardial edema or fibrosis was visually revealed by T2WI or LGE in our study. Acute HIV infection could cause myocarditis, revealed by obvious high intensity on T2WI and related symptoms, including chest pain, dyspnea, and palpitation [31]. Negative cardiac-related symptoms and myocardial injury biomarkers were consistent with the negative findings in T2WI images in our study. In the abovementioned study with a longer disease duration, LGE was identified in 24.3% of AIDS patients, and AIDS patients with positive LGE were more likely to develop adverse cardiac events (46% vs. 18%, $p = 0.002$) [18]. The relatively shorter disease course in our study might be insufficient to cause myocardial necrosis. Recent studies applying mapping sequences in AIDS found subclinical myocardial edema and fibrosis with increases in the T2 value, native T1 value, and extracellular volume [9,32]. However, detailed information about disease duration was not reported in these studies [9,32]. A mapping sequence was not performed in our study, and further research is needed to determine whether subclinical myocardial edema or fibrosis already exists in patients with a short disease duration.

Dyslipidemia is extremely common in ART-treated AIDS patients. ART is also accompanied by lipodystrophy, redistribution, and insulin resistance. All these factors would increase cardiovascular risk in AIDS patients receiving ART [33,34]. Additionally, injury caused by HIV infection, persistent inflammatory stimulation, and immune activation increase the possibility of myocardial injury [35,36]. ART is initiated immediately after the diagnosis of AIDS, and this treatment will be administered throughout life. The corresponding negative impact of ART and HIV infection-related factors on the heart will lead to significant cardiac complications. Thus, cardiovascular system monitoring is indispensable in ART-treated AIDS patients. Finally, we found that there was no significant cardiac dysfunction in ART-treated males with AIDS within 3 years, which indicated that ART was relatively safe for the heart in AIDS patients with a short disease duration. However, previous studies found obvious myocardial injury in ART-treated patients with longer disease durations. Thus, further studies focusing on ART-treated AIDS patients with disease durations longer than 3 years are also needed.

One limitation of this study was that the sample size was too small. In the next study, we will expand the sample size and perform a follow-up CMR study. Furthermore, a mapping sequence was not applied in this study. We used only the sequences that are most commonly used in the clinic for evaluation. However, the technology we used in this study is mature in clinical applications and more accessible in medical institutions.

5. Conclusions

ART-treated males with AIDS with a short disease duration may not develop obvious cardiac dysfunction as evaluated by routine CMR, so it is reasonable to appropriately extend the interval between cardiovascular follow-ups to more than 3 years.

Author Contributions: All authors contributed to the protocol of this study and to this manuscript. K.H. and H.F. performed CMR image analysis. W.X., Y.G. (Yingkun Guo), L.X. (Liqiu Xie), T.Z., L.C., X.F., N.J. and M.J. recorded clinical information and laboratory tests. H.Z. and B.K. carried out telephone questionnaires interviews. J.H. and L.X. (Lei Xiong) designed the tables and figures. K.H., H.F. and W.X. performed the statistical analysis and manuscript writing. N.Z. and Y.G. (Yingkun Guo) helped to revise the manuscript. Y.G. (Yueqin Gao) is the guarantor. K.H. and H.F. contributed equally to this manuscript and should be considered co-first authors. All authors have read and agreed to the published version of the manuscript.

Funding: This work was supported by the Sichuan Medical Scientific Research Project Plan (S19001), the Chengdu Health Commission Medical Scientific Research Project (2021024), the Sichuan Provincial Health Commission Scientific Research Project (21JP155), the National Natural Science Foundation of China (82120108015, 82102020, 82071874, 81971586), and the Sichuan Science and Technology Program (2020YJ0029).

Institutional Review Board Statement: The study was conducted according to the guidelines of the Declaration of Helsinki and was approved by the Institutional Research Ethics Committee of Public Health Clinical Center of Chengdu (PJ-K2020-29-01).

Informed Consent Statement: Informed consent was obtained from all subjects involved in the study. Written informed consent has been obtained from the patients to publish this paper.

Data Availability Statement: All data generated or analyzed during the study are included in the published paper.

Conflicts of Interest: The authors declare that they have no competing interests.

References

1. De Filippi, C.R.; Grinspoon, S.K. Myocardial Dysfunction With Contemporary Management of HIV: Prevalence, Pathophysiology, and Opportunities for Prevention. *JACC Heart Fail.* **2019**, *7*, 109–111. [CrossRef]
2. Anderson, D.W.; Virmani, R.; Reilly, J.M.; O'Leary, T.; Cunnion, R.E.; Robinowitz, M.; Macher, A.M.; Punja, U.; Villaflor, S.T.; Parrillo, J.E. Prevalent myocarditis at necropsy in the acquired immunodeficiency syndrome. *J. Am. Coll. Cardiol.* **1988**, *11*, 792–799. [CrossRef]
3. D'Amati, G.; di Gioia, C.R.; Gallo, P. Pathological findings of HIV-associated cardiovascular disease. *Ann. N. Y. Acad. Sci.* **2001**, *946*, 23–45. [CrossRef]
4. Lewis, W. AIDS: Cardiac findings from 115 autopsies. *Prog. Cardiovasc. Dis.* **1989**, *32*, 207–215. [CrossRef]
5. Islam, F.M.; Wu, J.; Jansson, J.; Wilson, D.P. Relative risk of cardiovascular disease among people living with HIV: A systematic review and meta-analysis. *HIV Med.* **2012**, *13*, 453–468. [CrossRef]
6. Shah, A.S.; Stelzle, D.; Lee, K.K.; Beck, E.J.; Alam, S.; Clifford, S.; Longenecker, C.T.; Strachan, F.; Bagchi, S.; Whiteley, W.; et al. Global burden of atherosclerotic cardiovascular disease in people living with HIV. *Circulation* **2018**, *138*, 1100–1112. [CrossRef] [PubMed]
7. Wu, K.C.; Woldu, B.; Post, W.S.; Hays, A.G. Prevention of heart failure, tachyarrhythmias and sudden cardiac death in HIV. *Curr. Opin. HIV AIDS* **2022**, *17*, 261–269. [CrossRef] [PubMed]

8. BMJ Publishing Group. Heart disease is the next hurdle for HIV positive Africans surviving concurrent infection. *Sex. Transm. Infect.* **2003**, *79*, 219. [CrossRef]

9. Ntusi, N.; O'Dwyer, E.; Dorrell, L.; Wainwright, E.; Piechnik, S.; Clutton, G.; Hancock, G.; Ferreira, V.; Cox, P.; Badri, M.; et al. HIV-1-Related Cardiovascular Disease Is Associated With Chronic Inflammation, Frequent Pericardial Effusions, and Probable Myocardial Edema. *Circ. Cardiovasc. Imaging* **2016**, *9*, e004430. [CrossRef]

10. Holloway, C.J.; Ntusi, N.; Suttie, J.; Mahmod, M.; Wainwright, E.; Clutton, G.; Hancock, G.; Beak, P.; Tajar, A.; Piechnik, S.K.; et al. Comprehensive cardiac magnetic resonance imaging and spectroscopy reveal a high burden of myocardial disease in HIV patients. *Circulation* **2013**, *128*, 814–822. [CrossRef]

11. Soon, G.G.; Min, M.; Struble, K.A.; Chan-Tack, K.M.; Hammerstrom, T.; Qi, K.; Zhou, S.; Bhore, R.; Murray, J.S.; Birnkrant, D.B. Meta-analysis of gender differences in efficacy outcomes for HIV-positive subjects in randomized controlled clinical trials of antiretroviral therapy (2000–2008). *AIDS Patient Care STDS* **2012**, *26*, 444–453. [CrossRef] [PubMed]

12. O'Neil, A.; Scovelle, A.J.; Milner, A.J.; Kavanagh, A. Gender/Sex as a Social Determinant of Cardiovascular Risk. *Circulation* **2018**, *137*, 854–864. [CrossRef] [PubMed]

13. Hariri, S.; McKenna, M.T. Epidemiology of human immunodeficiency virus in the United States. *Clin. Microbiol. Rev.* **2007**, *20*, 478–488. [CrossRef]

14. NCAIDS; NCSTD; China CDC. Update on the AIDS/STD epidemic in China in December 2017. *Chin. J. AIDS STD* **2018**, *24*, 111. [CrossRef]

15. Reinsch, N.; Buhr, C.; Krings, P.; Kaelsch, H.; Kahlert, P.; Konorza, T.; Neumann, T.; Erbel, R.; Competence Network of Heart Failure. Effect of gender and highly active antiretroviral therapy on HIV-related pulmonary arterial hypertension: Results of the HIV-HEART Study. *HIV Med.* **2008**, *9*, 550–556. [CrossRef] [PubMed]

16. Ten Freyhaus, H.; Vogel, D.; Lehmann, C.; Kümmerle, T.; Wyen, C.; Fätkenheuer, G.; Rosenkranz, S. Echocardiographic screening for pulmonary arterial hypertension in HIV-positive patients. *Infection* **2014**, *42*, 737–741. [CrossRef]

17. Snopková, S.; Husa, P. Metabolický syndrom u onemocniní HIV/AIDS [Metabolic syndrome and HIV/AIDS disorder]. *Klin. Mikrobiol. A Infekcni Lek.* **2006**, *12*, 108–116.

18. De Leuw, P.; Arendt, C.T.; Haberl, A.E.; Froadinadl, D.; Kann, G.; Wolf, T.; Stephan, C.; Schuettfort, G.; Vasquez, M.; Arcari, L.; et al. Myocardial Fibrosis and Inflammation by CMR Predict Cardiovascular Outcome in People Living With HIV. *JACC Cardiovasc. Imaging* **2021**, *14*, 1548–1557. [CrossRef]

19. Kalam, K.; Otahal, P.; Marwick, T.H. Prognostic implications of global LV dysfunction: A systematic review and meta-analysis of global longitudinal strain and ejection fraction. *Heart* **2014**, *100*, 1673–1680. [CrossRef] [PubMed]

20. Andre, F.; Steen, H.; Matheis, P.; Westkott, M.; Breuninger, K.; Sander, Y.; Kammerer, R.; Galuschky, C.; Giannitsis, E.; Korosoglou, G.; et al. Age- and gender-related normal left ventricular deformation assessed by cardiovascular magnetic resonance feature tracking. *J. Cardiovasc. Magn. Reson.* **2015**, *17*, 25. [CrossRef]

21. Luetkens, J.A.; Doerner, J.; Schwarze-Zander, C.; Wasmuth, J.C.; Boesecke, C.; Sprinkart, A.M.; Schmeel, F.C.; Homsi, R.; Gieseke, J.; Schild, H.H.; et al. Cardiac Magnetic Resonance Reveals Signs of Subclinical Myocardial Inflammation in Asymptomatic HIV-Infected Patients. *Circ. Cardiovasc. Imaging* **2016**, *9*, e004091. [CrossRef] [PubMed]

22. McCrary, A.W.; Nyandiko, W.M.; Ellis, A.M.; Chakraborty, H.; Muehlbauer, M.J.; Koech, M.M.; Daud, I.; Birgen, E.; Thielman, N.M.; Kisslo, J.A.; et al. Early cardiac dysfunction in children and young adults with perinatally acquired HIV. *AIDS* **2020**, *34*, 539–548. [CrossRef] [PubMed]

23. Hansen, L.; Parker, I.; Roberts, L.M.; Sutliff, R.L.; Platt, M.O.; Gleason, R.L., Jr. Azidothymidine (AZT) leads to arterial stiffening and intima-media thickening in mice. *J. Biomech.* **2013**, *46*, 1540–1547. [CrossRef] [PubMed]

24. Ho, J.E.; Deeks, S.G.; Hecht, F.M.; Xie, Y.; Schnell, A.; Martin, J.N.; Ganz, P.; Hsue, P.Y. Initiation of antiretroviral therapy at higher nadir CD4$^+$ T-cell counts is associated with reduced arterial stiffness in HIV-infected individuals. *AIDS* **2010**, *24*, 1897–1905. [CrossRef]

25. Volpe, G.E.; Tang, A.M.; Polak, J.F.; Mangili, A.; Skinner, S.C.; Wanke, C.A. Progression of carotid intima-media thickness and coronary artery calcium over 6 years in an HIV-infected cohort. *J. Acquir. Immune Defic. Syndr.* **2013**, *64*, 51–57. [CrossRef]

26. Maloberti, A.; Giannattasio, C.; Dozio, D.; Betelli, M.; Villa, P.; Nava, S.; Cesana, F.; Facchetti, R.; Giupponi, L.; Castagna, F.; et al. Metabolic syndrome in human immunodeficiency virus-positive subjects: Prevalence, phenotype, and related alterations in arterial structure and function. *Metab. Syndr. Relat. Disord.* **2013**, *11*, 403–411. [CrossRef]

27. Hansen, L.; Parker, I.; Sutliff, R.L.; Platt, M.O.; Gleason, R.L., Jr. Endothelial dysfunction, arterial stiffening, and intima-media thickening in large arteries from HIV-1 transgenic mice. *Ann. Biomed. Eng.* **2013**, *41*, 682–693. [CrossRef] [PubMed]

28. Calza, L.; Manfredi, R.; Colangeli, V.; Trapani, F.F.; Salvadori, C.; Magistrelli, E.; Danese, I.; Verucchi, G.; Serra, C.; Viale, P. Two-year treatment with rosuvastatin reduces carotid intima-media thickness in HIV type 1-infected patients receiving highly active antiretroviral therapy with asymptomatic atherosclerosis and moderate cardiovascular risk. *AIDS Res. Hum. Retrovir.* **2013**, *29*, 547–556. [CrossRef] [PubMed]

29. Ferraioli, G.; Tinelli, C.; Maggi, P.; Gervasoni, C.; Grima, P.; Viskovic, K.; Carerj, S.; Filice, G.; Filice, C.; Arterial Stiffness Evaluation in HIV-Infected Subjects Study Group. Arterial stiffness evaluation in HIV-positive patients: A multicenter matched control study. *AJR Am. J. Roentgenol.* **2011**, *197*, 1258–1262. [CrossRef]

30. Cristofaro, M.; Cicalini, S.; Busi Rizzi, E.; Schininà, V.; Petrosillo, N.; Bibbolino, C. Ultrasonography in lesions of the carotid vessels in HIV positive patients. *Radiol. Med.* **2011**, *116*, 61–70. [CrossRef] [PubMed]

31. Stöbe, S.; Tayal, B.; Tünnemann-Tarr, A.; Hagendorff, A. Dynamics in myocardial deformation as an indirect marker of myocardial involvement in acute myocarditis due to HIV infection: A case report. *Eur. Heart J. Case Rep.* **2021**, *5*, ytaa511. [CrossRef] [PubMed]

32. Yan, C.; Li, R.; Guo, X.; Yu, H.; Li, W.; Li, W.; Ren, M.; Yang, M.; Li, H. Cardiac Involvement in Human Immunodeficiency Virus Infected Patients: An Observational Cardiac Magnetic Resonance Study. *Front. Cardiovasc. Med.* **2021**, *8*, 756162. [CrossRef] [PubMed]

33. DAD Study Group; Friis-Møller, N.; SReiss, P.; VSabin, C.A.; VWeber, R.; Monforte Ad El-Sadr, W.; Thiébaut, R.; VDe Wit, S.; Kirk, O.; Fontas, E.; et al. Class of antiretroviral drugs and the risk of myocardial infarction. *N. Engl. J. Med.* **2007**, *356*, 1723–1735. [CrossRef]

34. Blanco, F.; San Román, J.; Vispo, E.; López, M.; Salto, A.; Abad, V.; Soriano, V. Management of metabolic complications and cardiovascular risk in HIV-infected patients. *AIDS Rev.* **2010**, *12*, 231–241.

35. Barbaro, G.; Fisher, S.D.; Lipshultz, S.E. Pathogenesis of HIV-associated cardiovascular complications. *Lancet Infect Dis.* **2001**, *1*, 115–124, Erratum in *Lancet Infect Dis.* **2004**, *4*, 533. [CrossRef]

36. Chaves, A.A.; Mihm, M.J.; Schanbacher, B.L.; Basuray, A.; Liu, C.; Ayers, L.W.; Bauer, J.A. Cardiomyopathy in a murine model of AIDS: Evidence of reactive nitrogen species and corroboration in human HIV/AIDS cardiac tissues. *Cardiovasc. Res.* **2003**, *60*, 108–118. [CrossRef]

Article

Differential Expression of microRNAs in Hypertrophied Myocardium and Their Relationship to Late Gadolinium Enhancement, Left Ventricular Hypertrophy and Remodeling in Hypertrophic Cardiomyopathy

Chen Zhang [1,†], Hongbo Zhang [1,2,†], Lei Zhao [2], Zhipeng Wei [3], Yongqiang Lai [3,*] and Xiaohai Ma [1,*]

[1] Department of Interventional Diagnosis and Treatment, Beijing Anzhen Hospital, Capital Medical University, 2nd Anzhen Road, Chaoyang District, Beijing 100020, China
[2] Department of Radiology, Beijing Anzhen Hospital, Capital Medical University, 2nd Anzhen Road, Chaoyang District, Beijing 100020, China
[3] Department of Cardiac Surgery, Beijing Anzhen Hospital, Capital Medical University, 2nd Anzhen Road, Chaoyang District, Beijing 100020, China
* Correspondence: yongqianglai2022@126.com (Y.L.); maxi8238@gmail.com (X.M.)
† These authors contributed equally to this work.

Citation: Zhang, C.; Zhang, H.; Zhao, L.; Wei, Z.; Lai, Y.; Ma, X. Differential Expression of microRNAs in Hypertrophied Myocardium and Their Relationship to Late Gadolinium Enhancement, Left Ventricular Hypertrophy and Remodeling in Hypertrophic Cardiomyopathy. *Diagnostics* **2022**, *12*, 1978. https://doi.org/10.3390/diagnostics12081978

Academic Editors: Minjie Lu and Arlene Sirajuddin

Received: 29 July 2022
Accepted: 11 August 2022
Published: 16 August 2022

Publisher's Note: MDPI stays neutral with regard to jurisdictional claims in published maps and institutional affiliations.

Abstract: Background: Differential expression has been found in a variety of circulating miRNAs in patients with hypertrophic cardiomyopathy (HCM). However, study on myocardial miRNAs is limited and a lot of miRNAs were not studied in previous studies. **Methods:** Twenty-one HCM patients and four patients who died from non-cardiovascular diseases were prospectively recruited for our study. A total of 26 myocardial tissues were collected, which were stored in liquid nitrogen immediately for miRNA detection using the Agilent Human miRNA Microarray Kit. All HCM patients underwent cardiovascular magnetic resonance (CMR) examination before surgery and cvi42 software was used to analyze cardiac function and myocardial fibrosis. **Results:** Compared with the control group, the expression of 22 miRNAs was found to be significantly increased in the HCM group, while 46 miRNAs were found to be significantly decreased in the HCM group. The expression levels of hsa-miR-3960 and hsa-miR-652-3p were significantly correlated with left ventricular mass index (r = 0.449 and 0.474, respectively). Meanwhile, Hsa-miR-642a-3p expression was positively correlated to the quantification of late gadolinium enhancement (r = 0.467). **Conclusions:** Our study found that 68 myocardial miRNAs were significantly increased or decreased in the HCM group. Myocardial miRNA levels could be used as potential biomarkers for LV hypertrophy, fibrosis and remodeling.

Keywords: hypertrophic cardiomyopathy; microRNA; cardiovascular magnetic resonance; myocardial fibrosis; cardiac remodeling

1. Introduction

As the most common monogenic cardiovascular disorder, hypertrophic cardiomyopathy (HCM) is associated with mutations in 11 or more genes [1]. MicroRNAs (miRNAs) are short RNA molecules that regulate the post-transcriptional silencing of target genes. A single miRNA can target hundreds of mRNAs and influence the expression of numerous genes [2]. A variety of miRNAs have been found as important regulators of multiple phases in cardiac development [3–5]. Previous studies reported that many miRNAs play roles in the pathogenic mechanisms of heart failure, such as hypertrophy, remodeling, hypoxia and apoptosis [6,7]. In the study by Roberta et al. [8], circulating miR-29a was found to be significantly up-regulated in HCM patients and correlated with fibrosis and left ventricular (LV) hypertrophy. However, circulating miRNAs did not only originate in the myocardial tissue but in other involved organs. Assessment of myocardial miRNAs could better reflect the miRNA expression in hypertrophied myocardium. Moreover, only dozens of miRNAs were analyzed in previous studies [8,9]; other miRNAs may also show differential expression in HCM patients.

Cardiovascular magnetic resonance (CMR) plays an important role in clinical work, which provides a mechanism to assess vascular or cardiac function, anatomy, tissue characteristics and perfusion in a highly reproducible manner [10–13]. CMR measurements of LV myocardial thickening, myocardial mass, or infarct size were confirmed with high reproducibility and low variance in repeated samples [14]. Late gadolinium enhancement (LGE) images derived from CMR could be used for identifying the location and extent of myocardial necrosis. The extent of LGE closely mirrors the distribution of myocyte necrosis at early periods [15], which may be used as a valuable tool to predict major adverse cardiac events (MACE) and cardiac mortality [16].

In this study, we mainly aimed to characterize the myocardial miRNA profile of HCM, and investigate miRNAs that showed differential expression in hypertrophied myocardium. Then, the correlation between myocardial miRNA levels and CMR variables was assessed to explore potential myocardial miRNA biomarkers of myocardial fibrosis, LV hypertrophy and remodeling.

2. Materials and Methods

2.1. Study Population

This study was approved by the Ethics Committee of Beijing Anzhen hospital and written informed consent was obtained from all the subjects. Twenty-one patients diagnosed with HCM were recruited for our study and all patients underwent transaortic extended septal myectomy in Beijing Anzhen hospital from November 2019 to December 2020. The diagnosis of HCM was based on echocardiographic or CMR demonstration of a hypertrophied but nondilated LV (with maximal wall thickness >15 mm at end diastolic) in the absence of any other systemic or cardiac disorder causing a similar grade of hypertrophy [17]. The indications for surgical myectomy included patients with obstructive HCM who remain severely symptomatic; symptomatic patients with obstructive HCM who have associated cardiac disease requiring surgical treatment and patients' voluntary acceptance [17]. Four normal myocardium specimens were used as controls, which were obtained from patients who died from non-cardiovascular diseases in Beijing Anzhen hospital from November 2019 to December 2020. Their myocardial tissues were stored in liquid nitrogen immediately for miRNA detection. A total of 21 patients with HCM (8 men and 13 women) and 4 healthy controls (2 men and 2 women) were included in this study. Patient's clinical characteristics were listed in Table 1.

Table 1. Clinical characteristics of HCM patients and healthy controls.

	HCM Patients		**Controls**		*p* **Value**
	n	**Value**	*n*	**Value**	
Age (years)	21	55 (41,60)	4	51 (32,59)	0.738
Male	8	38%	2	50%	1.000
Significant LVOT gradient (>30 mm Hg)	19	90%	0	0%	-
Family history of HCM	4	19%	0	0%	-
History of syncope	5	24%	0	0%	-
Nonsustained tachycardia	1	5%	0	0%	-
Maximal EDWT > 30 mm	2	10%	0	0%	-
Mitral regurgitation			0	0%	-
Mild	2	10%	0	0%	-
Moderate	7	33%	0	0%	-
Severe	12	57%	0	0%	-

HCM, hypertrophic cardiomyopathy; LVOT, left ventricular outflow tract; EDWT, end diastolic left ventricular wall thickness. Quantitative data were expressed as the median and interquartile range (IQR).

2.2. CMR Image Acquisition

CMR images were acquired with a 32-channel surface phased array cardiac coil in two different pieces of equipment (Ingenia 3.0T, Philips Healthcare, Best, Netherlands; Discovery MR750 3.0T, GE Medical Systems, Milwaukee, WI, USA) following routine scan protocol. Cardiac cine images were collected using balanced steady-state free precession (bFISP) sequences with retrospective cardiac gating. Images covered the whole LV from the base to the apex and each cardiac cycle had 25 phases. Parameters of cine images for GE Discovery MR750 were: TR/TE = 3.6/1.4 ms, FA = 60°, FOV = 380 × 380 mm^2; for Philips Ingenia: TR/TE = 3.0/1.52 ms, FA = 45°, FOV = 270 × 270 mm^2. LGE images were collected using a breath-hold 2D phase-sensitive inversion-recovery (PSIR) segmented gradient echo sequence 10 minutes after the contrast agent (Gadopentetate dimeglumine, Bayer Healthcare) was intravenously administered at a dose of 0.2 mmol/kg body weight. Parameters of LGE images for GE Discovery MR750 were: TR/TE = 6.2/2.9 ms, FA = 25°, FOV = 380 × 380 mm^2; for Philips Ingenia: TR/TE = 6.1/3.0 ms, FA = 25°, FOV = 350 × 350 mm^2. All slice thicknesses were 5 mm for long axis images and 8 mm for short axis images with no interval between slice locations.

2.3. CMR Image Analysis

Endocardial and epicardial borders of LV from the base to the apex were automatically delineated and manually adjusted using commercial software (cvi42, version 5.11.2, Circle Cardiovascular Imaging Inc., Calgary, AB, Canada) in both end systolic and end diastolic phase. Then, the left ventricular end systolic volume (LVESV), left ventricular end diastolic volume (LVEDV), left ventricular mass (LVM) and LVEF were calculated by the software. In addition, we further standardized LVESV, LVEDV and LVM by body surface area (BSA). We also manually measured maximal LV end diastolic wall thickness (EDWT) by an experienced observer (L.Z., with 10 years of CMR experience). After epicardial and endocardial borders were manually delineated by an experienced observer (L.Z., with 10 years of CMR experience) in the LGE images, a visually normal appearing area of myocardium without hyperenhancement was manually selected as a normal myocardial region of interest. We defined LGE as myocardium 6 standard deviations (SD) above the mean signal intensity.

2.4. Microarray Infomation

The Agilent Human miRNA Microarray Kit, Release 21.0, 8 × 60K (DesignID:070156) experiment and data analysis of the 26 samples were conducted by LB Technology Co., Ltd. (Beijing, China). The microarray contains 2570 probes for mature miRNA.

2.5. Experiment

Total RNA was quantified by the NanoDrop ND-2000 (Thermo Scientific, Waltham, MA, USA) and the RNA integrity (RIN) was assessed using the Agilent Bioanalyzer 2100 (Agilent Technologies, Santa Clara, CA, USA). The sample labeling, microarray hybridization and washing were performed based on the manufacturer's standard protocols. Briefly, total RNA was dephosphorylated, denatured and then labeled with Cyanine-3-CTP. After purification, the labeled RNAs were hybridized onto the microarray. After washing, the arrays were scanned with the Agilent Scanner G2505C (Agilent Technologies).

2.6. Experimental Data Analysis

Feature extraction software (version10.7.1.1, Agilent Technologies) was used to analyze array images to get raw data. Next, the raw data were normalized with the quantile algorithm. The probes detected with at least 75.0 percent in any group were chosen for further data analysis. Differentially expressed miRNAs were then identified through fold change and the p-value was calculated using a t-test. The threshold set for up- and down-regulated genes was a fold change $\geq$ 2.0 and a p-value $\leq$ 0.05. Target genes of differentially expressed miRNAs were the intersection predicted with 2 databases (miRDB, miRWalk). GO analysis and KEGG analysis were applied to determine the roles of these target genes. Hierarchical clustering was performed to show the distinguishable miRNAs expression pattern among samples. The miRNAs extraction and screening process are shown in Figure 1.

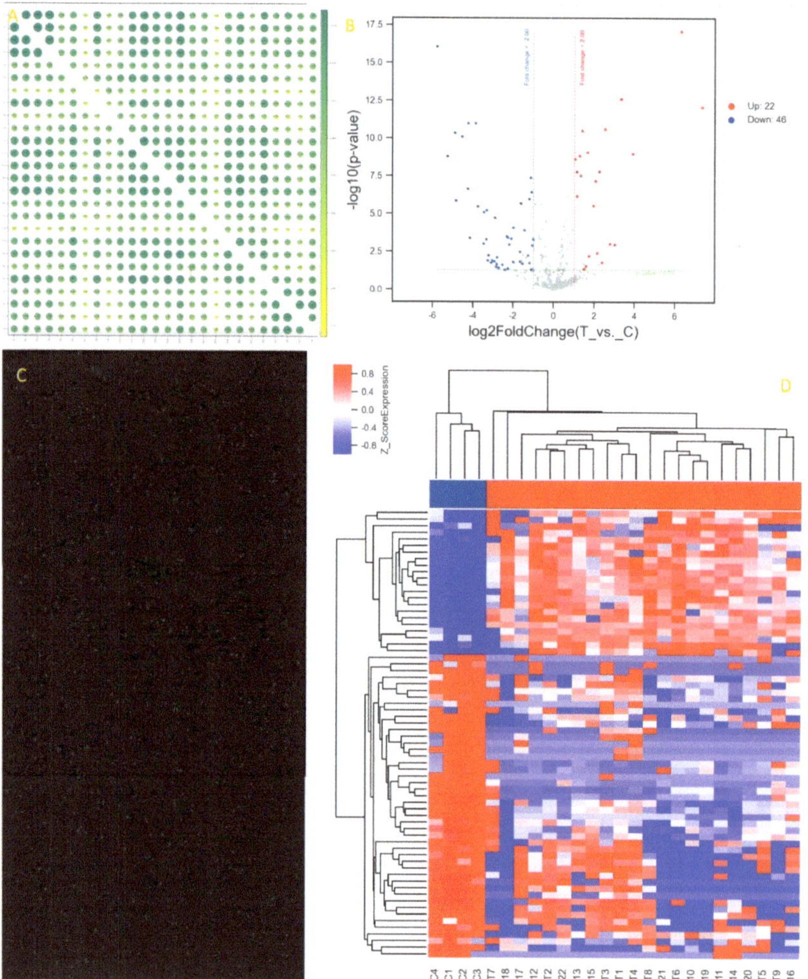

Figure 1. MiRNAs extraction and screening process. (**A**) Correlation of gene expression levels among samples using Pearson correlation methods. (**B**) Volcano map, in which each point represents a miRNA, the abscissa represents the logarithm of the difference multiple of the expression of a certain miRNA in the two groups of samples (log2fc), and the ordinate represents the negative logarithm ($-\log10$ (p-value)) of the statistical significance of the change of miRNA expression. The greater the absolute value of abscissa, the greater the difference multiple between the two groups; The larger the ordinate is, the more significant the differential expression is, and the more reliable the differentially expressed miRNA screened is. Red dots indicate up-regulation, blue dots indicate down-regulation, and gray dots indicate that at least 75% of the samples in one group are marked as ' detected ' or non-differentially expressed miRNA. (**C**) Raw data scanning diagram of a chip, each point on the diagram corresponds to a probe (only shows a partial probe). (**D**) Clustering relationship between samples, which could distinguish two or more groups of samples.

2.7. Statistical Analysis

Statistical analyses were undertaken with the SPSS software (IBM SPSS Statistics for Windows, Version 23.0; IBM Corp., Armonk, NY, USA) and R programming language (version 3.4.2, Available online: http://www.r-project.org (accessed on 19 January 2022). Quantitative data were expressed as the median and interquartile range (IQR), and categorical variables were presented as frequencies or percentages. Spearman correlation analyses were used to evaluate the potential correlation between miRNA levels and CMR variables. The Mann–Whitney U test was used to compare the difference in quantitative data in two different groups, and the chi-square test or Fisher's exact test were used to compare the difference of categorical variables between two different groups.

3. Results

3.1. Patients' CMR and Echocardiography Findings

Patients' CMR and echocardiography findings are listed in Table 2, including LVEF, LVEDVi, LVESVi, EDWT, quantification of LGE, LVMi, LV outflow tract gradient (LVOTO) and left atrial volume index (LAVi).

Table 2. CMR and echocardiography findings of HCM patients.

CMR Parameters	HCM Patients
Left atrial volume index (mL/m^2)	53 (44,68)
Left ventricular ejection fraction (%)	66 (59,73)
Left ventricular end diastolic volume index (mm/m^2)	75 (57,88)
Left ventricular end systolic volume index (mm/m^2)	24 (19,35)
Left ventricular outflow tract gradient (mm Hg)	78 (54,101)
Maximal wall thickness (mm)	22 (18,28)
Left ventricular mass index (mm/m^2)	95 (79,126)
Quantification of late gadolinium enhancement (%)	6 (2,20)

CMR, cardiac magnetic resonance; HCM, hypertrophic cardiomyopathy. Quantitative data were expressed as the median and interquartile range (IQR).

3.2. Up-Regulated and Down-Regulated miRNAs in HCM Patients

Compared with the control group, the expression of 22 miRNAs was found significantly increased in the HCM group, while 46 miRNAs were found significantly decreased in the HCM group (Table 3). The expression of myocardial miR-208b-3p, -221-3p, -224-3p were extremely up-regulated (with foldchange >10) and miR-218-5p, -4741, -5787, -208a-3p, -551b-3p, -4788, -575, -4466, -1246, -7150, -204-5p, -208a-5p, -6850-5p, -7847-3p were extremely down-regulated (with foldchange < −10).

Table 3. Up-regulated and down-regulated miRNAs in HCM patients.

miRNAs	*p*-Value	FoldChange	log2FoldChange	Regulation
hsa-miR-15a-5p	1.67×10^{-8}	2.190249138	1.131094984	Up
hsa-miR-24-1-5p	6.63×10^{-7}	2.185726725	1.128113036	Up
hsa-miR-95-3p	0.00107674	8.082612468	3.014821677	Up
hsa-miR-208a-3p	1.37×10^{-6}	−28.85041914	−4.850520374	Down
hsa-miR-148a-3p	0.007665064	−7.04991722	−2.817606318	Down
hsa-miR-10a-5p	0.023519485	−5.823609987	−2.54191374	Down
hsa-miR-181a-5p	2.42×10^{-9}	2.046487301	1.033149714	Up
hsa-miR-181b-5p	1.61×10^{-8}	4.767522463	2.253239736	Up
hsa-miR-204-5p	0.000980838	−11.19625089	−3.484943816	Down
hsa-miR-218-5p	8.97×10^{-17}	−54.93956941	−5.779773698	Down
hsa-miR-221-3p	8.08×10^{-18}	77.96215989	6.284702155	Up
hsa-miR-142-3p	0.00037835	−4.857959151	−2.280350359	Down
hsa-miR-149-5p	2.81×10^{-6}	3.888492702	1.959211031	Up
hsa-miR-188-5p	0.001074418	−4.594043979	−2.199764668	Down
hsa-miR-378a-5p	2.43×10^{-13}	9.967233981	3.317193196	Up
hsa-miR-378a-3p	1.45×10^{-9}	2.409517431	1.268744238	Up
hsa-miR-424-5p	0.015718634	5.24821437	2.39182665	Up

Table 3. *Cont.*

miRNAs	*p*-Value	FoldChange	log2FoldChange	Regulation
hsa-miR-451a	0.048520532	−2.158469131	−1.110008461	Down
hsa-miR-486-5p	3.01×10^{-8}	2.491376359	1.316942978	Up
hsa-miR-499a-5p	8.04×10^{-10}	3.125868037	1.644256874	Up
hsa-miR-551b-3p	7.91×10^{-11}	−23.50718535	−4.555029902	Down
hsa-miR-575	1.02×10^{-11}	−18.94850888	−4.244012418	Down
hsa-miR-652-3p	0.003807107	4.396032747	2.136202133	Up
hsa-miR-23b-5p	0.006018149	3.310903689	1.727225045	Up
hsa-miR-125a-3p	0.003260678	−3.153527041	−1.656966304	Down
hsa-miR-486-3p	0.000950208	6.877931643	2.781974778	Up
hsa-miR-490-5p	0.005608572	−2.417505315	−1.273519062	Down
hsa-miR-455-3p	6.87×10^{-8}	4.151633047	2.053678933	Up
hsa-miR-208b-3p	8.33×10^{-13}	159.1822033	7.31453524	Up
hsa-miR-1225-5p	1.02×10^{-6}	−2.301696596	−1.202697674	Down
hsa-miR-1246	1.03×10^{-11}	−14.79899608	−3.887427406	Down
hsa-miR-1915-3p	0.012200862	−8.341626335	−3.060328688	Down
hsa-miR-224-3p	1.04×10^{-9}	14.97603971	3.904584261	Up
hsa-miR-3141	0.015232493	−4.006507145	−2.00234505	Down
hsa-miR-4298	0.042552949	−4.810758055	−2.266264244	Down
hsa-miR-4270	0.012522098	−9.686961125	−3.276044152	Down
hsa-miR-4291	0.029289007	2.996508744	1.583282584	Up
hsa-miR-3679-5p	0.014999028	−3.151810102	−1.656180614	Down
hsa-miR-378d	2.36×10^{-11}	5.814021124	2.539536313	Up
hsa-miR-4442	0.014525527	−7.714730539	−2.947615767	Down
hsa-miR-4466	0.000397024	−17.79445123	−4.153355537	Down
hsa-miR-4530	3.77×10^{-7}	−2.141285992	−1.098477496	Down
hsa-miR-378i	3.05×10^{-11}	2.634035175	1.397274612	Up
hsa-miR-3960	0.000482361	−2.028841036	−1.020655831	Down
hsa-miR-4634	1.85×10^{-5}	−7.491169168	−2.905190902	Down
hsa-miR-4669	0.032262486	−7.26027626	−2.860024445	Down
hsa-miR-4687-3p	2.06×10^{-7}	−3.042668733	−1.605337271	Down
hsa-miR-4741	1.67×10^{-9}	−38.44059936	−5.264558926	Down
hsa-miR-4787-5p	0.000435723	−4.295676251	−2.102885267	Down
hsa-miR-4788	2.32×10^{-7}	−19.06965697	−4.253206987	Down
hsa-miR-642a-3p	0.049097861	−5.377734753	−2.426998599	Down
hsa-miR-5001-5p	8.45×10^{-5}	−3.991371678	−1.996884629	Down
hsa-miR-1229-5p	0.005670955	−9.395082827	−3.231905881	Down
hsa-miR-5787	4.37×10^{-11}	−29.79856364	−4.897170886	Down
hsa-miR-6088	0.00125777	−2.075538432	−1.053485646	Down
hsa-miR-6090	0.000126781	−2.72669402	−1.447152815	Down
hsa-miR-6510-5p	0.020496139	−2.921621676	−1.546769374	Down
hsa-miR-208a-5p	8.26×10^{-6}	−11.036198	−3.464171341	Down
hsa-miR-6727-5p	0.000316337	−5.062208252	−2.339766859	Down
hsa-miR-6739-5p	0.044423071	2.783974922	1.477146215	Up
hsa-miR-6800-5p	0.017351182	−8.6565924	−3.113799231	Down
hsa-miR-6850-5p	0.000487495	−10.15242969	−3.343753131	Down
hsa-miR-6891-5p	0.021973529	−7.101027763	−2.828027847	Down
hsa-miR-7107-5p	0.017734323	−2.296376215	−1.199359018	Down
hsa-miR-7110-5p	0.037958046	−6.662449596	−2.736052713	Down
hsa-miR-7150	3.40×10^{-6}	−13.56395008	−3.761705475	Down
hsa-miR-7847-3p	6.05×10^{-6}	−10.01635556	−3.324285775	Down
hsa-miR-8069	4.29×10^{-8}	−2.194231964	−1.133716049	Down

HCM, hypertrophic cardiomyopathy.

3.3. Correlation between miRNA Levels and CMR Variables

The expression levels of hsa-miR-3960 and hsa-miR-652-3p were significantly and positively correlated with LVMi (r = 0.449, *p* = 0.036; r = 0.474, *p* = 0.026, respectively). The expression levels of hsa-miR-3679-5p and hsa-miR-7107-5p were significantly and positively correlated with LVEF (r = 0.486, *p* = 0.022; r = 0.454, *p* = 0.034, respectively), while the expression level of hsa-miR-499a-5p was significantly and negatively correlated with LVEF (r = −0.571, *p* = 0.005). Meanwhile, hsa-miR-642a-3p expression was positively correlated to the quantification of LGE (r = 0.467, *p* = 0.028). Then, hsa-miR-3141 and hsa-miR-3679-5p expression were found to be negatively correlated to LVESVi (r = −0.556, *p* = 0.007; r = 0.459, *p* = 0.032, respectively) (Figures 2 and 3).

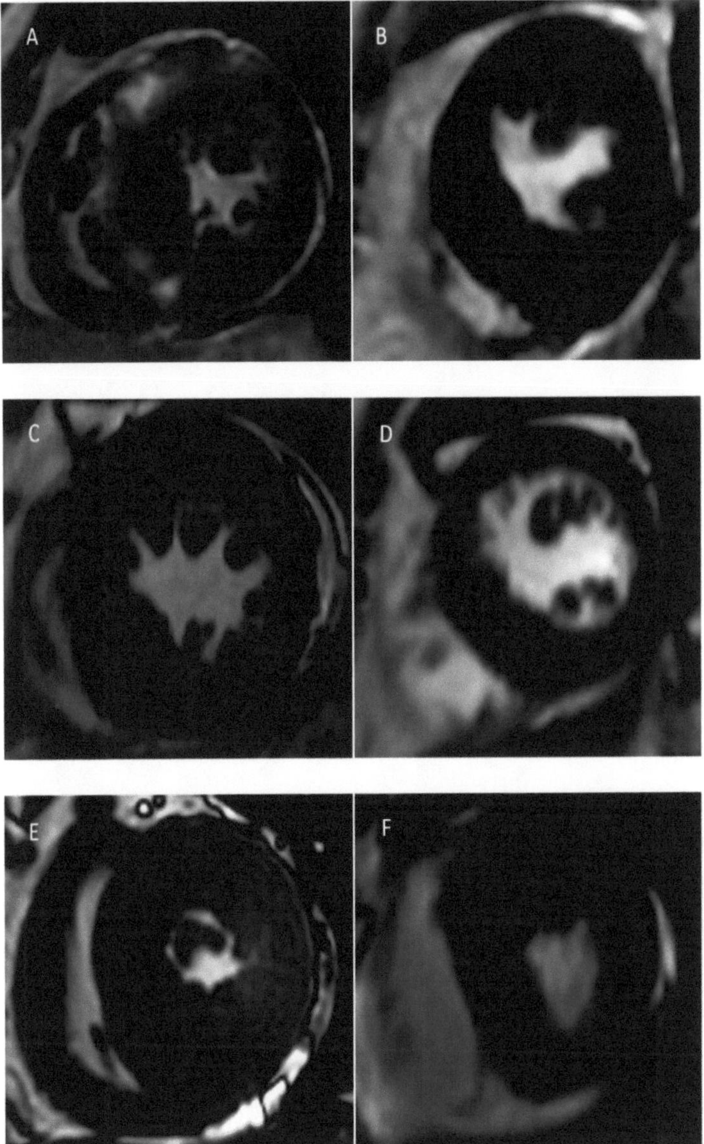

Figure 2. CMR findings in different patients with different myocardial miRNA levels. (**A**) Short axis LGE image in a patient with high expression of myocardial hsa-miR-642a-3p showed extensive LGE in the septal wall of LV. (**B**) Short axis LGE image in a patient with low expression of myocardial hsa-miR-642a-3p, no LGE was found in the LV myocardial. (**C**) Short axis cine image in a patient with high expression of myocardial hsa-miR-652-3p and hsa-miR-3960 showed significantly thickened myocardial. (**D**) Short axis cine image in a patient with low expression of myocardial hsa-miR-652-3p and hsa-miR-3960 showed myocardial thickening but was not as obvious as the previous one. (**E**) Short axis cine image in a patient with high expression of myocardial hsa-miR-3679-5p and hsa-miR-3141 showed low LVESVi. (**F**) Short axis cine image in a patient with low expression of myocardial hsa-miR-3679-5p and hsa-miR-3141 showed higher LVESVi than the previous one. **CMR**, cardiovascular magnetic resonance; **LGE**, late gadolinium enhancement; **LV**, left ventricle; **LVESVi**, left ventricular end systolic volume index.

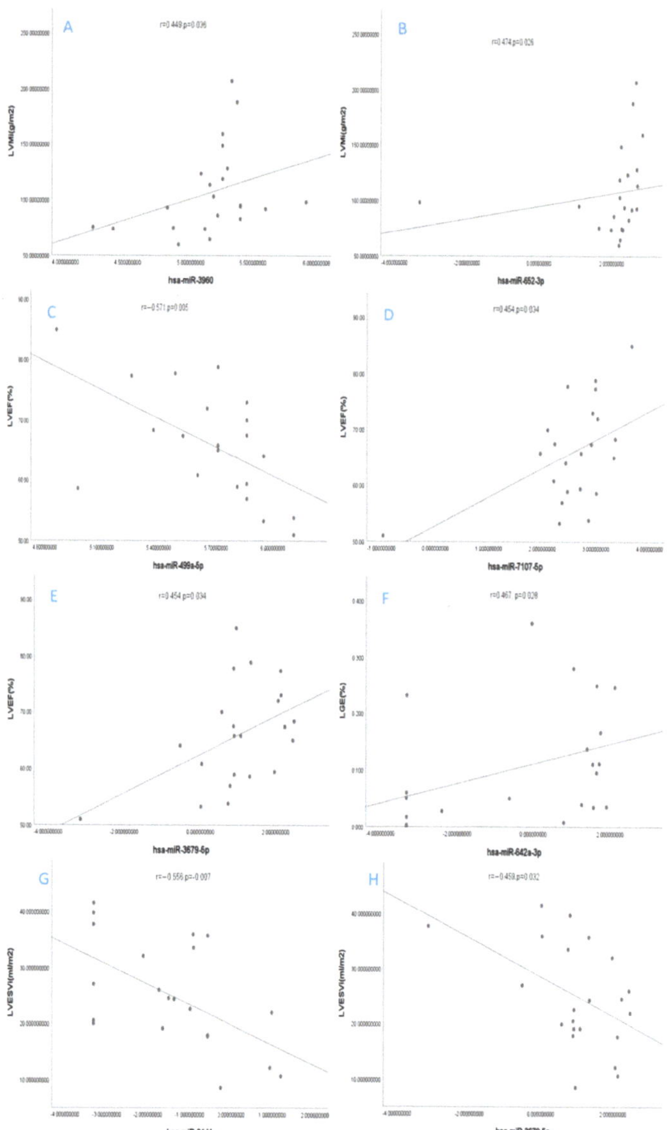

Figure 3. Correlation between miRNA levels and clinical variables. (**A**) The expression levels of hsa−miR−3960 were significantly and positively correlated with LVMi. (**B**) The expression levels of hsa−miR−652−3p were significantly and positively correlated with LVMi. (**C**) The expression lev of hsa−miR−499a−5p was significantly and negatively correlated with LVEF. (**D**) The expression levels of hsa−miR−7107−5p were significantly and positively correlated with LVEF. (**E**) The expression levels of hsa−miR−3679−5p were significantly and positively correlated with LVEF. (**F**) Hsa−miR−642a−3p expression was positively correlated to the quantification of LGE. (**G**) Hsa−miR−3141 expression was negatively correlated to LVESVi. (**H**) Hsa−miR−3679−5p expression was negatively correlated to LVESVi. **LVMi**, left ventricular mass index; **LVEF**, left ventricular ejection fraction; **LGE**, late gadolinium enhancement; **LVESVi**, left ventricular and systolic volume index.

4. Discussion

We evaluated the expression of 2570 miRNAs in each myocardial specimen and found that the expression of 22 miRNAs was significantly increased and 46 miRNAs were significantly decreased in the HCM group. Then, seven miRNAs were found to be significantly correlated to CMR parameters.

We evaluated the expression of myocardial miRNA levels rather than circulating miRNA levels, so the result in this study is different from previous studies [8,9]. In the study by Roberta et al. [8], the expression of circulating miR-27a, -199a-5p, -26a, -145, -133a, -143, -199a-3p, -126-3p, -29a, -155, -30a, and -21 were found to be increased in HCM patients. In the study by Derda [9] et al., circulating expression of miR-155 was significantly decreased in both obstructive and non-obstructive HCM patients. However, in this study, the expression of these miRNAs was not significantly up-regulated or down-regulated in HCM patients. This result means that the expression of miRNAs was different between myocardial and circulation.

In the preliminary study on the expression of myocardial miRNAs, Kuster [18] et al. found that the expression of miRNA-10b was down-regulated and miRNA-204, -497, -184, -222 and -34 were up-regulated. In the study by Song [19] et al., expression of miRNA-451 was significantly down-regulated, and overexpression of miR-451 in neonatal rat cardiomyocytes could reduce cell size. In the study by Huang [20] et al., the expression of miR-221, miR-222 and miR-433 was significantly up-regulated. The results in our study were partly similar to these studies as we also found up-regulated expression of miR-221 and down-regulated expression of miR-451, but no similar results were found in other miRNAs.

A number of circulating miRNAs were found significantly associated with LV hypertrophy and fibrosis [8,21,22]. In this study, we further evaluated the relationship between myocardial miRNAs and LVH, myocardial fibrosis and LV remodeling.

Previous studies focused on circulating miR-27a, miR-29a and miR-199a-5p, which were validated to be significantly correlated with LVH and LV fibrosis in various studies [8,22]. Our study found that myocardial hsa-miR-3960 and hsa-miR-652-3p were significantly associated with LV hypertrophy, and hsa-miR-642a-3p was significantly correlated with LV fibrosis; this means that myocardial miRNAs could also perform as biomarkers of LVH and LV fibrosis. In addition, we also found that hsa-miR-499a-5p, hsa-miR-7107-5p, hsa-miR-3141 and hsa-miR-3679-5p were significantly associated with the remodeling of the LV cavity. To our knowledge, this has not been studied yet.

MiR-499 was shown to regulate the cardiac β-MyHC/α-MyHC ratio, and β-MyHC expression was increased in human hearts of patients with ischemic cardiomyopathy and other heart diseases [23–26]. Circulating miR-499a-5p was proven as one of the dysregulated miRNAs in HCM, which expressed higher in HCM patients than in healthy controls, and carriers of P/LP variants in the MYH7 gene expressed higher levels than in controls [27,28]. In our study, we found that myocardial miR-499a-5p was also up-regulated, and higher expression of myocardial miR-499a-5p correlated with lower LVEF. However, the relationship between LV remodeling and miR-499a-5p needs further validation and mechanism exploration.

MiR-642-3p was revealed as an adipocyte-specific microRNA in a previous study [29], but the function of miR-642 families on myocytes has not been reported. So, although myocardial miR-642-3p was significantly down-regulated in hypertrophied myocardium and correlated with LV fibrosis in our study, further studies are needed to explore the mechanism of this relationship.

In the study by Eyyupkoca et al. [30], mir-652-3p was identified to be associated with adverse left ventricular remodeling (ALVR), which was defined as an increase in LVEDV and LVESV > 13% 6 months after acute myocardial infarction (AMI). Unfortunately, studies were limited in other miRNAs, such as miR-3960, miR-7107-5p, miR-3141 and miR-3679-5p in the cardiovascular field, so greater efforts are needed in this field to explore the mechanism of how these miRNAs work and to find potential therapeutic targets.

Several studies have suggested that endogenic factors such as myeloperoxidase (MPO) and nitric oxide synthase (NOS) may play important roles in the frail population [31,32]. However, the effect of endogenic factors derived from myocardial tissue on HCM patients was not investigated in this study, and further study is needed to explore the possible protective role or the unfavorable role of endogenic factors in the HCM population.

There are several limitations to our study. First, since the myocardial species were difficult to obtain, especially in patients without HCM, we only included 21 HCM patients and four healthy controls. So, our result needs further validation with large-scale studies. Then, due to insufficient image quality, T1 maps were not available for analysis in many patients, so we did not analyze the relationship between myocardial miRNAs and diffuse myocardial fibrosis. Next, we did not collect blood samples from these patients, so it is impossible to analyze the relationship between circulating miRNAs and myocardial miRNAs; this will be improved in future research. Finally, patients' follow-up was not conducted in this study, and the relationship between myocardial miRNAs and long-term outcomes was not discussed; this will be added in the next research.

5. Conclusions

In conclusion, our study found that 68 myocardial miRNAs were significantly increased or decreased in the HCM group. Myocardial miRNA levels could be used as potential biomarkers for LV hypertrophy, fibrosis and remodeling.

Author Contributions: Conceptualization, C.Z. and X.M.; Methodology, H.Z.; Software, H.Z.; Validation, X.M., Y.L. and L.Z.; Resources, H.Z. and Z.W.; Data Curation, C.Z. and H.Z.; Writing—Original Draft Preparation, H.Z.; Writing—Review & Editing, L.Z.; Supervision, X.M.; Project Administration, X.M and Y.L.; Funding Acquisition, X.M. All authors have read and agreed to the published version of the manuscript.

Funding: This work was supported by the Beijing Natural Science Foundation (7212025) and the Beijing Natural Science Foundation (7222302).

Institutional Review Board Statement: The study was conducted according to the guidelines of the Declaration of Helsinki, and approved by the Institutional Review Board (number:2022129X, approved date: 29 July 2022).

Informed Consent Statement: Written informed consent has been obtained from the patients.

Data Availability Statement: The data presented in this study are available on request from the corresponding author.

Conflicts of Interest: The authors declare no conflict of interest.

References

1. Maron, B.J. Clinical Course and Management of Hypertrophic Cardiomyopathy. *N. Engl. J. Med.* **2018**, *379*, 655–668. [CrossRef] [PubMed]
2. Lu, T.X.; Rothenberg, M.E. MicroRNA. *J. Allergy Clin. Immunol.* **2018**, *141*, 1202–1207. [CrossRef] [PubMed]
3. Zhao, Y.; Ransom, J.F.; Li, A.; Vedantham, V.; von Drehle, M.; Muth, A.N.; Tsuchihashi, T.; McManus, M.T.; Schwartz, R.J.; Srivastava, D. Dysregulation of cardiogenesis, cardiac conduction, and cell cycle in mice lacking miRNA-1-2. *Cell* **2007**, *129*, 303–317. [CrossRef] [PubMed]
4. Da Costa Martins, P.A.; Bourajjaj, M.; Gladka, M.; Kortland, M.; van Oort, R.J.; Pinto, Y.M.; Molkentin, J.D.; De Windt, L.J. Conditional dicer gene deletion in the postnatal myocardium provokes spontaneous cardiac remodeling. *Circulation* **2008**, *118*, 1567–1576. [CrossRef]
5. Chen, J.; Huang, Z.-P.; Seok, H.Y.; Ding, J.; Kataoka, M.; Zhang, Z.; Hu, X.; Wang, G.; Lin, Z.; Wang, S.; et al. mir-17-92 cluster is required for and sufficient to induce cardiomyocyte proliferation in postnatal and adult hearts. *Circ. Res.* **2013**, *112*, 1557–1566. [CrossRef]
6. Melman, Y.F.; Shah, R.; Das, S. MicroRNAs in heart failure: Is the picture becoming less miRky? *Circ. Heart Fail.* **2014**, *7*, 203–214. [CrossRef]
7. Tijsen, A.J.; Pinto, Y.M.; Creemers, E.E. Non cardiomyocyte microRNAs in heart failure. *Cardiovasc. Res.* **2012**, *93*, 573–582. [CrossRef]

8. Roncarati, R.; Anselmi, C.V.; Losi, M.A.; Papa, L.; Cavarretta, E.; Da Costa Martins, P.; Contaldi, C.; Jotti, G.S.; Franzone, A.; Galastri, L.; et al. Circulating miR-29a, among other up-regulated microRNAs, is the only biomarker for both hypertrophy and fibrosis in patients with hypertrophic cardiomyopathy. *J. Am. Coll. Cardiol.* **2014**, *63*, 920–927. [CrossRef]
9. Derda, A.A.; Thum, S.; Lorenzen, J.M.; Bavendiek, U.; Heineke, J.; Keyser, B.; Stuhrmann, M.; Givens, R.C.; Kennel, P.J.; Schulze, P.C.; et al. Blood-based microRNA signatures differentiate various forms of cardiac hypertrophy. *Int. J. Cardiol.* **2015**, *196*, 115–122. [CrossRef]
10. Hundley, W.G.; Bluemke, D.A.; Finn, J.P.; Flamm, S.D.; Fogel, M.A.; Friedrich, M.G.; Ho, V.B.; Jerosch-Herold, M.; Kramer, C.M.; Manning, W.J.; et al. ACCF/ACR/AHA/NASCI/SCMR 2010 expert consensus document on cardiovascular magnetic resonance: A report of the American College of Cardiology Foundation Task Force on Expert Consensus Documents. *J. Am. Coll. Cardiol.* **2010**, *55*, 2614–2662. [CrossRef]
11. Wolk, M.J.; Bailey, S.R.; Doherty, J.U.; Douglas, P.S.; Hendel, R.C.; Kramer, C.M.; Min, J.K.; Patel, M.R.; Rosenbaum, L.; Shaw, L.J.; et al. ACCF/AHA/ASE/ASNC/HFSA/HRS/SCAI/SCCT/SCMR/STS 2013 multimodality appropriate use criteria for the detection and risk assessment of stable ischemic heart disease: A report of the American College of Cardiology Foundation Appropriate Use Criteria Task Force, American Heart Association, American Society of Echocardiography, American Society of Nuclear Cardiology, Heart Failure Society of America, Heart Rhythm Society, Society for Cardiovascular Angiography and Interventions, Society of Cardiovascular Computed Tomography, Society for Cardiovascular Magnetic Resonance, and Society of Thoracic Surgeons. *J. Am. Coll. Cardiol.* **2014**, *63*, 380–406.
12. Patel, M.R.; White, R.D.; Abbara, S.; Bluemke, D.A.; Herfkens, R.J.; Picard, M.; Shaw, L.J.; Silver, M.; Stillman, A.E.; Udelson, J.; et al. 2013 ACCF/ACR/ASE/ASNC/SCCT/SCMR appropriate utilization of cardiovascular imaging in heart failure: A joint report of the American College of Radiology Appropriateness Criteria Committee and the American College of Cardiology Foundation Appropriate Use Criteria Task Force. *J. Am. Coll. Cardiol.* **2013**, *61*, 2207–2231.
13. Petersen, S.E.; Khanji, M.Y.; Plein, S.; Lancellotti, P.; Bucciarelli-Ducci, C. European Association of Cardiovascular Imaging expert consensus paper: A comprehensive review of cardiovascular magnetic resonance normal values of cardiac chamber size and aortic root in adults and recommendations for grading severity. *Eur. Heart J.-Cardiovasc. Imaging* **2019**, *20*, 1321–1331. [CrossRef]
14. Gandy, S.J.; Waugh, S.A.; Nicholas, R.S.; Simpson, H.J.; Milne, W.; Houston, J.G. Comparison of the reproducibility of quantitative cardiac left ventricular assessments in healthy volunteers using different MRI scanners: A multicenter simulation. *J. Magn. Reson. Imaging* **2008**, *28*, 359–365. [CrossRef]
15. Kim, R.J.; Lima, J.A.C.; Chen, E.-L.; Reeder, S.B.; Klocke, F.J.; Zerhouni, E.A.; Judd, R.M. Fast ^{23}Na magnetic resonance imaging of acute reperfused myocardial infarction. Potential to assess myocardial viability. *Circulation* **1997**, *95*, 1877–1885. [CrossRef]
16. Kwong, R.Y.; Chan, A.K.; Brown, K.A.; Chan, C.W.; Reynolds, H.G.; Tsang, S.; Davis, R.B. Impact of unrecognized myocardial scar detected by cardiac magnetic resonance imaging on event-free survival in patients presenting with signs or symptoms of coronary artery disease. *Circulation* **2006**, *113*, 2733–2743. [CrossRef]
17. Ommen, S.R.; Mital, S.; Burke, M.A.; Day, S.M.; Deswal, A.; Elliott, P.; Evanovich, L.L.; Hung, J.; Joglar, J.A.; Kantor, P.; et al. 2020 AHA/ACC Guideline for the Diagnosis and Treatment of Patients With Hypertrophic Cardiomyopathy: A Report of the American College of Cardiology/American Heart Association Joint Committee on Clinical Practice Guidelines. *J. Am. Coll. Cardiol.* **2020**, *142*, e558–e631.
18. Kuster, D.W.; Mulders, J.; Cate, F.J.T.; Michels, M.; dos Remedios, C.G.; da Costa Martins, P.A.; van der Velden, J.; Oudejans, C.B. MicroRNA transcriptome profiling in cardiac tissue of hypertrophic cardiomyopathy patients with MYBPC3 mutations. *J. Mol. Cell. Cardiol.* **2013**, *65*, 59–66. [CrossRef]
19. Song, L.; Su, M.; Wang, S.; Zou, Y.; Wang, X.; Wang, Y.; Cui, H.; Zhao, P.; Hui, R.; Wang, J. MiR-451 is decreased in hypertrophic cardiomyopathy and regulates autophagy by targeting TSC1. *J. Cell. Mol. Med.* **2014**, *18*, 2266–2274. [CrossRef]
20. Huang, D.; Chen, Z.; Wang, J.; Chen, Y.; Liu, D.; Lin, K. MicroRNA-221 is a potential biomarker of myocardial hypertrophy and fibrosis in hypertrophic obstructive cardiomyopathy. *Biosci. Rep.* **2020**, *40*, BSR20191234. [CrossRef]
21. Zhou, J.; Zhou, Y.; Wang, C.X. LncRNA-MIAT regulates fibrosis in hypertrophic cardiomyopathy (HCM) by mediating the expression of miR-29a-3p. *J. Cell. Biochem.* **2019**, *120*, 7265–7275. [CrossRef] [PubMed]
22. Fang, L.; Ellims, A.H.; Moore, X.-L.; White, D.A.; Taylor, A.J.; Chin-Dusting, J.; Dart, A.M. Circulating microRNAs as biomarkers for diffuse myocardial fibrosis in patients with hypertrophic cardiomyopathy. *J. Transl. Med.* **2015**, *13*, 314. [CrossRef] [PubMed]
23. Broadwell, L.J.; Smallegan, M.J.; Rigby, K.M.; Navarro-Arriola, J.S.; Montgomery, R.L.; Rinn, J.L.; Leinwand, L.A. Myosin 7b is a regulatory long noncoding RNA (lncMYH7b) in the human heart. *J. Biol. Chem.* **2021**, *296*, 100694. [CrossRef] [PubMed]
24. Nakao, K.; Minobe, W.; Roden, R.; Bristow, M.R.; Leinwand, L.A. Myosin heavy chain gene expression in human heart failure. *J. Clin. Investig.* **1997**, *100*, 2362–2370. [CrossRef] [PubMed]
25. Miyata, S.; Minobe, W.; Bristow, M.R.; Leinwand, L.A. Myosin heavy chain isoform expression in the failing and nonfailing human heart. *Circ. Res.* **2000**, *86*, 386–390. [CrossRef] [PubMed]
26. Lowes, B.D.; Gilbert, E.M.; Abraham, W.T.; Minobe, W.A.; Larrabee, P.; Ferguson, D.; Wolfel, E.E.; Lindenfeld, J.; Tsvetkova, T.; Robertson, A.D.; et al. Myocardial gene expression in dilated cardiomyopathy treated with beta-blocking agents. *N. Engl. J. Med.* **2002**, *346*, 1357–1365. [CrossRef]
27. Baulina, N.; Pisklova, M.; Kiselev, I.; Chumakova, O.; Zateyshchikov, D.; Favorova, O. Circulating miR-499a-5p Is a Potential Biomarker of *MYH7*-Associated Hypertrophic Cardiomyopathy. *Int. J. Mol. Sci.* **2022**, *23*, 3791. [CrossRef]

28. Thottakara, T.; Lund, N.; Krämer, E.; Kirchhof, P.; Carrier, L.; Patten, M. A Novel miRNA Screen Identifies miRNA-4454 as a Candidate Biomarker for Ventricular Fibrosis in Patients with Hypertrophic Cardiomyopathy. *Biomolecules* **2021**, *11*, 1718. [CrossRef]

29. Zaragosi, L.-E.; Wdziekonski, B.; Le Brigand, K.; Villageois, P.; Mari, B.; Waldmann, R.; Dani, C.; Barbry, P. Small RNA sequencing reveals miR-642a-3p as a novel adipocyte-specific microRNA and miR-30 as a key regulator of human adipogenesis. *Genome Biol.* **2011**, *12*, R64. [CrossRef]

30. Eyyupkoca, F.; Ercan, K.; Kiziltunc, E.; Ugurlu, I.B.; Kocak, A.; Eyerci, N. Determination of microRNAs associated with adverse left ventricular remodeling after myocardial infarction. *Mol. Cell. Biochem.* **2022**, *477*, 781–791. [CrossRef]

31. Marzetti, E.; Calvani, R.; DuPree, J.; Lees, H.A.; Giovannini, S.; Seo, D.-O.; Buford, T.W.; Sweet, K.; Morgan, D.; Strehler, K.Y.E.; et al. Late-life enalapril administration induces nitric oxide-dependent and independent metabolic adaptations in the rat skeletal muscle. *AGE* **2013**, *35*, 1061–1075. [CrossRef]

32. Giovannini, S.; Onder, G.; Leeuwenburgh, C.; Carter, C.; Marzetti, E.; Russo, A.; Capoluongo, E.; Pahor, M.; Bernabei, R.; Landi, F. Myeloperoxidase levels and mortality in frail community-living elderly individuals. *J. Gerontol. Ser. A* **2010**, *65*, 369–376. [CrossRef]

Article

The Variation in the Diastolic Period with Interventricular Septal Displacement and Its Relation to the Right Ventricular Function in Pulmonary Hypertension: A Preliminary Cardiac Magnetic Resonance Study

Fan Yang [1,†], Wen Ren [2,†], Dan Wang [3], Yan Yan [1], Yuan-Lin Deng [1], Zhen-Wen Yang [4], Tie-Lian Yu [1], Dong Li [1,*] and Zhang Zhang [1,*]

[1] Department of Radiology, Tianjin Medical University General Hospital, Tianjin 300052, China
[2] Department of Radiology, The Affiliated Suzhou Hospital of Nanjing Medical University, Suzhou 215008, China
[3] Department of Ultrasonography, Shanxi Bethune Hospital, Taiyuan 030032, China
[4] Department of Cardiology, Tianjin Medical University General Hospital, Tianjin 300052, China
* Correspondence: dr_lidong@163.com (D.L.); filea1249@sina.com (Z.Z.)
† These authors contributed equally to this work.

Citation: Yang, F.; Ren, W.; Wang, D.; Yan, Y.; Deng, Y.-L.; Yang, Z.-W.; Yu, T.-L.; Li, D.; Zhang, Z. The Variation in the Diastolic Period with Interventricular Septal Displacement and Its Relation to the Right Ventricular Function in Pulmonary Hypertension: A Preliminary Cardiac Magnetic Resonance Study. *Diagnostics* **2022**, *12*, 1970. https://doi.org/10.3390/diagnostics12081970

Academic Editor: Andrea D. Annoni

Received: 30 June 2022
Accepted: 10 August 2022
Published: 15 August 2022

Publisher's Note: MDPI stays neutral with regard to jurisdictional claims in published maps and institutional affiliations.

Abstract: Background: Pulmonary hypertension (PH) is known to alter the biventricular shape and temporal phases of the cardiac cycle. The presence of interventricular septal (IVS) displacement has been associated with the severity of PH. There has been limited cardiac magnetic resonance (CMR) data regarding the temporal parameters of the cardiac cycle in PH. This study aimed to quantify the temporal changes in the cardiac cycle derived from CMR in PH patients with and without IVS displacement and sought to understand the mechanism of cardiac dysfunction in the cardiac cycle. Methods: Patients with PH who had CMR and right heart catheterization (RHC) examinations were included retrospectively. Patients were divided into an IVS non-displacement (IVS_{ND}) group and an IVS displacement (IVS_D) group according to IVS morphology, as observed on short-axis cine CMR images. Additionally, age-matched healthy volunteers were included as the health control (HC). Temporal parameters, IVS displacement, ventricular volume and functional parameters were obtained by CMR, and pulmonary hemodynamics were obtained by RHC. The risk stratification of the PH patients was also graded according to the guidelines. Results: A total of 70 subjects were included, consisting of 33 IVS_D patients, 15 IVS_{ND} patients, and 22 HC patients. In the IVS_{ND} group, only the right ventricle ejection fraction (RVEF) was decreased in the ventricular function, and no temporal change in the cardiac cycle was found. A prolonged isovolumetric relaxation time (IRT) and shortened filling time (FT) in both ventricles, along with biventricular dysfunction, were detected in the IVS_D group ($p < 0.001$). The IRT of the right ventricle (IRT_{RV}) and FT of the right ventricle (FT_{RV}) in the PH patients were associated with pulmonary vascular resistance, right cardiac index, and IVS curvature, and the IRT_{RV} was also associated with the RVEF in a multivariate regression analysis. A total of 90% of the PH patients in the IVS_D group were stratified into intermediate- and high-risk categories, and they showed a prolonged IRT_{RV} and a shortened FT_{RV}. The IRT_{RV} was also the predictor of the major cardiovascular events. Conclusions: The temporal changes in the cardiac cycle were related to IVS displacement and mainly impacted the diastolic period of the two ventricles in the PH patients. The IRT and FT changes may provide useful pathophysiological information on the progression of PH.

Keywords: pulmonary hypertension; magnetic resonance imaging; cine; cardiovascular physiological phenomena; cardiac cycle; ventricular function

1. Introduction

The right ventricle (RV) and left ventricle (LV) work within a distensible pericardium and are connected to each other through the interventricular septum (IVS), which shares myocardial fibers with both ventricles and accommodates the interactions between the two ventricles [1]. Therefore, the overloading of the RV pressure and volume affects not only the RV morphology and function, but also the LV morphology and function, in both the systolic and diastolic phases [2].

Increased pulmonary arterial pressure (PAP) can lead to prolonged and inhomogeneous RV contraction and has been associated with negative ventricular–ventricular interactions in pulmonary hypertension (PH) [3,4]. There is also evidence that the mechanism of RV-induced LV discoordination involves a combination of delayed early systolic electromechanical activation, late-systolic IVS shift, and prolonged post-systolic IVS thickening [5]. IVS displacement is defined as a flattening or bowing toward the LV [6], and it is observed in patients with PH. The presence of IVS displacement has been associated with the severity of PH [7] and may impair the LV filling dynamics [8]. IVS displacement is considered to be a consequence of the prolonged contraction of the RV free wall relative to that of the IVS and the LV free wall, causing interventricular relaxation dyssynchrony [9]. Therefore, IVS displacement causes the ventricular interdependency to become visible in PH. The simulations using the computational model have shown that the altered duration of the RV free wall contraction and profound IVS dyskinesia are associated with interventricular mechanical discoordination and decreased early LV filling in PH [10]. Nevertheless, the computational simulations of PH cannot explain the cases of all PH patients, especially in regard to the severe symptomatic PH patients.

Cardiac magnetic resonance (CMR) imaging is a non-invasive, robust diagnostic follow-up tool used for PH patients [11]. Cine CMR imaging is the reference standard for the evaluation of the morphology, volume, and function of both the LV and RV [12], and its value in the evaluation of patients with PH is increasingly recognized [12]. Cine CMR is able to describe the morphological changes in IVS to give a detailed assessment of the severity of PH [13]. It can display the opening and closure of the aortic valves, pulmonary arterial valves, mitral valves, and tricuspid valves distinctly [14], so that the temporal parameters of the cardiac cycle can be derived accurately. However, limited CMR data are available regarding the changes in the cardiac cycle in PH patients.

The goal of this study was to quantify the temporal changes in the cardiac cycle in PH patients with different IVS shapes and their relationship with the cardiac function using CMR, and to understand the mechanism of cardiac dysfunction in the cardiac cycle.

2. Materials and Methods

2.1. Study Design and Patient Enrollment

Data were obtained from the records of adult patients who were diagnosed with or suspected of having PH from May 2012 to August 2018, who had been examined by CMR. The eligible patients retrospectively included in this study were those who were diagnosed with PH by right heart catheterization (RHC), as defined by the European Society of Cardiology [15], but who had not received any treatment. The interval between the CMR and the RHC examinations was less than a week. All the patients were in sinus rhythm. All patients whose etiology was not precapillary PH were excluded because of their different hemodynamic conditions. Precapillary PH was defined as a mean pulmonary arterial pressure (mPAP) of $\geq$25 mmHg with a pulmonary artery wedge pressure (PAWP) of $\leq$15 mmHg and pulmonary vascular resistance (PVR) of >3 Wood units, measured by RHC [15]. Patients who were aged below 18 years or those whose CMR image qualities did not meet the post-processing requirements were excluded. All the PH patients were divided into two groups according to IVS morphology, observed on short-axis cine CMR images. The PH patients with IVS flattening or even bowing toward the LV were classified as the IVS displacement (IVS$_D$) group, while the PH patients without IVS displacement were classified as the IVS non-displacement (IVS$_{ND}$) group. Additionally, 22 age-matched healthy volunteers with no evidence of any heart diseases also underwent CMR examination and were included as the health control (HC) group (Figure 1). The study was approved by the hospital research ethics committee and conducted in compliance with all the clinical practice requirements as prescribed by the committee. The requirement of informed consent was waived.

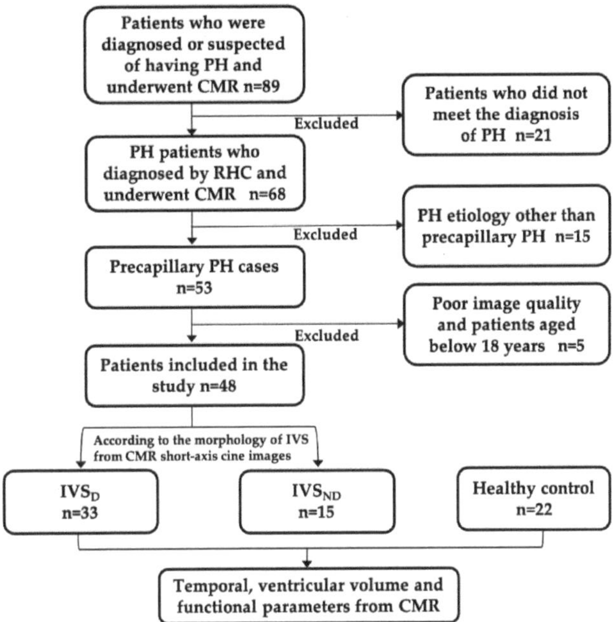

Figure 1. Patient flowchart. PH, pulmonary hypertension; CMR, cardiac magnetic resonance; RHC, right heart catheterization; IVS_{ND}, pulmonary hypertension patients without interventricular septum displacement; IVS_D, pulmonary hypertension patients with interventricular septum displacement.

2.2. CMR Examination

CMR was performed on a 3.0 T MR scanner (GE Healthcare, Discovery MR 750, Milwaukee, WI, USA) with an 8-channel cardiac coil, using a vector-cardiographic method for electrocardiogram gating. The short-axis cine CMR images (slice thickness = 8 mm), 4-chamber cine images (slice thickness = 6 mm), and LV and RV outflow tracts (slice thickness = 6 mm) were acquired using fast imaging employing steady-state acquisition (FIESTA) during breath-holds. The acquisition parameters were as follows: 20 frames per cardiac cycle, repetition time 3.40 to 3.60 ms, echo time 1.50 to 1.60 ms, flip angle = 45°, bandwidth = 125 KHz/pixel, field of view = 35 cm × 35 cm, matrix = 224 × 224, and NEX = 1.

2.3. CMR Image Analysis

The CMR image analysis was performed with Report Card 3.7 on GE Advantage Workstation 4.6. The analysis of the IVS morphology was performed using the short-axis cine CMR images, as previously described in the literature [12–14], including the interventricular septal curvature (C_{IVS}) and curvature ratio (CR). The RV and LV function analysis was also performed using the short-axis cine CMR images. The RV and LV endocardial and epicardial borders were automatically traced with manual adjustments to obtain the end-diastolic volume (EDV), end-systolic volume (ESV), stroke volume (SV) and ejection fraction (EF). The papillary muscles and trabeculae were included in the ventricular cavity volume. The myocardial mass (MM) was calculated by multiplying the volume of the ventricular myocardium in the end-diastolic phase with its density ($1.05 \text{ g}/\text{cm}^3$). The ventricular mass index (VMI) was calculated by dividing the RVMM by the LVMM. All the cardiac functional parameters were divided by the specific body surface area (BSA) for normalization and recorded as the end-diastolic volume index (EDVI), end-systolic volume index (ESVI), stroke volume index (SVI) and myocardial mass index (MMI), respectively.

The opening and closure of the aortic valves, pulmonary valves, mitral valves and tricuspid valves were reviewed using the cine CMR images. The aortic valves and pulmonary valves were reviewed using the outflow tract cine images of the LV and RV, respectively. The mitral valves and tricuspid valves were observed using the 4-chamber cine view. The opening time and closure time of the valves were normalized and recorded as the percentage of the R-R interval. The cardiac cycle was composed of an isovolumetric relaxation phase, a filling phase, an isovolumetric contraction phase and an ejection phase. The durations of the four phases were calculated as follows (isovolumetric relaxation time, IRT; filling time, FT; isovolumetric contraction time, ICT; ejection time, ET; tricuspid valve, T; pulmonary artery valve, P; mitral valve, M; aortic valve, A; open, o; closure, c):

$$IRT_{RV} = T_O - P_C$$

$$FT_{RV} = T_C - T_O$$

$$ICT_{RV} = P_O - T_C$$

$$ET_{RV} = P_C - P_O$$

$$IRT_{LV} = M_O - A_C$$

$$FT_{LV} = M_C - M_O$$

$$ICT_{LV} = A_O - M_C$$

$$ET_{LV} = A_C - A_O$$

To assess the inter-observer and intra-observer reproducibility of the temporal parameters of the cardiac cycle, 18 (26%) PH patients were selected randomly and re-examined independently by the two readers (FY, 6 years CMR experience and DW, 3 years CMR experience). For the intra-observer reproducibility, one observer (FY) repeated the measurements four weeks later. The intra-class correlation coefficient was used to assess the reproducibility.

2.4. Risk Stratification, Follow-Up, and Study Endpoint

In line with the risk assessment instrument from the abbreviated version of the 2015 European Society of Cardiology (ESC)/European Respiratory Society (ERS) risk stratification strategy [15], all the PH patients were graded according to the World Health Organization functional classification (WHO FC), including 6 min walking distance (6 MWD), brain natriuretic peptide (BNP), N-terminal prohormone of the brain natriuretic peptide (NT-proBNP), mean right atrial pressure (mRAP), cardiac index (CI) and mixed venous oxygen saturation (SvO2). For each patient, the sum of all grades was divided by the number of available variables and rounded to the next integer to define the risk group. The cut-off values proposed in the guidelines were graded from 1–3 (1: low risk, 2: intermediate risk, 3: high risk). The overall treatment goal for patients with PH is to achieve a low-risk status; thus, the PH patients were divided into the low-risk group and the intermediate- and high-risk groups.

All the PH patients were followed up with a census date of 12 January 2021. The designed primary endpoint was defined as major cardiovascular events (MACE), which included hospitalization for heart failure, lung transplantation, malignant ventricular arrhythmia and death.

2.5. Statistical Analyses

SPSS 23.0 was used for all the statistical analyses. All the continuous variables were presented as medians and interquartile ranges (IQR) when the variables were not normally distributed. The Kruskal–Wallis one-way ANOVA test, with Bonferroni correction post hoc analysis, was used to compare the continuous variables among the three groups of HC, IVS_{ND} and IVS_{D}. The categorical variables were presented as frequencies (%) and compared using Fisher's exact test. The correlation between the parameters was calculated with Spearman's correlation coefficient. Multivariable stepwise regression analysis was used to explore the factors associated with RVEF. Each variable with a significant association ($p < 0.05$) in the univariate analysis was introduced into the regression model. The differences in the ventricular function and the temporal parameters between the low-risk group and the intermediate- and high-risk groups were tested by the Mann–Whitney U test. The univariable and multivariable Cox regression models included demographic factors, clinical factors, laboratory tests, and RHC and CMR variables (variables whose p-value was <0.15 in the univariable Cox regression analysis were included in the multivariable Cox regression analysis). All the analyses were two-sided, and p-values of <0.05 were considered to be statistically significant.

3. Results

3.1. Patient Characteristics

The clinical characteristics of the subjects are presented in Table 1. A total of 48 consecutive precapillary PH patients confirmed by RHC were enrolled, including 29 patients with PAH, 18 patients with chronic thromboembolic pulmonary hypertension (CTEPH), and one patient who had PH with unclear and/or multifactorial mechanisms. There were no significant differences in age, sex or BSA among the IVS_{ND} group, the IVS_{D} group and the HC group. The IVS_{D} group had significantly decreased 6MWD and increased NT-proBNP and WHO FC levels. The HR was significantly faster in the IVS_{D} group than in the IVS_{ND} group ($p < 0.05$) and in the HC group ($p < 0.001$). No significant differences between the IVS_{ND} group and the IVS_{D} group in the proportion of PH subsets ($p > 0.05$) were observed. The mPAP, PVR, CI and mRAP were higher in the IVS_{D} group than in the IVS_{ND} group ($p < 0.05$).

Table 1. Demographic, clinical and RHC characteristics in the health control and PH patients.

	HC (n = 22)	IVS_{ND} (n = 15)	IVS_{D} (n = 33)
Demographics			
Female, n (%)	20 (90.9)	14 (93.3)	27 (81.8)
Age, years	43 (35–49)	52 (37–62)	40 (32–62)
PAH/CTEPH	–	8/21	7/11
WHO FC I/II/III/IV	–	0/14/1/0	0/13/17/3 [###]
6MWD, m	–	444 (302–463)	235 (157–339) [###]
NT-proBNP, pg/mL	–	97 (56–245)	1678 (634–1961) [###]
HR, bpm	68 (63–73)	71 (66–79)	84 (78–93) [*** #]
BSA, m^2	1.65 (1.58–1.77)	1.66 (1.51–1.75)	1.69 (1.50–1.76)
Low-risk/intermediate and high risk	–	9/6	3/30
RHC measurements			
mPAP, mmHg	–	39 (33–46)	54 (41–62) [##]
PVR, Wood	–	10 (7–14)	16 (11–21) [##]
CI, L/min/m^2	–	2.3 (2.1–3.1)	1.9 (1.5–2.5) [#]
mRAP, mmHg	–	5 (3–6)	6 (5–9) [#]
PAWP, mmHg	–	10 (7–12)	9 (6–10)

[***] $p < 0.001$: versus HC; [#] $p < 0.05$, [##] $p < 0.01$, [###] $p < 0.001$: versus IVS_{ND}. HC, health control; IVS_{ND}, pulmonary hypertension patient without interventricular septum displacement; IVS_{D}, pulmonary hypertension patient with interventricular septum displacement; HR, heart rate; BSA, body surface area; WHO FC, World Health Organization functional classification; 6MWD, 6 min walking distance; NT-proBNP, N-terminal prohormone of the brain natriuretic peptide; PAH, pulmonary arterial hypertension; CTEPH, chronic thromboembolic pulmonary hypertension; RHC, right heart catheterization; mPAP, mean pulmonary arterial pressure; PVR, pulmonary vascular resistance; CI, cardiac index; mRAP, mean right atrium pressure; PAWP, pulmonary arterial wedge pressure.

3.2. IVS Morphology and Cardiac Functional Parameters

The IVS_D group had significantly decreased C_{IVS} and CR when compared with the IVS_{ND} group and the HC group (Table 2).

Table 2. CMR-derived morphologic and functional parameter characteristics in the health control and PH patients.

	HC (*n* = 22)	IVS_{ND} (*n* = 15)	IVS_D (*n* = 33)
RVEDVI, mL/m^2	81 (71–87)	68 (61–93)	117 (104–140) *** ###
RVESVI, mL/m^2	37 (32–43)	43 (36–49)	84 (66–95) *** ###
RVSVI, mL/m^2	42 (38–49)	35 (27–42)	37 (29–47)
RVEF, %	52 (50–60)	45 (38–50) *	31 (25–40) *** ##
RVMMI, g/m^2	13 (10–17)	17 (13–19)	27 (20–31) *** ##
LVEDVI, mL/m^2	80 (72–90)	68 (53–78)	54 (42–48) ***
LVESVI, mL/m^2	33 (29–41)	24 (21–35)	26 (17–29) ***
LVSVI, mL/m^2	45 (42–52)	44 (32–50)	30 (23–39) *** #
LVEF, %	57 (52–67)	61 (58–64)	57 (53–61)
LVMMI, g/m^2	40 (37–48)	41 (36–47)	42 (36–48)
VMI	0.34 (0.27–0.42)	0.42 (0.30–0.46)	0.61 (0.48–0.76) *** ##
C_{IVS}	0.07 (0.06–0.07)	0.06 (0.04–0.07)	0.01 (−0.01–0.02) *** ###
CR	1.00 (0.95–1.04)	0.88 (0.71–0.92)	0.24 (−0.23–0.45) *** ###

* $p < 0.05$, *** $p < 0.001$: versus HC; # $p < 0.05$, ## $p < 0.01$, ### $p < 0.001$: versus IVS_{ND}; CMR, cardiac magnetic resonance; HC, health control; IVS_{ND}, pulmonary hypertension patient without interventricular septum displacement; IVS_D, pulmonary hypertension patient with interventricular septum displacement; RV, right ventricle; LV, left ventricle; EDVI, end-diastolic volume index; ESVI, end-systolic volume index; SVI, stroke volume index; EF, ejection fraction; MMI, myocardial mass index; VMI, ventricular mass index; C_{IVS}, interventricular septal curvature; CR, curvature ratio.

Figure 2 shows the trend of the percentage changes in the RV (a) and LV (b) function from the HC to the IVS_D group. The detailed differences in the above functional parameters among the groups are shown in the Figure S1. For the RV parameters (Figure 2a), the IVS_D patients had increased RVEDVI and RVESVI, and decreased RVEF when compared with the IVS_{ND} group and the HC ($p < 0.01$). The RVEF was the only decreased parameter in the IVS_{ND} group compared with the HC ($p < 0.05$). While there was no significant difference in the RVSVI among the three groups ($p > 0.05$), among the LV parameters (Figure 2b), the LVEDVI, LVESVI, and LVSVI were decreased in the IVS_D patients ($p < 0.001$). There was no significant difference in the LVEF among the three groups ($p > 0.05$) (Table 2). The RVMMI and VMI of the IVS_D group were increased ($p < 0.01$), but the LVMMI showed no difference.

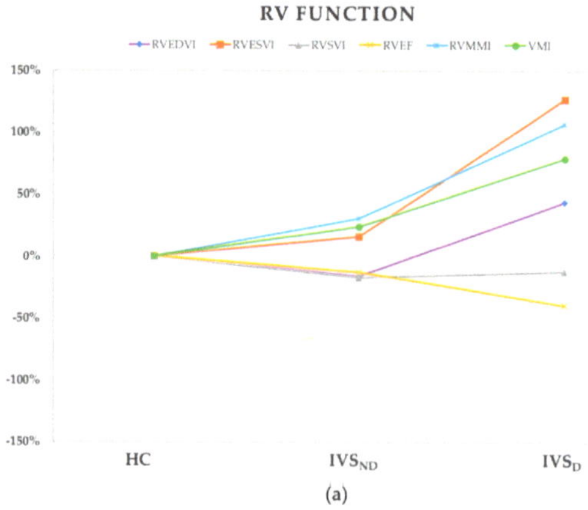

Figure 2. *Cont.*

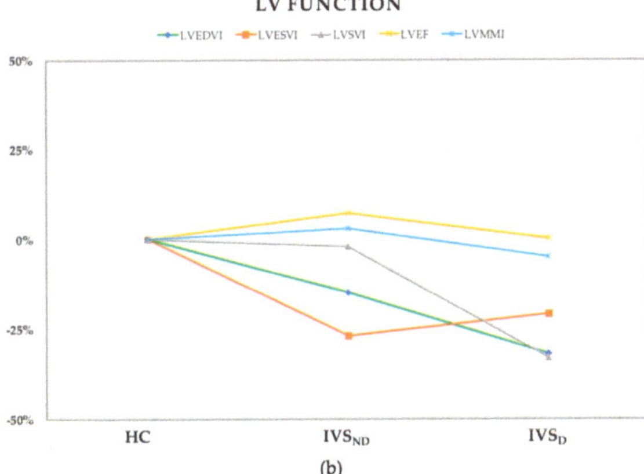

(b)

Figure 2. The trend of the percentage changes in the RV (**a**) and LV (**b**) function from the HC to IVS_D group. All of the values of the coordinate points were calculated from the percentage change of the median of the parameters in the IVS_{ND} and IVS_D groups using the HC as the baseline. HC, health control; IVS_{ND}, pulmonary hypertension patients without interventricular septum displacement; IVS_D, pulmonary hypertension patients with interventricular septum displacement; RV, right ventricle; LV, left ventricle; EDVI, end-diastolic volume index; ESVI, end-systolic volume index; SVI, stroke volume index; EF, ejection fraction; MMI, myocardial mass index; VMI, ventricular mass index.

3.3. Temporal Parameters in the Cardiac Cycle

Table 3 and Figure 3 present the characteristics of the temporal parameters in the cardiac cycle. All the temporal parameters showed no difference between the IVS_{ND} and HC groups.

Table 3. CMR-derived temporal parameters of the three groups.

Temporal Parameters	HC ($n = 22$)	IVS_{ND} ($n = 15$)	IVS_D ($n = 33$)
P_O, %	6.1 (5.6–6.2)	6.0 (1.2–10.8)	1.3 (−3.4–6.2)
P_C, %	41.0 (36.2–41.3)	40.8 (36.1–41.3)	36.3 (33.2–38.9)
A_O, %	6.1 (4.8–6.2)	6.0 (1.1–10.7)	1.2 (−3.5–6.2) *
A_C, %	41.0 (36.1–41.2)	36.1 (36.0–41.0)	36.1 (31.5–36.3) **
M_O, %	46.1 (41.1–51.0)	46.2 (41.3–51.5)	46.4 (43.8–48.4)
M_C, %	101.0 (96.2–101.1)	101.0 (91.5–101.1)	91.6 (86.7–96.3) *** #
T_O, %	46.1 (41.1–51.0)	46.2 (45.9–51.2)	51.3 (46.5–54.0) *** #
T_C, %	101.1 (99.7–101.1)	101.0 (91.5–101.1)	96.2 (91.5–96.4) ***
ICT_{RV}, %	5.0 (4.9–5.3)	5.00 (4.9–10.0)	5.0 (4.9–10.0)
ET_{RV}, %	35.0 (32.4–38.1)	30.0 (30.0–40.0)	35.0 (30.1–40.0)
IRT_{RV}, %	5.0 (4.9–5.3)	9.9 (5.1–15.0)	15.0 (10.1–19.9) *** #
FT_{RV}, %	55.0 (50.0–56.3)	50.0 (45.0–55.0)	44.2 (35.7–49.9) *** #
ICT_{LV}, %	5.0 (5.0–10.0)	5.0 (4.9–10.0)	9.9 (5.0–10.1)
ET_{LV}, %	35.0 (30.0–35.6)	30.0 (30.0–35.0)	35.0 (30.0–35.1)
IRT_{LV}, %	5.1 (5.0–10.0)	9.9 (5.1–15.0)	13.4 (10.0–15.0) ***
FT_{LV}, %	52.5 (49.9–55.1)	50.1 (39.9–60.0)	45.1 (40.1–50.0) **

* $p < 0.05$, ** $p < 0.01$, *** $p < 0.001$: versus HC; # $p < 0.05$: versus IVS_{ND}; HC, health control; IVS_{ND}, pulmonary hypertension patient without interventricular septum displacement; IVS_D, pulmonary hypertension patient with interventricular septum displacement; RV, right ventricle; LV, left ventricle; P, pulmonary artery valve; A, aortic valve; M, mitral valve; T, tricuspid valve; ICT, isovolumetric contraction time; IRT, isovolumetric relaxation time; ET, ejection time; FT, filling time.

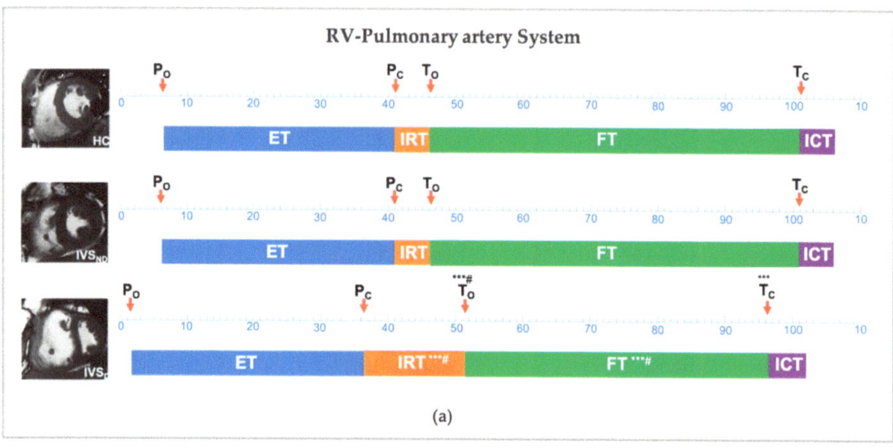

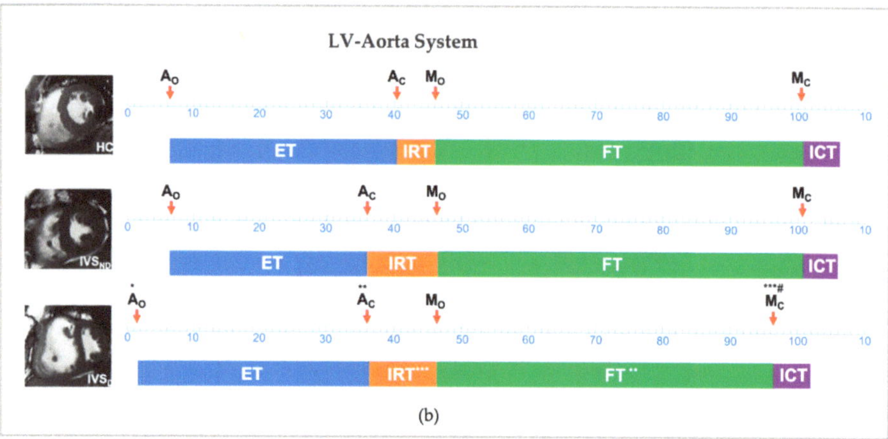

Figure 3. Temporal changes between the HC, IVS_{ND} and IVS_D in the right ventricle-pulmonary artery system (**a**) and left ventricle-aorta system (**b**). * $p < 0.05$, ** $p < 0.01$, *** $p < 0.001$: versus HC; # $p < 0.05$: versus IVS_{ND}; P, pulmonary artery valve; A, aortic valve; M, mitral valve; T, tricuspid valve; ICT, isovolumetric contraction time; IRT, isovolumetric relaxation time; ET, ejection time; FT, filling time; HC, health control; IVS_{ND}, pulmonary hypertension patients without interventricular septum displacement; IVS_D, pulmonary hypertension patients with interventricular septum displacement.

Compared with the HC group, the IVS_D showed no difference in the P_O, P_C, ET_{RV} and ICT_{RV} in terms of the RV-pulmonary artery system, while this group showed a delayed T_O, an advanced T_C, a longer IRT_{RV} and a shorter FT_{RV} ($p < 0.001$). With regards to the LV-aorta system, although both the A_O and the A_C were advanced, the ET_{LV} and the ICT_{LV} showed no differences in the IVS_D group. The IVS_D group showed no differences in the M_O, while showing an advanced M_C, a longer IRT_{LV} and a shorter FT_{LV} ($p < 0.001$).

Comparing the two groups of IVS_D and IVS_{ND}, a longer IRT_{RV} and shorter FT_{RV} were detected along with a delayed T_O in the IVS_D group ($p < 0.05$) (Table 3). However, in the temporal parameters of the LV, no differences were detected, except an advanced M_C ($p < 0.05$).

The intra- and inter-observer reproducibility results using intra-class correlation coefficients are shown in the Supplementary Materials, Table S1. The inter-observer and intra-observer variability results for all the temporal parameters were very low.

3.4. Correlation of the Temporal Parameters with the RHC and CMR Data

The IRT_{RV} was significantly correlated with the PVR, CI and C_{IVS} ($r = 0.38$, $r = -0.34$, $r = -0.49$, $p < 0.05$). The FT_{RV} was significantly correlated with the mPAP, PVR, CI and C_{IVS} ($r = -0.46$, $r = -0.52$, $r = 0.47$, $r = 0.55$, $p < 0.01$). There were significant correlations between some ventricular functional parameters and mPAP, PVR, CI and C_{IVS}, respectively ($p < 0.05$) (Table 4).

Table 4. Spearman correlations *r* between the CMR data and RHC data.

	mPAP	PVR	CI	Civs
RVEDVI	0.19	0.36 *	−0.35 *	−0.72 ***
RVESVI	0.34 *	0.53 **	−0.50 ***	−0.79 ***
RVEF	−0.50 **	−0.56 **	0.52 ***	0.67 ***
RVMMI	0.25	0.43 **	−0.41 **	−0.67 **
LVEDVI	−0.50 **	−0.41 **	0.40 **	0.44 ***
LVESVI	−0.39 **	−0.32 *	0.35 *	0.29 ***
LVSVI	−0.51 **	−0.43 **	0.39 **	0.48 *
VMI	0.41 **	0.46 **	−0.44 **	−0.66 ***
IRT$_{RV}$	0.27	0.38 **	−0.34 *	−0.49 ***
FT$_{RV}$	−0.46 **	−0.52 ***	0.47 **	0.55 ***

* $p < 0.05$, ** $p < 0.01$, *** $p < 0.001$. CMR, cardiac magnetic resonance; RHC, right heart catheterization; RV, right ventricle; LV, left ventricle; EDVI, end-diastolic volume index; ESVI, end-systolic volume index; SVI, stroke volume index; EF, ejection fraction; MMI, myocardial mass index; VMI, ventricular mass index; mPAP, mean pulmonary arterial pressure; PVR, pulmonary vascular resistance; CI, cardiac index; C_{IVS}, interventricular septal curvature. IRT, isovolumetric relaxation time; FT, filling time.

The multivariate linear regression included age, sex, 6MWD, NT-proBNP, and RHC and CMR variables and showed that the IRT_{RV}, LVEF and PVR were significantly associated with the RVEF (Table 5).

Table 5. Multiple linear regression analysis for RVEF.

Variates	Unstandardized Coefficients		Standardized Coefficients	*t*	*p*-Value
	B (95% CI)	Std. Error	Beta		
IRT$_{RV}$	−0.456 (−0.990−−0.246)	0.188	−0.293	−2.426	0.019
LVEF	0.490 (0.128–0.852)	0.180	0.308	2.728	0.009
PVR	−0.618 (−0.835−−0.077)	0.185	−0.403	−3.350	0.002

The correlation coefficient between the result of the model and the CMR-derived RVEF was as follow, R = 0.667, R^2 = 0.444, adjusted R^2 = 407, F = 11.733, $p < 0.001$. RVEF, right ventricular ejection fraction; LVEF, left ventricular ejection fraction; RV, right ventricle; IRT, isovolumetric relaxation time.

3.5. Differences in the CMR and RHC Characteristics Based on Risk Stratification

Table 1 shows that 40% of the IVS_{ND} group and 90% of the IVS_D group were in the intermediate- and high-risk categories. Patients in the intermediate- and high-risk groups had a significantly increased RVEDVI, RVESVI and RVMASSI VMI, and a lower RVEF, LVEDVI, LVSVI and CR compared to the low-risk group ($p < 0.05$) (Table 6). Patients in the intermediate- and high-risk groups displayed a delayed T_O, a longer IRT_{RV} and a shortened FT_{RV} ($p < 0.01$).

Table 6. Comparison of the CMR-derived indices and RHC characteristics of the PH patients based on risk stratification.

Variates	Low Risk (n = 12)	Intermediate and High Risk (n = 36)	p-Value
RVEDVI, mL/m^2	89 (66–93)	112 (91–135)	0.008 *
RVESVI, mL/m^2	47 (36–53)	80 (57–94)	0.001 *
RVEF, %	46 (40–50)	32 (27–40)	<0.001 *
RVMASSI, g/m^2	17 (14–23)	23 (19–30)	0.012 *
LVEDVI, mL/m^2	70 (56–81)	55 (43–70)	0.034 *
LVSVI, mL/m^2	41 (31–53)	32 (23–41)	0.024 *
VMI	0.45 (0.32–0.50)	0.60 (0.44–0.72)	0.022 *
CR	0.78 (0.34–0.91)	0.29 (−0.21–0.69)	0.017 *
T_O, %	46.2 (42.5–51.1)	51.3 (46.4–55.2)	0.005 *
IRT$_{RV}$, %	9.9 (5.1–13.8)	15.0 (10.0–19.7)	0.009 *
FT$_{RV}$, %	50.0 (49.9–55.0)	43.6 (35.0–49.9)	0.001 *

* $p < 0.05$.

3.6. Survival Analysis

Twelve patients (25%) with PH died during the median follow-up period of 62 months (interquartile range: 38–67 months). The univariate Cox proportional hazards regression analysis of all the PH patients showed that age, WHO FC, 6MWD, IRT$_{RV}$, CR and mPAP were associated with the MACE. The multivariable analysis showed that the IRT$_{RV}$ and mPAP were the significant predictors of the MACE (Table 7).

Table 7. Univariable and multivariable Cox proportional hazard analysis for the MACE.

Variates	Univariate		Multivariate	
	HR (95% CI)	p-Value	HR (95% CI)	p-Value
age	0.971 (0.940–1.004)	0.08		
WHO FC	2.227 (1.039–4.774)	0.04		
6MWD	0.996 (0.991–1.000)	0.05		
IRT$_{RV}$	0.946 (0.881–1.017)	0.13	0.930 (0.867–0.997)	0.04
CR	0.421 (0.152–1.164)	0.10		
mPAP	1.037 (1.004–1.072)	0.03	1.046 (1.012–1.081)	0.007

MACE, major cardiovascular event; HR, hazard ratio; CI, confidence interval.

4. Discussion

This cardiac MR study has provided detailed information about the temporal variations in the cardiac cycle in the progression of PH patients. The main variations were in the diastolic period, and the PH patients with IVS displacement had a longer IRT and shorter FT for both the RV and LV, and a prolonged IRT and shortened FT of the RV were associated with the RV afterload. Additionally, the IRT$_{RV}$ may have the potential to predict the prognosis for PH patients. However, the PH patients without IVS displacement showed no temporal variations in the cardiac cycle.

Increased PVR and elevated PAP are the main factors responsible for PH, leading to vascular remodeling and cardiac remodeling [16]. During the compensation stage, the PH patients showed a mildly decreased RV contractile function and sustained LV function. The filling pressure and interventricular pressure gradients were still normal [10,17]. Previous studies showed that RV contractile dysfunction was significantly associated with the severity of PH and the curvature of the IVS showed a strong correlation with the sPAP [6,18]. The current findings showed that the only impaired cardiac functional parameter was the RVEF in the PH patients without IVS displacement.

Although increased mPAP and PVR reflect the severity of PH, increased mPAP and PVR do not reflect an adverse ventricular–ventricular interaction, interventricular pressure gradient and interventricular septal motion directly [10]. The increased afterload and impaired ventricular interaction deteriorate both ventricles to the decompensation state.

The RV is remodeled both concentrically and eccentrically with the RV dilatation and myocardial hypertrophy [19]. The IVS becomes flattened, and even bows to the LV. As the results of this study showed, the C_{IVS} and CR were both reduced in the IVS_D patients. The RVEDVI, RVESVI and RVMMI were significantly increased and the RVEF was further decreased in the IVS_D patients compared with those of the IVS_{ND} patients and HC patients. The LVEDVI, LVESVI and LVSVI were decreased with the IVS displacement and LV compression.

Furthermore, RV pressure loading influences not only the IVS morphology but also the temporal phases of the cardiac cycle [2]. In general, the opening and closure of the atrioventricular valves are mainly dependent on the different pressure gradients between the atrium and ventricle. When the intraventricular pressure drops below the intra-atrial pressure, the atrioventricular valve opens, while the atrioventricular valve closes when the intraventricular pressure rises above the intra-atrial pressure [14].

With respect to the pulmonary artery-RV system, our study identified delayed To, advanced Tc, prolonged IRT_{RV} and shortened FT_{RV} in the IVS_D patients. In PH patients with IVS displacement, the increased RV pressure and inefficient RV diastolic function could result in a prolonged IRT_{RV}, which could in turn decrease the intraventricular pressure. Moreover, increased RV pressure would promote the closure of the tricuspid valve and shorten the FT_{RV} accordingly. An increased afterload would reduce the myocyte velocity and prolong the myocyte shortening. Thus, RV actin-myosin cross-bridge cycling may result in a stiffer myocardium near the end of the ventricular ejection, prolonging the IRT and hampering the early relaxation [20]. The IRT_{RV} was prolonged in PH patients when compared with the controls, resulting in a noticeable delay in atrioventricular opening [21]. The shortened FT_{RV} then impedes its filling, which has an essential impact on the RV systolic function [2].

With respect to the LV-aorta system, the IVS_D patients displayed advanced Mc, prolonged IRT_{LV} and shortened FT_{LV}. This was linked to IVS displacement. The increased RV pressure causing the RV dilatation and IVS flattening or even bowing to the LV would result in a smaller LV cavity, which leads to the LV underfilling and reduction in the LVSV. RV dilatation and IVS displacement had a further deleterious effect on ventricular interactions [4]. Under these circumstances, a prolonged IRT and shortened FT of the LV and RV could cause ventricular underfilling and decrease the systolic function of both ventricles [10,19,21–23].

Our findings also showed that IRT_{RV}, FT_{RV} and some of the cardiac functional parameters (such as RVEDVI, RVESVI, RVEF and VMI) were associated with PVR, CI and C_{IVS}. Additionally, the IRT_{RV}, LVEF and PVR were associated with the RVEF in the multivariate regression analysis. All these findings indicate that the IVS displacement and increased RV afterload were not the only factors related to ventricular dysfunction. The temporal variations of both the IRT_{RV} and FT_{RV} in the cardiac cycle may also be associated with an IVS morphological abnormality and impaired ventricular function, especially in the case of the IRT_{RV}.

Guidelines on the diagnosis and treatment of PH, published by the European Society of Cardiology and European Respiratory Society, state that the overall treatment goal for patients with PH is to achieve a low-risk status, which is usually associated with good exercise capacity, good quality of life, good RV function and a low-mortality risk [15]. The main indicators for the risk stratification in PH are WHO FC, 6MWD, NT-proBNP/BNP and CI [11], and the corresponding results of this study were consistent with the guidelines. Most of the PH patients with IVS displacement were stratified into the intermediate- and high-risk categories, reaching as high as 90% in this study, and they showed delayed To, prolonged IRT_{RV} and shortened FT_{RV}. These findings indicate that the temporal variations of the IRT_{RV} and FT_{RV} in cardiac cycle could also inform the risk stratification of PH patients. Moreover, the multivariable Cox proportional hazards regression analysis showed that the IRT_{RV} was the significant predictor of the MACE. This indicates that the IRT_{RV} might serve as a potential prognostic factor for PH patients.

By applying the current methods in the clinics, we can not only evaluate the function and morphology of the ventricles, but also provide extra temporal information for PH patients, without introducing additional CMR sequences and scanning times. This pathophysiological information may give the doctors a profound understanding of both the progression of PH and the interaction between the ventricles, offering help in the guidance of the treatment strategy for individuals with PH. IVS displacement, prolonged IRT_{RV}, and shortened FT_{RV} indicate RV failure in patients with precapillary-PH, and, more seriously, predict clinical worsening [24]. On the contrary, the reversible displacement of IVS is useful in predicting the alleviation of PH [25]. The temporal parameters of the diastole may have prognostic value and may act as indicators of effective treatment. Cardiac resynchronization therapy is a potential strategy for patients with severe symptomatic PH [26]. The findings regarding the changes in the temporal parameters might provide additional implications for intervention. Large and long-term prospective studies on IVS displacement and the associated temporal changes are needed to confirm this.

This study has several limitations. Firstly, this is a single-center study with a small sample size. The current results and discoveries should be further validated using a large PH population. Next, although the four phases in the cardiac cycle were clearly set out in this study, the pilot temporal parameter method, which used CMR cine data with images including 20 frames from multiple cardiac cycles, might have caused sampling errors or inaccuracies. All the patients were in sinus rhythm, and this could minimize the sampling errors. Finally, the heterogeneous etiology causing precapillary-PH might have led to other uncontrolled factors in the analysis. Further exploration is needed.

With the progression of PH, the diastolic period variations in the RV and LV were related to IVS displacement and ventricular dysfunction. The temporal variations in the cardiac cycle may also be possible parameters, which can explain the severity of the disease and provide useful information about the pathophysiological mechanism of the ventricular dysfunction and indicate the prognosis for PH patients.

Supplementary Materials: The following supporting information can be downloaded at: https://www.mdpi.com/article/10.3390/diagnostics12081970/s1, Figure S1: The differences in RV and LV function among the HC, IVS_{ND}, and IVS_D groups; Table S1: Reproducibility using intra-class correlation coefficients of temporal parameters.

Author Contributions: Conceptualization, D.L. and F.Y.; methodology, F.Y. and W.R.; software, Y.Y.; validation, Y.-L.D., D.W. and Y.Y.; formal analysis, F.Y.; resources, Z.-W.Y. and T.-L.Y.; writing—original draft preparation, F.Y.; writing—review & editing, D.L. and Z.Z.; visualization, F.Y.; project administration, D.L. and Z.Z. All authors have read and agreed to the published version of the manuscript.

Funding: This work was supported by the National Natural Science Foundation of China (82071907), the China International Medical Foundation Sky Imaging Research Fund (Z-2014-07-2003-05), the Natural Science Foundation of Tianjin (18JCYBJC25100), the Health Science and Technology Project of Tianjin (MS20022), the Wu Jieping Medical Foundation Special Fund for Clinical Research (320.6750.2022-3-5), the Tianjin Key Medical Discipline (Specialty) Construction Project (TJYXZDXK-001A), and the Suzhou "Science and Education Revitalize Health" Youth Science and Technology Project (KJXW2021042).

Institutional Review Board Statement: The study was conducted according to the guidelines of the Declaration of Helsinki and approved by the ethics committee of Tianjin Medical University General Hospital (IRB2022-KY-182, February 2022).

Informed Consent Statement: Patient consent was waived due to the fact that our study was retrospective and approved by the ethics committee.

Data Availability Statement: Data available on request.

Acknowledgments: The authors wish to thank Jean Glover for the help on the English grammar and expression.

Conflicts of Interest: The authors declare no conflict of interest.

References

1. Buckberg, G.D.; Hoffman, J.I.E.; Cecil Coghlan, H.; Nanda, N.C. Ventricular structure-function relations in health and disease: Part I. The normal heart. *Eur. J. Cardio-Thoracic Surg.* **2015**, *47*, 587–601. [CrossRef] [PubMed]
2. Naeije, R.; Badagliacca, R. The overloaded right heart and ventricular interdependence. *Cardiovasc. Res.* **2017**, *113*, 1474–1485. [CrossRef] [PubMed]
3. Gillebert, T.C.; Sys, S.U.; Brutsaert, D.L. Influence of loading patterns on peak length-tension relation and on relaxation in cardiac muscle. *J. Am. Coll. Cardiol.* **1989**, *13*, 483–490. [CrossRef]
4. Marcus, J.T.; Gan, C.T.J.; Zwanenburg, J.J.M.; Boonstra, A.; Allaart, C.P.; Götte, M.J.W.; Vonk-Noordegraaf, A. Interventricular Mechanical Asynchrony in Pulmonary Arterial Hypertension. Left-to-Right Delay in Peak Shortening Is Related to Right Ventricular Overload and Left Ventricular Underfilling. *J. Am. Coll. Cardiol.* **2008**, *51*, 750–757. [CrossRef]
5. Frank, B.S.; Schäfer, M.; Douwes, J.M.; Ivy, D.D.; Abman, S.H.; Davidson, J.A.; Burzlaff, S.; Mitchell, M.B.; Morgan, G.J.; Browne, L.P.; et al. Novel measures of left ventricular electromechanical discoordination predict clinical outcomes in children with pulmonary arterial hypertension. *Am. J. Physiol. Heart Circ. Physiol.* **2020**, *318*, H401–H412. [CrossRef]
6. Alunni, J.P.; Degano, B.; Arnaud, C.; Tétu, L.; Blot-Soulétie, N.; Didier, A.; Otal, P.; Rousseau, H.; Chabbert, V. Cardiac MRI in pulmonary artery hypertension: Correlations between morphological and functional parameters and invasive measurements. *Eur. Radiol.* **2010**, *20*, 1149–1159. [CrossRef]
7. López-Candales, A. Determinants of an abnormal septal curvature in chronic pulmonary hypertension. *Echocardiography* **2015**, *32*, 49–55. [CrossRef]
8. Gurudevan, S.V.; Malouf, P.J.; Auger, W.R.; Waltman, T.J.; Madani, M.; Raisinghani, A.B.; DeMaria, A.N.; Blanchard, D.G. Abnormal Left Ventricular Diastolic Filling in Chronic Thromboembolic Pulmonary Hypertension. True Diastolic Dysfunction or Left Ventricular Underfilling? *J. Am. Coll. Cardiol.* **2007**, *49*, 1334–1339. [CrossRef]
9. Haddad, F.; Guihaire, J.; Skhiri, M.; Denault, A.Y.; Mercier, O.; Al-Halabi, S.; Vrtovec, B.; Fadel, E.; Zamanian, R.T.; Schnittger, I. Septal curvature is marker of hemodynamic, anatomical, and electromechanical ventricular interdependence in patients with pulmonary arterial hypertension. *Echocardiography* **2014**, *31*, 699–707. [CrossRef]
10. Palau-Caballero, G.; Walmsley, J.; Van Empel, V.; Lumens, J.; Delhaas, T. Why septal motion is a marker of right ventricular failure in pulmonary arterial hypertension: Mechanistic analysis using a computer model. *Am. J. Physiol.-Hear. Circ. Physiol.* **2017**, *312*, H691–H700. [CrossRef]
11. Leuchte, H.H.; ten Freyhaus, H.; Gall, H.; Halank, M.; Hoeper, M.M.; Kaemmerer, H.; Kähler, C.; Riemekasten, G.; Ulrich, S.; Schwaiblmair, M.; et al. Risk stratification strategy and assessment of disease progression in patients with pulmonary arterial hypertension: Updated Recommendations from the Cologne Consensus Conference 2018. *Int. J. Cardiol.* **2018**, *272*, 20–29. [CrossRef] [PubMed]
12. Ibanez, B.; Aletras, A.H.; Arai, A.E.; Arheden, H.; Bax, J.; Berry, C.; Bucciarelli-Ducci, C.; Croisille, P.; Dall'Armellina, E.; Dharmakumar, R.; et al. Cardiac MRI Endpoints in Myocardial Infarction Experimental and Clinical Trials: JACC Scientific Expert Panel. *J. Am. Coll. Cardiol.* **2019**, *74*, 238–256. [CrossRef] [PubMed]
13. Dellegrottaglie, S.; Sanz, J.; Poon, M.; Viles-Gonzalez, J.F.; Sulica, R.; Goyenechea, M.; Macaluso, F.; Fuster, V.; Rajagopalan, S. Pulmonary hypertension: Accuracy of detection with left ventricular septal-to-free wall curvature ratio measured at cardiac MR. *Radiology* **2007**, *243*, 63–69. [CrossRef]
14. Sheth, P.J.; Danton, G.H.; Siegel, Y.; Kardon, R.E.; Infante, J.C.; Ghersin, E.; Fishman, J.E. Cardiac physiology for radiologists: Review of relevant physiology for interpretation of cardiac MR imaging and CT. *Radiographics* **2015**, *35*, 1335–1351. [CrossRef]
15. Galiè, N.; Humbert, M.; Vachiery, J.L.; Gibbs, S.; Lang, I.; Torbicki, A.; Simonneau, G.; Peacock, A.; Vonk Noordegraaf, A.; Beghetti, M.; et al. 2015 ESC/ERS Guidelines for the diagnosis and treatment of pulmonary hypertension. *Eur. Heart J.* **2016**, *37*, 67–119. [CrossRef] [PubMed]
16. Broncano, J.; Bhalla, S.; Gutierrez, F.R.; Vargas, D.; Williamson, E.E.; Makan, M.; Luna, A. Cardiac MRI in pulmonary hyper-tension: From magnet to bedside. *Radiographics* **2020**, *40*, 982–1002. [CrossRef]
17. Vonk Noordegraaf, A.; Chin, K.M.; Haddad, F.; Hassoun, P.M.; Hemnes, A.R.; Hopkins, S.R.; Kawut, S.M.; Langleben, D.; Lumens, J.; Naeije, R. Pathophysiology of the right ventricle and of the pulmonary circulation in pulmonary hypertension: An update. *Eur. Respir. J.* **2019**, *53*, 1801900. [CrossRef]
18. Roeleveld, R.J.; Marcus, T.J.; Faes, T.J.C.; Gan, T.J.; Boonstra, A.; Postmus, P.E.; Vonk-Noordegraaf, A. Interventricular septal configuration at MR imaging and pulmonary arterial pressure in pulmonary hypertension. *Radiology* **2005**, *234*, 710–717. [CrossRef] [PubMed]
19. Sanz, J.; Sánchez-Quintana, D.; Bossone, E.; Bogaard, H.J.; Naeije, R. Anatomy, Function, and Dysfunction of the Right Ventricle: JACC State-of-the-Art Review. *J. Am. Coll. Cardiol.* **2019**, *73*, 1463–1482. [CrossRef]
20. Chen, H.; Man, R.Y.K.; Leung, S.W.S. Ppar-α agonists acutely inhibit Ca^{2+}-independent PLA_2 to reduce H_2O_2- induced contractions in aortae of spontaneously hypertensive rats. *Am. J. Physiol.-Hear. Circ. Physiol.* **2018**, *314*, H681–H691. [CrossRef]
21. Driessen, M.M.P.; Hui, W.; Bijnens, B.H.; Dragulescu, A.; Mertens, L.; Meijboom, F.J.; Friedberg, M.K. Adverse ventricular–ventricular interactions in right ventricular pressure load: Insights from pediatric pulmonary hypertension versus pulmonary stenosis. *Physiol. Rep.* **2016**, *4*, e12833. [CrossRef] [PubMed]
22. López-Candales, A.; Shaver, J.; Edelman, K.; Candales, M.D. Temporal differences in ejection between right and left ventricles in chronic pulmonary hypertension: A pulsed doppler study. *Int. J. Cardiovasc. Imaging* **2012**, *28*, 1943–1950. [CrossRef]

23. Koestenberger, M.; Sallmon, H.; Avian, A.; Cantinotti, M.; Gamillscheg, A.; Kurath-Koller, S.; Schweintzger, S.; Hansmann, G. Ventricular–ventricular interaction variables correlate with surrogate variables of clinical outcome in children with pulmonary hypertension. *Pulm. Circ.* **2019**, *9*, 1–9. [CrossRef] [PubMed]
24. Critser, P.J.; Higano, N.S.; Lang, S.M.; Kingma, P.S.; Fleck, R.J.; Hirsch, R.; Taylor, M.D.; Woods, J.C. Cardiovascular magnetic resonance imaging derived septal curvature in neonates with bronchopulmonary dysplasia associated pulmonary hypertension. *J. Cardiovasc. Magn. Reson.* **2020**, *22*, 50. [CrossRef]
25. Nishina, Y.; Inami, T.; Kataoka, M.; Kariyasu, T.; Shimura, N.; Ishiguro, H.; Yokoyama, K.; Yoshino, H.; Satoh, T. Evaluation of Right Ventricular Function on Cardiac Magnetic Resonance Imaging and Correlation with Hemodynamics in Patients with Chronic Thromboembolic Pulmonary Hypertension. *Circ. Rep.* **2020**, *2*, 174–181. [CrossRef]
26. Westerhof, B.E.; Saouti, N.; Van Der Laarse, W.J.; Westerhof, N.; Vonk Noordegraaf, A. Treatment strategies for the right heart in pulmonary hypertension. *Cardiovasc. Res.* **2017**, *113*, 1465–1473. [CrossRef] [PubMed]

diagnostics

The Role of Circulating Collagen Turnover Biomarkers and Late Gadolinium Enhancement in Patients with Non-Ischemic Dilated Cardiomyopathy

Radu Revnic [1], Bianca Olivia Cojan-Minzat [1], Alexandru Zlibut [2,*], Rares-Ilie Orzan [2], Renata Agoston [3], Ioana Danuta Muresan [2], Dalma Horvat [2], Carmen Cionca [4], Bogdan Chis [2,5] and Lucia Agoston-Coldea [2,5]

[1] Department of Family Medicine, Iuliu Hatieganu University of Medicine and Pharmacy, 400012 Cluj-Napoca, Romania; radu_revnic@yahoo.com (R.R.); cojanminzat.bianca@yahoo.com (B.O.C.-M.)
[2] Department of Internal Medicine, Iuliu Hatieganu University of Medicine and Pharmacy, 400012 Cluj-Napoca, Romania; orzanrares@gmail.com (R.-I.O.); ioanamuresandanuta@yahoo.com (I.D.M.); hdalma92@yahoo.com (D.H.); bogdan_a_chis@yahoo.com (B.C.); luciacoldea@yahoo.com (L.A.-C.)
[3] Faculty of Medicine, Iuliu Hatieganu University of Medicine and Pharmacy, 400012 Cluj-Napoca, Romania; renata.agoston@yahoo.com
[4] Department of Radiology, Affidea Hiperdia Diagnostic Imaging Centre, 400487 Cluj-Napoca, Romania; carmen.cionca@gmail.com
[5] Department of Internal Medicine, Emergency County Hospital, 400347 Cluj-Napoca, Romania
* Correspondence: alex.zlibut@yahoo.com; Tel.: +40-264-591942; Fax: +40-264-599817

Citation: Revnic, R.; Cojan-Minzat, B.O.; Zlibut, A.; Orzan, R.-I.; Agoston, R.; Muresan, I.D.; Horvat, D.; Cionca, C.; Chis, B.; Agoston-Coldea, L. The Role of Circulating Collagen Turnover Biomarkers and Late Gadolinium Enhancement in Patients with Non-Ischemic Dilated Cardiomyopathy. *Diagnostics* **2022**, *12*, 1435. https://doi.org/10.3390/diagnostics12061435

Academic Editors: Minjie Lu and Arlene Sirajuddin

Received: 8 May 2022
Accepted: 8 June 2022
Published: 10 June 2022

Publisher's Note: MDPI stays neutral with regard to jurisdictional claims in published maps and institutional affiliations.

Abstract: Background: Myocardial scarring is a primary pathogenetic process in nonischemic dilated cardiomyopathy (NIDCM) that is responsible for progressive cardiac remodeling and heart failure, severely impacting the survival of these patients. Although several collagen turnover biomarkers have been associated with myocardial fibrosis, their clinical utility is still limited. Late gadolinium enhancement (LGE) determined by cardiac magnetic resonance imaging (CMR) has become a feasible method to detect myocardial replacement fibrosis. We sought to evaluate the association between collagen turnover biomarkers and replacement myocardial scarring by CMR and, also, to test their ability to predict outcome in conjunction with LGE in patients with NIDCM. Method: We conducted a prospective study on 194 patients (48.7 $\pm$ 14.3 years of age; 74% male gender) with NIDCM. The inclusion criteria were similar to those for the definition of NIDCM, performed exclusively by CMR: (1) LV dilation with an LV end-diastolic volume (LVEDV) of over 97 mL/m^2; (2) global LV dysfunction, expressed as a decreased LVEF of under 45%. CMR was used to determine the presence and extent of LGE. Several collagen turnover biomarkers were determined at diagnosis, comprising galectin-3 (Gal3), procollagen type I carboxy-terminal pro-peptide (PICP) and N-terminal pro-peptide of procollagen type III (PIIINP). A composite outcome (all-cause mortality, ventricular tachyarrhythmias, heart failure hospitalization) was ascertained over a median of 26 months. Results: Gal3, PICP and PIIINP were considerably increased in those with LGE+ ($p < 0.001$), also being directly correlated with LGE mass ($r^2 = 0.42$; $r^2 = 0.44$; $r^2 = 0.31$; all $p < 0.001$). Receiver operating characteristic (ROC) analysis revealed a significant ability to diagnose LGE, with an area under the ROC of 0.816 for Gal3, 0.705 for PICP, and 0.757 for PIIINP (all $p < 0.0001$). Kaplan–Meier analysis showed that at a threshold of >13.8 ng/dL for Gal3 and >97 ng/dL for PICP, they were able to significantly predict outcome (HR = 2.66, $p < 0.001$; HR = 1.93, $p < 0.002$). Of all patients, 17% ($n = 33$) reached the outcome. In multivariate analysis, after adjustment for covariates, only LGE+ and Gal3+ remained independent predictors for outcome ($p = 0.008$; $p = 0.04$). Nonetheless, collagen turnover biomarkers were closely related to HF severity, providing incremental predictive value for severely decreased LVEF of under 30% in patients with NIDCM, beyond that with LGE alone. Conclusions: In patients with NIDCM, circulating collagen turnover biomarkers such as Gal3, PICP and PIIINP are closely related to the presence and extent of LGE and can significantly predict cardiovascular outcome. The joint use of LGE with Gal3 and PICP significantly improved outcome prediction.

Keywords: cardiac magnetic resonance imaging; galectin-3; procollagen type I carboxy-terminal pro-peptide; N-terminal pro-peptide of procollagen type III

1. Introduction

Despite recent therapeutic developments, non-ischemic dilated cardiomyopathy (NIDCM) remains a primary cause of progressive cardiac remodeling and heart failure (HF) that leads to frequent hospitalization and increased mortality. These patients have an increased risk of developing myocardial fibrosis, which, in turn, plays a central role in the progression of HF [1]. Cardiac magnetic resonance imaging (CMR) with late gadolinium enhancement (LGE) detects focal replacement myocardial fibrosis in up to 30% of patients with NIDCM and provides an incremental predictive value for cardiovascular risk stratification [2]. However, using this technique, diffuse interstitial fibrosis remains undetected [3].

Molecular markers of fibrosis such as galectin-3 (Gal3), procollagen type I carboxy-terminal pro-peptide (PICP) and N-terminal pro-peptide of procollagen type III (PIIINP) are in direct relationship with myocardial collagen turnover, and thus might aid in the prediction of major adverse cardiovascular events (MACEs) [4]. Gal3 binds to a specific beta-galactosidase which is overexpressed by phagocytic macrophages, and thus endorses the proliferation of myofibroblasts with myocardial collagen deposition—leading to the progression of myocardial fibrosis, inflammation, fibrosis and cardiac remodeling [5–10]. Moreover, increased serum levels of Gal3 have been identified in patients with chronic HF [5,11–13] and also predicts cardiac remodeling and mortality in this category of patients [7,8,14]. Additionally, in one recently published study, it has been shown that sera levels of Gal3 were closely associated with the extent of LGE in patients with NIDCM [15].

To date, PICP and PIIINP are the only proven peptides that can be identified in the bloodstream that are considerably correlated to histologically proven myocardial fibrosis [4]. These molecules have been observed to be considerably correlated to the progression of myocardial fibrosis in patients with ischemic heart disease and NIDCM [7,8]. Additionally, their increased sera levels are able to predict MACEs in patients with HF and preserved left ventricle (LV) ejection fractions (LVEFs) [4,16]. However, their role in patients with NIDCM is not entirely clear.

The aim of this study was to evaluate the link between circulating collagen turnover biomarkers and myocardial replacement fibrosis determined by CMR, and also to test their ability to predict outcome in conjunction with LGE.

2. Materials and Methods

2.1. Study Design and Patient Characteristics

We conducted an observational, prospective study on patients recently diagnosed with NIDCM who were examined in the Department of Internal Medicine, Iuliu Hatieganu University of Medicine and Pharmacy of Cluj-Napoca, between October 2017 and November 2020. The inclusion criteria were similar to those for the definition of NIDCM, performed exclusively by CMR: (1) LV dilation with an LV end-diastolic volume (LVEDV) of over 97 mL/m^2; (2) global LV dysfunction, expressed as a decreased LVEF of under 45% [17]. The exclusion criteria are presented in Figure 1. The current study was conducted in accordance with the Declaration of Helsinki and received approval from the Ethics Committee of Iuliu Hatieganu University of Medicine and Pharmacy of Cluj-Napoca, Romania. All patients were fully informed about the study protocol and provided written consent.

All patients underwent a similar investigation protocol, which included demographic and clinical data and biological sampling, along with standard cardiovascular evaluation and CMR.

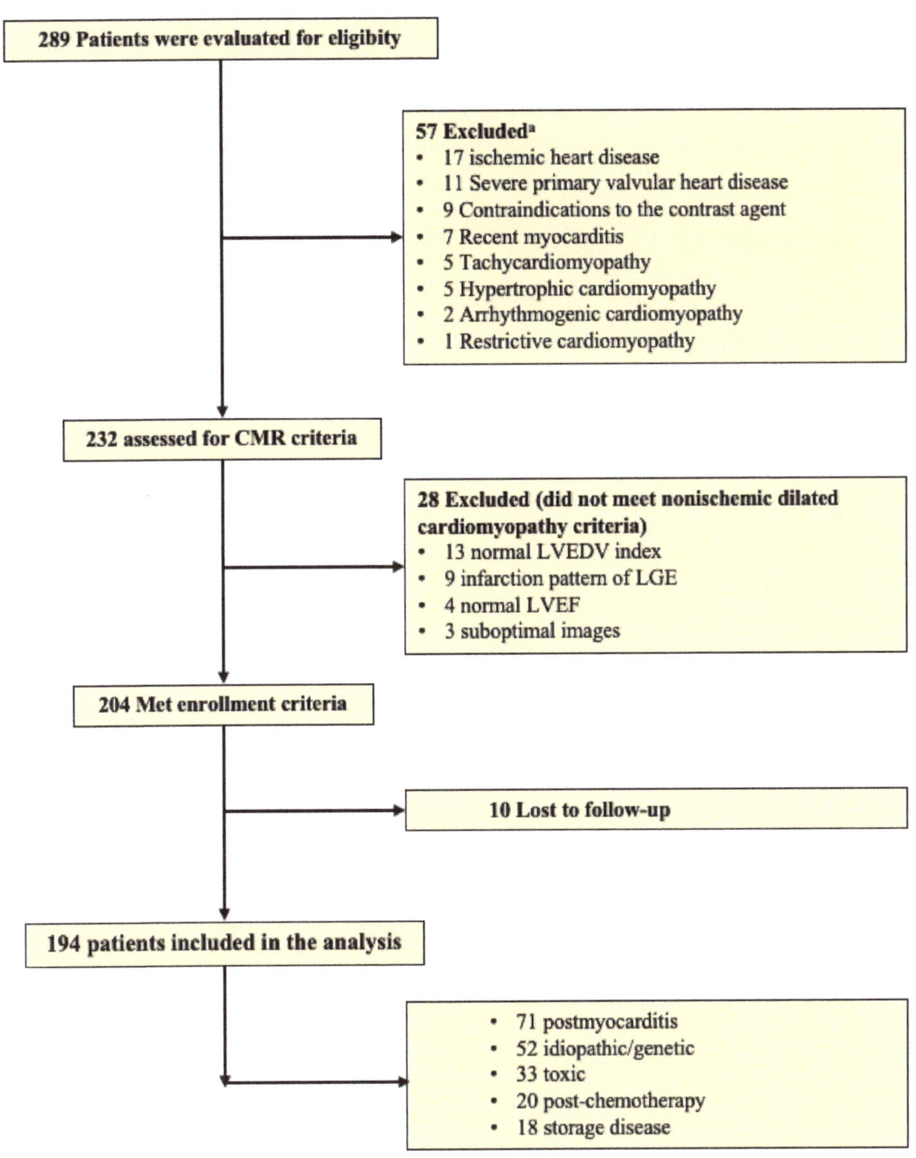

[a]coronary artery disease was defined as >50% angiographical stenosis in any epicardial coronary artery. Primary valvular disease was defined as moderate or higher valvular stenosis or regurgitation, with the exception of the functional ones. Functional mitral regurgitation was defined as mitral regurgitation secondary to left ventricular remodeling resulting in failure of leaflet coadaptation, in the setting of normal mitral valve anatomy, on echocardiography and cardiovascular magnetic resonance imaging.

Figure 1. Flowchart of the study.

2.2. Circulating Collagen Turnover Biomarkers

Biochemical workups were performed in the Clinical Biochemistry Laboratory from the 2nd Medical Clinic of the Cluj Emergency County Hospital. Two peripheral venous blood samples were harvested, centrifuged immediately after, and stored in special vials with ethylenediaminetetraacetic acid. Sera glucose and creatinine were determined from one vial using a Konelab-31 analyzer, while the other vial was stored at $-70\ ^{\circ}$C until the end of the study and used to determine the serum levels of cardiac biomarkers. PICP, PIIINP, copeptin (CPP) and N-terminal pro-Brain Natriuretic Peptide (NT-proBNP) were determined by the Sandwich ELISA technique according to the manufacturer's recommendations using Elabscience Biotechnology Co., Ltd, Wuhan, China. The inferior thresholds and variabilities for these markers were as follows: for PICP, a cut-off value of 0.13 ng/mL and inter-/intra-test variability <10%/<12%; for PIIINP, a cut-off value of 0.14 ng/mL and inter-/intra-test variability <10%; for CPP, a cut-off value of 0.18 ng/mL and inter-/intra-test variability <10%; for NT-proBNP, a cut-off value of 0.38 ng/mL and inter-/intra-test variability <10%. Serum levels of Gal3 were measured using an enzyme-linked immunosorbent assay (Human Galectin-3—Quantikine ELISA Kit, R&D Systems), with an inferior cut-off value of 0.016 ng/mL, without crosslinked reactivity with other galectin or collagen molecules. Intra-test and inter-test plasma variations for Gal3 were 3.5–4.3%, and 5.8–6%, respectively. Renal function was evaluated using the estimated glomerular filtration rate (eGFR) and renal impairment was considered as an eGFR of under 60 mL/min/1.73 m^2.

2.3. CMR Measurements

CMR images were appraised using a 1.5 T Open Bore system MR scanner (Magnetom Altea, Siemens Medical Solutions, Erlangen, Germany) in complete apnea by two level-III experienced operators who were blinded to all clinical and imaging data, in line with current international guidelines [18]. The acquisition of steady-state free precession (SSFP) CMR sequences was performed to detect ventricular function and mass using standard long- and short-axes (two-chamber, three-chamber, and four-chamber) to enclose both ventricles were covered from the base to the apex. Cine-SSFP parameters were as follows: repetition time (TR) 3.6 ms; echo time (TE) 1.8 ms; flip angle 60°; slice thickness 6 mm; field-of-view 360 mm; image matrix of 192 $\times$ 192 pixels; voxel size 1.9 $\times$ 1.9 $\times$ 6 mm; 25–40 ms temporal resolution reconstructed to 25 cardiac phases.

Focal myocardial fibrosis was evaluated by LGE detected at 10 min after intravenous infusion of 0.2 mmol/kg gadoxetic acid (Clariscan, GH Healthcare AS, Oslo, Norway) using long- and short axis-views, using a segmented inversion-recovery gradient-echo sequence. LGE acquisition parameters were TR 4.8 ms, TE 1.3 ms and inversion time 200 to 300 ms. Inversion time was adjusted to optimize nulling of normal myocardium. Brachial blood pressure was monitored during SSFP-CMR acquisitions.

LVEDV, LV end-systolic volume (LVESV), LVEF and end-diastolic LV mass (LVM) were measured on short-axis cine-SSFP images. Epicardial and endocardial borders were traced semi-automatically at end-diastole and end-systole using a Syngo Virtual Cockpit. All volumes were indexed to body surface area (BSA). Besides this, for a more accurate assessment of LV function, we assessed the LV longitudinal-axis strain (LV-LAS; the difference in mitral annular displacement at end-systole vs. end-diastole expressed as a percentage) and LV sphericity index (LVSI). The LVSI was calculated by dividing LVEDV by the volume of a sphere, whose LV length (L) was measured at the end-diastole: LVSI = LVEDV/(π/6 $\times$ (L)^3) [19,20].

The presence and distribution of LGE in the LV were assessed from short-axis images using the 17-segments model, as recommended by the American Heart Association [21], and quantified using a signal intensity threshold of >5 standard deviations (SDs) above a remote reference of the normal myocardium. This threshold proved to be in the best agreement with visual assessments and had the best reproducibility among the different technique thresholds [22]. The Full Width at Half Maximum (FWHM) technique was used to quantify LGE. The reference region was defined as an area that included the maximum intensity of the visually appreciated LGE signal on each slice. The maximum signal strength threshold was recorded to define LGE. Additionally, the total LGE was determined as the sum of all LGE areas for each slice, multiplied by the slice thickness. LGE quantification was performed by 2 independent observers. Inter-observer reproducibility was 0.91 95% CI (0.882–0.934) and intra-observer reproducibility was 0.93 95% CI (0.902–0.947). Specific LGE distribution patterns were accounted for: mid-wall or subepicardial, and focal or diffuse. The LGE mass was automatically quantified from short-axis LGE images, using cvi42, Circle Cardiovascular Imaging Inc., Calgary, CA, Canada. The extent of LGE was expressed in grams (g) and percentage of LV mass.

2.4. Clinical Outcome

The clinical follow-up was obtained by completing a questionnaire either during hospital visits, telephone house-calls, or both—aiming to delineate the occurrence of clinical outcomes, which corresponded to the first event occurring in each patient among the following MACEs: death or aborted death from cardiac causes, sustained ventricular tachyarrhythmia (beats with ventricular origin that last >30 s and have a rate greater than >100 beats/min), and HF requiring hospitalization—defined according to current international guidelines. Hospitalization due to non-cardiac causes was not counted as an event. Survival analysis was performed for the clinical outcomes. The median follow-up was 26 months and maximum follow-up reached 41 months.

2.5. Statistical Analysis

Initially, the Kolmogorov–Smirnov test was used to assess data normality. Continuous data were presented as median (inter-quartile range (IQR)) and mean $\pm$ standard deviation (SD). Discrete data were reported as percentages and frequencies. The distribution of variables was accounted for after logarithmic transformation. Comparisons between groups were approached using Hi^2 and Fischer tests for qualitative data and ANOVA or Kruskal–Wallis H tests for continuous data. Pearson's coefficient of correlation was used to examine the relationship between data. Furthermore, for specific descriptive analyses, the studied population was dichotomized according to LGE presence (LGE+) or absence (LGE−), and also with respect to higher than median levels of PICP, PIIINP and Gal3 (PICP+, PIIINP+, Gal3+) and lower than median levels of these biomarkers (PICP−, PIIINP−, Gal3−)—thus resulting in six specific groups. Also, logistic regression was used to evaluate the incremental ability of these markers.

Kaplan–Meier survival curves were created and differences between groups were assessed using log-rank tests. Unadjusted and adjusted Cox regression analysis was performed to determine hazard rates (HRs) and 95% confidence intervals (CIs). Furthermore, adjustment regression models were used to test if the biomarkers of cardiac fibrosis did or did not respect a linear trend. For the adjustment, specific covariates which are known to significantly predict outcome in patients with NIDCM such as LVEF, eGFR, body-mass index (BMI), NT-proBNP, diabetes, gender and age were used. Moreover, ROC analyses were used to calculate the cut-off values of circulating biomarkers for predicting MACEs.

Additionally, inter- and intra-observer Kappa Cohen coefficients were calculated. Retrospective calculus of statistical test powers and prospective dimensions of the sample were estimated using type I and type II variations, based on sample size. The statistical analysis was performed using statistical software MedCalc (Version 19.1.7, MedCalc Software, Ostend, Belgium).

3. Results

3.1. Baseline Characteristics

A total of 194 patients with NIDCM were enrolled in the study and their main characteristics are presented in Table 1. They were divided into two groups based on the presence and absence of LGE: 73 (37.7%) patients were LGE+ and 121 (62.3%) were LGE−. Those in the LGE+ group had significantly increased sera levels of circulating collagen turnover biomarkers—namely Gal3, PICP and PIIINP—compared to the others: 17.7 ng/mL vs. 9.1 ng/mL, $p < 0.001$; 156 ng/mL vs. 74 ng/mL, $p < 0.001$; and 5.1 ng/mL vs. 3.5 ng/mL, $p < 0.001$, respectively. Moreover, patients with LGE+ had significantly modified LVEDV (142.4 mL/m^2 vs. 124.2 mL/m^2, $p < 0.001$), LVESV (101.2 mL/m^2 vs. 78.1 mL/m^2, $p < 0.001$), LVSI (0.44 vs. 0.38, $p < 0.001$) and LV-LAS (−8.5 vs. −10.5%, $p < 0.001$), and significantly lower LVEF (30.3% vs. 38.2%, $p < 0.001$).

Table 1. Baseline characteristics.

Data	All Patients (n = 194)	LGE− (n = 121)	LGE+ (n = 73)	p
Clinical features				
Age, mean (SD), years	48.7 (14.3)	47.9 (14.7)	50.0 (13.6)	NS
Masculine gender, n (%)	144 (74.2)	88 (72.7)	56 (76.7)	NS
BMI, Kg/m^2	27.4 (4.7)	27.2 (4.5)	27.6 (5.1)	NS
HR, mean (SD), bpm	73 (16.0)	70 (14.2)	76 (17.9)	NS
SBP, mean (SD), mmHg	134 (19.1)	135 (18.5)	131 (19.7)	NS
AHT, n (%)	102 (52.5)	72 (59.5)	30 (41.0)	<0.05
Diabetes mellitus, n (%)	63 (32.5)	43 (35.5)	20 (27.4)	<0.05
Dyslipidemia, n (%)	111 (57.2)	70 (57.8)	41 (56.2)	NS
Smokers, n (%)	65 (33.5)	41 (33.8)	24 (32.8)	NS
NYHA I/II/III class	30/97/37	20/59/23	10/38/14	<0.05
Medication				
Betablockers, n (%)	149 (76.8)	93 (76.8)	56 (76.7)	NS
ACEI or ARB2, n (%)	147 (75.7)	92 (76.0)	56 (76.7)	NS
Calcium channel blockers, n (%)	32 (16.5)	20 (16.5)	12 (16.4)	NS
Diuretics, n (%)	118 (60.8)	73 (60.3)	45 (61.4)	NS
Biomarkers				
NT-proBNP, median (IQR), ng/L	16,900 (8700–39,500)	16,200 (8700–36,200)	17,200 (10,600–39,500)	<0.001
CPP, median (IQR), ng/mL	12.7 (1.8–87)	8.2 (1.8–68.2)	17.1 (4.3–87)	<0.001
PICP, median (IQR), ng/mL	97 (23–347)	74 (23–344)	156 (38–347)	<0.001
PIIINP, median (IQR), ng/mL	4.1 (1.7–8.7)	3.5 (1.7–7.1)	5.1 (2.1–8.7)	<0.001
Gal3, median (IQR), ng/mL	13.8 (2.2–26.6)	9.1 (2.2–23.6)	17.7 (6.1–26.6)	<0.001
eGFR, mean (SD), mL/min/1.73 m^2	87.1 (21.2)	87.7 (20.4)	86.1 (22.6)	NS
CMR				
LVEDV indexed, median (SD), mL/m^2	131.1 (34.5)	124.2 (29.7)	142.4 (39.1)	<0.001
LVESV indexed, median (SD), mL/m^2	86.8 (33.7)	78.1 (28.4)	101.2 (36.9)	<0.001
LVM indexed, median (SD), g/m^2	86.1 (20.5)	83.3 (19.4)	90.5 (21.6)	<0.01
LVEF, median (SD), %	35.2 (9.6)	38.2 (7.8)	30.3 (9.3)	<0.001
LAV indexed, median (SD), mL/m^2	55.5 (21.4)	53.1 (20.4)	60.6 (22.2)	<0.05
LV-LAS, median (SD), %	−9.7 (5.3)	−10.5 (5.1)	−8.5 (5.4)	<0.001
LVSI, median (SD)	0.41 (0.13)	0.38 (0.15)	0.44 (0.12)	<0.001
LGE mass, median (IQR), g	-	-	31.2 (1–89)	N/A
LGE mass/LVM, median (IQR), %	-	-	18.4 (0.6–56)	N/A

Abbreviations: ACEI, angiotensin-converting enzyme inhibitors; AHT, arterial hypertension; ARB2, angiotensin II receptor blockers; BMI, body-mass index; CPP, copeptin; eGFR, estimated glomerular filtration rate; Gal3, Galectin-3; HR, heart rate; IQR, interquartile range; LAS, left ventricle long-axis strain; LAV, left atrial volume; LGE, late gadolinium enhancement; LVEDV, left ventricle end-diastolic volume; LVEF, left ventricle ejection fraction; LVESV, left ventricle end-systolic volume; LVM, left ventricle mass; LVSI, left ventricle sphericity index; NYHA, New York Heart Association; PICP, procollagen type I carboxy-terminal pro-peptide; PIIINP, N-terminal pro-peptide of procollagen type III; SBP, systolic blood pressure; SD, standard deviation.

3.2. Association between Circulating Collagen Turnover Biomarkers and LGE

Overall, sera levels of PICP, PIIINP and Gal-3 were significantly increased in patients with LGE (LGE+), as compared to those without LGE (Table 1). Furthermore, Gal3, PICP and PIIINP were positively associated with LGE mass, with significant Pearson's correlation coefficients of $r^2 = 0.372$, $p < 0.0001$; $r^2 = 0.379$, $p < 0.0001$; and $r^2 = 0.315$, $p < 0.0001$. ROC analysis demonstrated that specific cut-off values significantly identified the presence of LGE: an area under the ROC of 0.816 for Gal3 (95% CI: 0.754–0.868; $p < 0.0001$), 0.705 for PICP (95% CI: 0.636–0.769; $p < 0.0001$) and of 0.757 for PIIINP (95% CI: 0.690–0.816, $p < 0.0001$; Figure 2). Additionally, an 11 ng/mL threshold for Gal3 revealed the presence of LGE with a high sensitivity of 90.4%, a specificity of 66.1% and a negative predictive value of 92%, while PIIINP proved—for a cut-off value of 1.18 ng/mL—to have a sensitivity of

71.83%, a specificity of 83% and a negative predictive value of 83.6%, and PICP had—for a cut-off value of 44.4 ng/mL—a 77.5% sensitivity, 76% specificity and a negative predictive value of 85.5%.

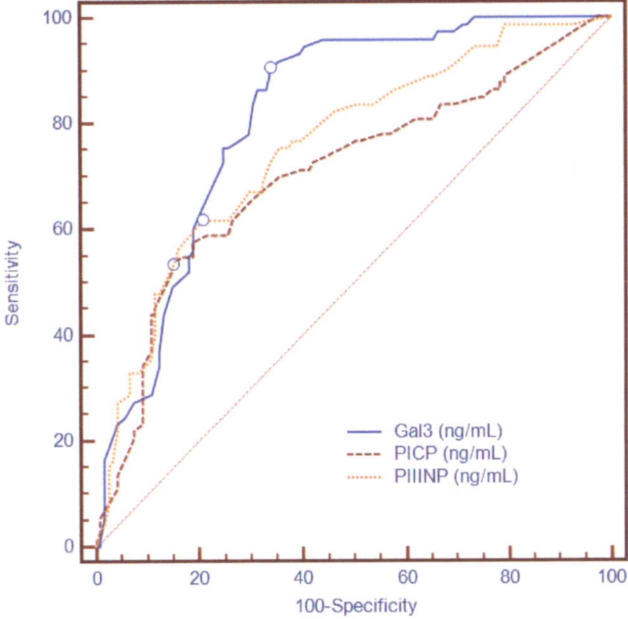

Figure 2. ROC analysis demonstrating the ability of Gal-3, PICP and PIIINP to identify the presence of LGE. Abbreviations: Gal-3, galectin-3; LGE, late gadolinium enhancement; PICP, procollagen type I carboxy-terminal pro-peptide; PIIINP, N-terminal pro-peptide of procollagen type III; ROC, receiver operating characteristics.

Regarding the relationship between circulating collagen turnover biomarkers and the severity of HF, Gal3 and PICP were inversely correlated with LVEF ($r^2 = -0.58$, $p < 0.0001$; $r^2 = -0.39$, $p < 0.0001$) and NYHA class $\geq$III ($r^2 = -0.56$, $p < 0.0001$; $r^2 = -0.39$, $p < 0.0001$), and directly correlated with NT-proBNP ($r^2 = 0.49$; $p < 0.0001$; $r^2 = 0.32$; $p < 0.0001$) and CCP ($r^2 = 0.41$; $p < 0.0001$; $r^2 = 0.38$; $p < 0.0001$) levels.

3.3. Characterization of Patients with DCM and Severely Decreased LV Function

As shown in Table 2, 31% ($n = 61$) of patients had a severely decreased LVEF of under 30%. These patients had significantly impaired LVEDV, LVESV, LVSI and LV-LAS (all $p < 0.001$) and considerably higher LGE mass (31.2 vs. 6.4, $p < 0.001$) and LGE mass/LV mass ratios (18.2 vs. 4.5, $p < 0.001$), as compared to those with LVEFs over 30%.

Sera markers of HF were notably increased in those with LVEF < 30%: NT-proBNP: 17,500 ng/L vs. 16,200 ng/L, $p < 0.001$ and CPP: 17.5 ng/mL vs. 9.5 ng/mL, $p < 0.001$, respectively.

A stepwise logistic regression proportional-hazard model analysis was deployed to test if collagen turnover biomarkers are useful in the risk stratification of patients with NIDCM and severely decreased LVEF, beyond LGE. LGE alone significantly predicted the presence of LVEF <30% in patients with NIDCM (Chi-square = 55.72, $p < 0.0001$). The addition of Gal3 to LGE significantly increased the diagnosis power (Chi-square = 69.69, $p < 0.0001$), while further adding PICP increased their identification ability even more (Chi-square = 79.31, $p < 0.0001$). Lastly, the association of Gal3, PICP and PIIINP with LGE provided a significant incremental value for predicting decreased LVEFs of <30% in patients with NIDCM (Chi-square = 86.09, $p < 0.0001$; Figure 3).

Table 2. Comparison between NIDCM patients with severely and non-severely decreased LVEF.

Data	All Patients (n = 194)	LVEF 31–45% (n = 131)	LVEF < 30% (n = 61)	p
NT-proBNP, median (IQR), ng/L	16,900 (8700–39,500)	16,200 (8700–36,200)	17,500 (10,500–39,500)	<0.01
CPP, median (IQR), ng/mL	12.7 (1.8–87)	9.5 (1.8–68.2)	17.5 (3.2–87)	<0.001
PICP, median (IQR), ng/mL	97 (23–347)	79 (23–344)	147 (32–347)	<0.001
PIIINP, median (IQR), ng/mL	4.1 (1.7–8.7)	3.9 (1.7–8.7)	4.5 (1.9–8.7)	<0.001
Gal3, median (IQR), ng/mL	13.8 (2.2–26.6)	9.6 (2.2–26.6)	17.7 (3.1–23.6)	<0.001
eGFR, mean (SD), mL/min/1.73 m^2	87.1 (21.2)	87.7 (20.4)	86.1 (22.6)	NS
LVEDV indexed, median (SD), mL/m^2	131.1 (34.5)	117.4 (21.6)	160.7 (38.8)	<0.001
LVESV indexed, median (SD), mL/m^2	86.8 (33.7)	69.9 (15.7)	124.2 (32.3)	<0.001
LVM indexed, median (SD), g/m^2	86.1 (20.5)	80.9 (17.7)	97.1 (21.9)	<0.01
LAV indexed, median (SD), mL/m^2	55.5 (21.4)	51.7 (19.5)	63.7 (22.8)	<0.01
LAS, median (SD), %	−9.7 (5.3)	−11.6 (5.1)	−5.7 (2.5)	<0.001
LVSI, median (SD)	0.41 (0.13)	0.37 (0.09)	0.46 (0.13)	<0.001
LGE mass, median (IQR), g	14.2 (0.9–88)	6.4 (0.9–71.1)	31.2 (1–88)	<0.001
LGE mass/LVM, median (IQR), %	8.8 (0.6–64.2)	4.5 (0.6–44.7)	18.4 (16.9–64.2)	N/A

Abbreviations: ACEI, angiotensin-converting enzyme inhibitors; AHT, arterial hypertension; ARB2, angiotensin II receptor blockers; BMI, body-mass index; CPP, copeptin; eGFR, estimated glomerular filtration rate; Gal3, Galectin-3; HR, heart rate; IQR, interquartile range; LAS, left ventricle long-axis strain; LAV, left atrial volume; LGE, late gadolinium enhancement; LVEDV, left ventricle end-diastolic volume; LVEF, left ventricle ejection fraction; LVESV, left ventricle end-systolic volume; LVM, left ventricle mass; LVSI, left ventricle sphericity index; NYHA, New York Heart Association; PICP, procollagen type I carboxy-terminal pro-peptide; PIIINP, N-terminal pro-peptide of procollagen type III; SBP, systolic blood pressure; SD, standard deviation.

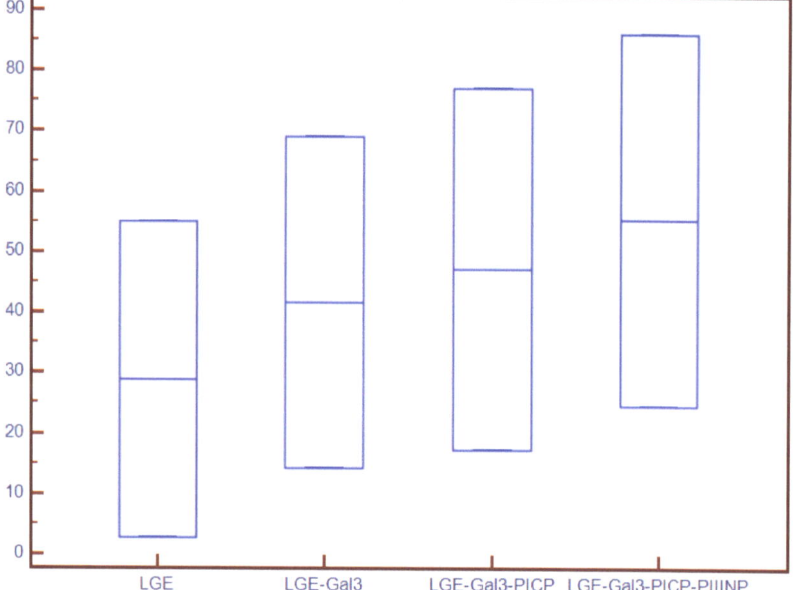

Figure 3. Incremental ability of LGE, LGE stepwise added to Gal-3, PICP and PIIINP for identifying patients with NIDCM and severely decreased LVEF. Abbreviations: Gal-3, galectin-3; LGE, late gadolinium enhancement; PICP, procollagen type I carboxy-terminal pro-peptide; PIIINP, N-terminal pro-peptide of procollagen type III; ROC, receiver operating characteristics.

3.4. Univariate and Multivariate Cox Analysis and Time-To-Event Analysis of LGE and Circulating Collagen Turnover Biomarkers

Patients were followed up for 26 months. Of them, 17% (n = 33) of patients reached the outcome: all-cause mortality (n = 6), malignant ventricular tachyarrhythmia (n = 14) and HF hospitalization (n = 13; Table 3).

Table 3. Comparison between patients with NIDCM who reached MACEs and the others.

Data	All Patients (*n* = 194)	MACEs− (*n* = 161)	MACEs+ (*n* = 33)	*p*
Clinical features				
Age, mean (SD), years	48.7 (14.3)	48.5 (13.6)	49.3 (17.7)	NS
Masculine gender, *n* (%)	144 (74.2)	121 (84.0)	23 (26.0)	NS
BMI, Kg/m^2	27.4 (4.7)	27.6 (4.7)	26.1 (4.4)	NS
HR, mean (SD), bpm	73 (16.0)	72 (15.4)	75 (18.2)	NS
SBP, mean (SD), mmHg	134 (19.1)	135 (19.2)	131 (17.9)	NS
AHT, *n* (%)	102 (52.5)	87 (85.2)	15 (14.8)	<0.001
Diabetes mellitus, *n* (%)	63 (32.5)	52 (82.5)	11 (17.5)	<0.001
Dyslipidemia, *n* (%)	111 (57.2)	91 (81.9)	20 (18.1)	<0.001
Smokers, *n* (%)	65 (33.5)	57 (87.7)	8 (12.3)	<0.001
NYHA I/II/III class	30/97/37	21/91/31	9/6/6	<0.05
Medication				
Betablockers, *n* (%)	149 (76.8)	124 (83.2)	25 (16.8)	<0.001
ACEI or ARB2, *n* (%)	147 (75.7)	125 (85.0)	22 (15.0)	<0.001
Calcium channel blockers, *n* (%)	32 (16.5)	22 (68.7)	8 (31.3)	<0.001
Diuretics, *n* (%)	118 (60.8)	93 (78.8)	25 (21.2)	<0.001
Biomarkers				
NT-proBNP, median (IQR), ng/L	16,900 (8700–39,500)	14,000 (8700–36,600)	19,300 (10,200–39,500)	<0.001
CPP, median (IQR), ng/mL	12.7 (1.8–87)	9.9 (1.8–87)	16.2 (3.1–82.9)	<0.001
PICP, median (IQR), ng/mL	97 (23–347)	92 (23–347)	118 (32–338)	<0.001
PIIINP, median (IQR), ng/mL	4.1 (1.7–8.7)	4.0 (1.7–8.3)	4.5 (2.1–8.7)	<0.01
Gal3, median (IQR), ng/mL	13.8 (2.2–26.6)	11 (2.2–26.6)	17.2 (3.0–24.0)	0.001
eGFR, mean (SD), mL/min/1.73 m^2	87.1 (21.2)	86.1 (19.7)	89.6 (25.8)	NS
CMR				
LVEDV indexed, median (SD), mL/m^2	131.1 (34.5)	130.4 (35.0)	134.7 (32.7)	NS
LVESV indexed, median (SD), mL/m^2	86.8 (33.7)	85.4 (33.9)	93.7 (32.4)	NS
LVM indexed, median (SD), g/m^2	86.1 (20.5)	85.9 (20.5)	86.4 (20.6)	NS
LVEF, median (SD), %	35.2 (9.6)	35.9 (9.2)	31.7 (9.1)	<0.01
LAV indexed, median (SD), mL/m^2	55.5 (21.4)	54.2 (21.7)	61.7 (18.4)	NS
LV-LAS, median (SD), %	−9.7 (5.3)	−10.2 (5.5)	−7.8 (3.5)	<0.01
LVSI, median (SD)	0.41 (0.13)	0.38 (0.11)	0.47 (0.13)	<0.001
LGE mass, median (IQR), g	14.3 (0–89)	11.2 (0–86)	29.9 (23–89)	<0.001
LGE mass/LVM, median (IQR), %	8.4 (0–56)	6.6 (0–52.8)	19.4 (1.2–56)	<0.001

Abbreviations: ACEI, angiotensin-converting enzyme inhibitors; AHT, arterial hypertension; ARB2, angiotensin II receptor blockers; BMI, body-mass index; CPP, copeptin; eGFR, estimated glomerular filtration rate; Gal3, Galectin-3; HR, heart rate; IQR, interquartile range; LAS, left ventricle long-axis strain; LAV, left atrial volume; LGE, late gadolinium enhancement; LVEDV, left ventricle end-diastolic volume; LVEF, left ventricle ejection fraction; LVESV, left ventricle end-systolic volume; LVM, left ventricle mass; LVSI, left ventricle sphericity index; MACEs, major adverse cardiovascular events; NYHA, New York Heart Association; PICP, procollagen type I carboxy-terminal pro-peptide; PIIINP, N-terminal pro-peptide of procollagen type III; SBP, systolic blood pressure; SD, standard deviation.

In the univariate Cox analysis, LGE and all cardiac biomarkers of fibrosis (Gal3, PICP and PIIINP) were significantly associated with MACEs. However, following the multivariate analysis, after adjustment for covariates comprised of age, gender, LVEF, eGFR, BMI, NT-proBNP and diabetes mellitus, only LGE+ and Gal3 remained independent predictors for MACEs ($p = 0.008$; $p = 0.04$; Table 4).

Table 4. Univariate and multivariate Cox analysis for MACEs.

Parameters	Univariate Analysis		Multivariate Analysis	
	HR Unadjusted (95% CI)	*p*	HR Adjusted (95% CI)	*p*
LGE	4.06 (1.94–8.52)	0.0001	4.91 (2.06–11.6)	0.008
Gal3	2.67 (1.32–5.28)	0.008	1.11 (1.03–1.19)	0.04
PICP	1.04 (1.01–1.07)	0.001	1.00 (0.98–1.07)	NS
PIIINP	1.09 (1.02–1.11)	0.001	1.00 (0.98–1.03)	NS

Abbreviations: Gal3, galectin-3; LGE, late gadolinium enhancement; PICP, procollagen type I carboxy-terminal pro-peptide; PIIINP, N-terminal pro-peptide of procollagen type III. Multivariate analysis, after adjustment for covariates which comprised age, gender, LVEF, eGFR, BMI, NT-proBNP, and diabetes mellitus.

Furthermore, gradual logistic regression proportional-hazard models showed a significant incremental predictive ability by adding Gal3 to LGE used alone (from Chi-square = 16.49, $p < 0.0001$ to Chi-square = 21.11, $p < 0.0001$).

Moreover, Kaplan–Meier analysis was performed to test the predictive ability of LGE, Gal3, PICP and PIIINP to predict the composite outcome. Thus, for specific thresholds, circulating collagen turnover biomarkers significantly predicted MACEs: >13.8 ng/mL for Gal3 (HR = 2.66, 95% CI (1.34–5.27), $p < 0.001$; Figure 4), >98 ng/dL for PICP (HR = 1.93, 95% CI (1.17–3.87), $p < 0.002$; Figure 5) and >4.1 ng/dL for PIIINP (HR = 1.42, 95% CI (1.07–2.81), $p < 0.03$; Figure 6), while LGE was also associated with a considerably increased risk of MACEs (HR = 4.06, 95% CI (1.99–8.26), $p = 0.0001$; Figure 7).

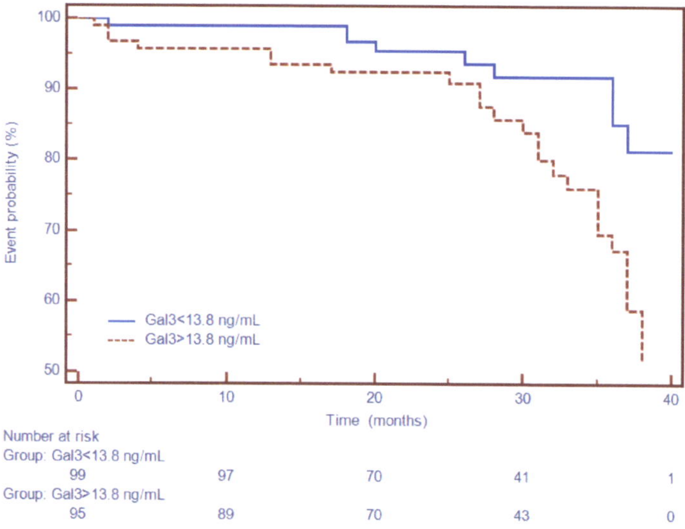

Figure 4. Kaplan–Meier analysis for the ability of Gal-3 to predict cardiovascular outcome. Abbreviations: Gal-3, galectin-3.

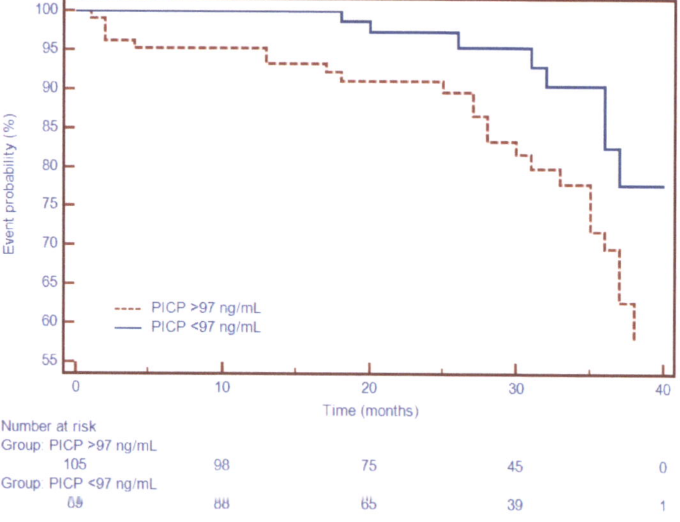

Figure 5. Kaplan–Meier analysis for the ability of PICP to predict cardiovascular outcome. Abbreviations: PICP, procollagen type I carboxy-terminal pro-peptide.

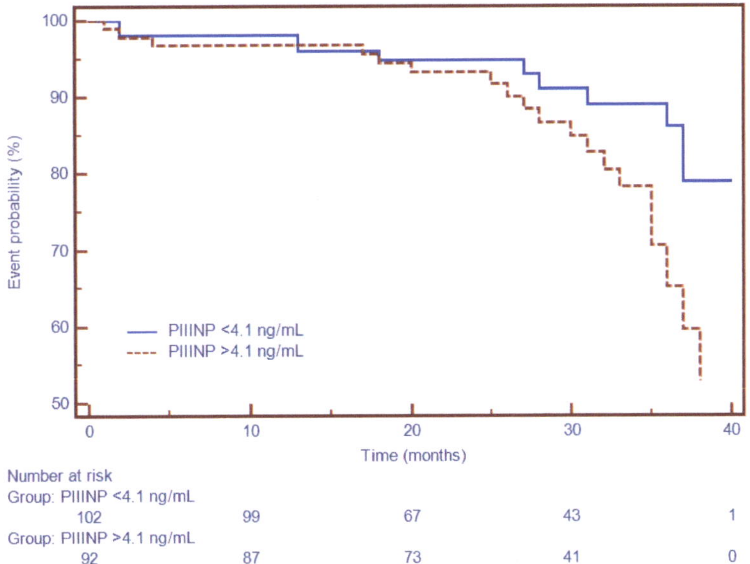

Figure 6. Kaplan–Meier analysis for the ability of PIIINP to predict cardiovascular outcome. Abbreviations: PIIINP, N-terminal pro-peptide of procollagen type III.

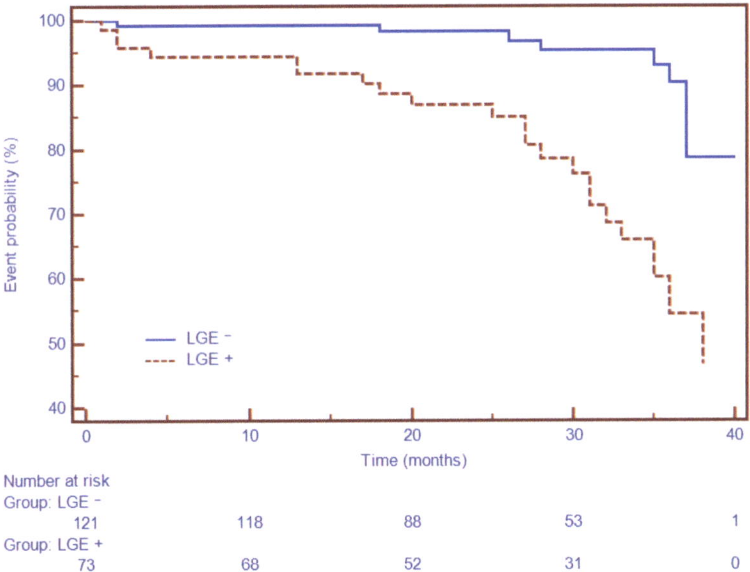

Figure 7. Kaplan–Meier analysis for the ability of LGE to predict cardiovascular outcome. Abbreviations: LGE, late gadolinium enhancement.

Moreover, a subgroup analysis that included patients with NIDCM and severely decreased LVEF (<30%) showed that, for similar cut-off values, Gal3 (Figure 8) and PICP (Figure 9) had an even higher predictive ability for outcome: HR = 4.27, 95% CI (2.58–7.06), $p < 0.0001$ and HR = 3.23, 95% CI (1.91–5.46), $p < 0.0001$.

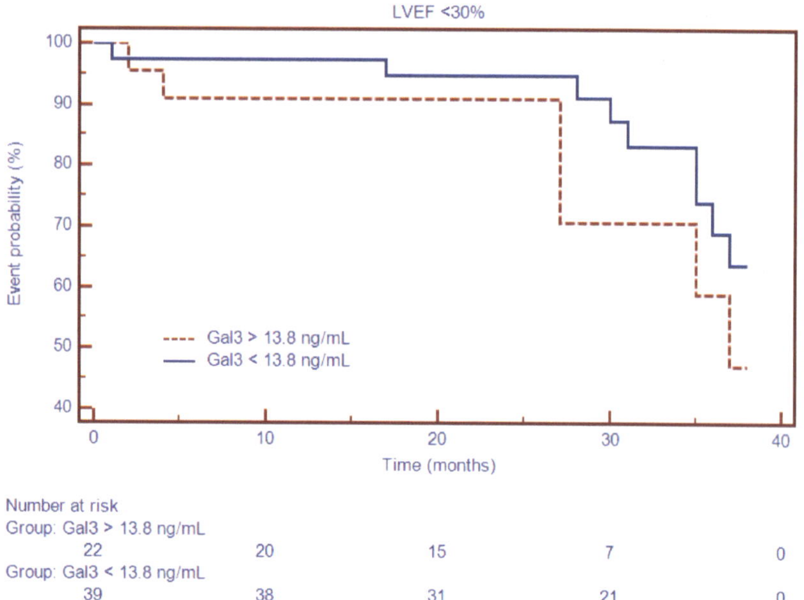

Figure 8. Kaplan–Meier analysis for the ability of Gal-3 to predict cardiovascular outcome in patients with NIDCM and severely decreased LVEF <30%. Abbreviations: Gal-3, galectin-3.

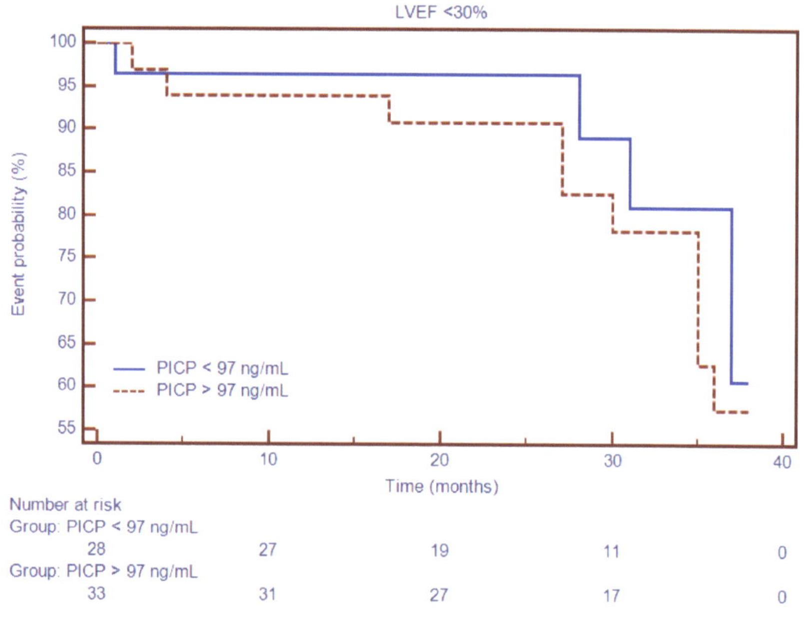

Figure 9. Kaplan–Meier analysis for the ability of PICP to predict cardiovascular outcome in patients with NIDCM and severely decreased LVEF <30%. Abbreviations: PICP, procollagen type I carboxy-terminal pro-peptide.

4. Discussion

In this study, we evaluated the association of circulating collagen turnover biomarkers with replacement myocardial fibrosis and with severely decreased LV systolic function, both determined by CMR, in patients with NIDCM. The main findings of our research article comprise: (1) sera levels of collagen turnover biomarkers, namely Gal3, PICP and PIIINP, were closely associated with CMR parameters of LV systolic dysfunction, such as LVEDV, LVESV, LVSI, LV-LAS—also being notably correlated with markers of HF severity, namely NYHA class $\geq$ III, NT-proBNP and CPP levels; (2) Gal3, PICP and PIIINP were directly associated with the mass of replacement myocardial fibrosis, represented as LGE mass and the LGE mass/LV mass ratio; (3) along with LGE, Gal3 and PICP were the most notable independent predictors of cardiovascular outcome; (4) the addition of Gal3, PICP and PIIINP provided an incremental ability to diagnose severely decreased LV systolic function in this category of patients.

Cardiac fibrosis is frequently found in patients with NIDCM and is associated with a more aggressive disease phenotype, being more difficult to treat. At the root of these findings stands accelerated progression of LV dysfunction, congestive HF, and increased risk of sudden cardiac death [1,23]. Previously published studies have shown that in patients with NIDCM, the presence and extent of LGE were independently associated with HF, malignant ventricular tachyarrhythmias, cardiac death and all-cause mortality [1,5,24,25]. Nonetheless, LGE has several limitations in detecting myocardial scarring; thus, by corroborating CMR with sera biomarkers, it might increase diagnostic accuracy [23]. In our current study, we have shown that the combined use of LGE with collagen turnover biomarkers significantly increased prognosis prediction and risk stratification in patients with NIDCM.

Amongst all the markers, Gal3 had the strongest predictive ability, being an independent predictor for outcome, together with LGE—even after the adjustment for standard covariates such as age, gender, LVEF, NYHA class, renal function, and NT-proBNP. Besides this, in our study, we have shown that circulant Gal3 was independently associated with myocardial fibrosis, quantified as LGE by CMR in patients with NIDCM—this being another innovative aspect of our research. Similarly, Vergaro et al. have shown that plasmatic levels of Gal3 are closely associated with LGE in patients with NIDCM [15]. Additionally, a recently published murine study has shown that the suppression of Gal3 has beneficial effects on the regression of NIDCM [26].

Sera PICP was reported to be an important circulant marker of type I collagen turnover, being significantly associated with myocardial fibrosis [27] and with an increased risk of MACEs in patients with NIDCM. Nonetheless, in formerly published studies, the prognostic ability of myocardial fibrosis is rather questionable due to their contradictory results [28,29]. In our study, PICP and PIIINP were closely associated with LGE mass and proved to have a significant ability to predict the occurrence of MACEs; however, in the Cox analysis—after the adjustment for confounders—none of them remained independent predictors for outcome.

Furthermore, we evaluated the profile of these biomarkers in patients with NIDCM and severely decreased LVEF of under 30%. All of them had significantly increased sera levels and were even closely related to HF parameters. Moreover, the stepwise addition of these biomarkers to LGE proved to increase their association with decreased LVEF, beyond that of each parameter used alone. Thus, our study suggests the utility of these sera biomarkers even in the risk stratification of these patients.

Furthermore, the joint use of circulating biomarkers and LGE might become useful in monitoring disease progression and also in identifying patients who would benefit from implantable cardioverter devices or cardiac resynchronization therapy [30,31], but these things are only in their infancy.

Withal, an important issue that needs to be considered is that these biomarkers reflect the systemic metabolism of collagen, and not only in cardiac collagen; thus, this is the reason why these markers could become useful in heart diseases only when they can be combined with cardiovascular imaging parameters [4]. Likewise, further research should focus on exploring if the combined use of CMR with circulant collagen turnover biomarkers might aid in therapeutic monitoring and cardiovascular risk stratification in patients with NIDCM.

Study Limitations

Firstly, being a single-center study represents a limitation by default. Secondly, T1-maps and extracellular volumes were not assessed in all patients since this technique was not available in our research facility at the beginning of the study. Lastly, the long recruitment period might have affected the sera samples from which the biomarkers were determined.

5. Conclusions

In patients with NIDCM, circulating collagen turnover biomarkers—namely Gal3, PICP and PIIINP—were independently associated with myocardial replacement fibrosis determined as LGE by CMR, being useful in the risk stratification of them. These markers were even higher in those with NIDCM and severely decreased LVEF. Moreover, they were useful in prognosis prediction; however, only Gal3 proved to be an independent predictor for cardiovascular outcome.

Author Contributions: R.R., B.O.C.-M. and L.A.-C., conception and design of the study; L.A.-C., R.A. and A.Z., acquisition, analysis, and interpretation of data; statistical methods and interpretation of results; project administration. R.A., I.D.M., C.C., D.H., R.-I.O. and B.C., acquisition, analysis, and interpretation of data; critical revision of the manuscript for intellectual content. R.R., A.Z. and L.A.-C., drafting of the manuscript. All authors have read and agreed to the published version of the manuscript.

Funding: This research received no external funding.

Institutional Review Board Statement: The current study was conducted in accordance with the Declaration of Helsinki and received approval from the Ethics Committee of Iuliu Hatieganu University of Medicine and Pharmacy of Cluj-Napoca, Romania.

Informed Consent Statement: All patients were fully informed about the study protocol and provided written consent.

Data Availability Statement: Not applicable.

Acknowledgments: This work was supported by an internal institutional doctoral fellowship from the Iuliu Hatieganu University of Medicine and Pharmacy of Cluj-Napoca.

Conflicts of Interest: The authors declare no conflict of interest.

References

1. Gulati, A.; Jabbour, A.; Ismail, T.F.; Guha, K.; Khwaja, J.; Raza, S.; Morarji, K.; Brown, T.D.H.; Ismail, N.A.; Dweck, M.R.; et al. Association of Fibrosis with Mortality and Sudden Cardiac Death in Patients with Nonischemic Dilated Cardiomyopathy. *JAMA* **2013**, *309*, 896. [CrossRef] [PubMed]
2. Shanbhag, S.M.; Greve, A.M.; Aspelund, T.; Schelbert, E.B.; Cao, J.J.; Danielsen, R.; Þorgeirsson, G.; Sigurðsson, S.; Eiríksdóttir, G.; Harris, T.B.; et al. Prevalence and prognosis of ischaemic and non-ischaemic myocardial fibrosis in older adults. *Eur. Heart J.* **2019**, *40*, 529–538. [CrossRef] [PubMed]
3. De Boer, R.A.; De Keulenaer, G.; Bauersachs, J.; Brutsaert, D.; Cleland, J.G.; Diez, J.; Du, X.J.; Ford, P.; Heinzel, F.R.; Lipson, K.E.; et al. Towards better definition, quantification and treatment of fibrosis in heart failure. A scientific roadmap by the Committee of Translational Research of the Heart Failure Association (HFA) of the European Society of Cardiology. *Eur. J. Heart Fail.* **2019**, *21*, 272–285. [CrossRef] [PubMed]
4. González, A.; Schelbert, E.B.; Díez, J.; Butler, J. Myocardial Interstitial Fibrosis in Heart Failure. *J. Am. Coll. Cardiol.* **2018**, *71*, 1696–1706. [CrossRef]
5. Weir, R.A.P.; Petrie, C.J.; Murphy, C.A.; Clements, S.; Steedman, T.; Miller, A.M.; McInnes, I.B.; Squire, I.B.; Ng, L.L.; Dargie, H.J.; et al. Galectin-3 and Cardiac Function in Survivors of Acute Myocardial Infarction. *Circ. Heart Fail.* **2013**, *6*, 492–498. [CrossRef]
6. Sharma, U.C.; Pokharel, S.; van Brakel, T.J.; van Berlo, J.H.; Cleutjens, J.P.M.; Schroen, B.; André, S.; Crijns, H.J.G.M.; Gabius, H.-J.; Maessen, J.; et al. Galectin-3 Marks Activated Macrophages in Failure-Prone Hypertrophied Hearts and Contributes to Cardiac Dysfunction. *Circulation* **2004**, *110*, 3121–3128. [CrossRef] [PubMed]
7. Lok, D.J.A.; Van Der Meer, P.; de la Porte, P.W.B.-A.; Lipsic, E.; Van Wijngaarden, J.; Hillege, H.L.; van Veldhuisen, D.J. Prognostic value of galectin-3, a novel marker of fibrosis, in patients with chronic heart failure: Data from the DEAL-HF study. *Clin. Res. Cardiol.* **2010**, *99*, 323–328. [CrossRef]

8. Lok, D.J.; Lok, S.I.; Bruggink-André de la Porte, P.W.; Badings, E.; Lipsic, E.; van Wijngaarden, J.; de Boer, R.A.; van Veldhuisen, D.J.; van der Meer, P. Galectin-3 is an independent marker for ventricular remodeling and mortality in patients with chronic heart failure. *Clin. Res. Cardiol.* **2013**, *102*, 103–110. [CrossRef]

9. Calvier, L.; Miana, M.; Reboul, P.; Cachofeiro, V.; Martinez-Martinez, E.; de Boer, R.A.; Poirier, F.; Lacolley, P.; Zannad, F.; Rossignol, P.; et al. Galectin-3 Mediates Aldosterone-Induced Vascular Fibrosis. *Arterioscler. Thromb. Vasc. Biol.* **2013**, *33*, 67–75. [CrossRef] [PubMed]

10. Yang, R.-Y.; Rabinovich, G.A.; Liu, F.-T. Galectins: Structure, function and therapeutic potential. *Expert Rev. Mol. Med.* **2008**, *10*, e17. [CrossRef] [PubMed]

11. Masci, P.G.; Doulaptsis, C.; Bertella, E.; Del Torto, A.; Symons, R.; Pontone, G.; Barison, A.; Droogné, W.; Andreini, D.; Lorenzoni, V.; et al. Incremental Prognostic Value of Myocardial Fibrosis in Patients With Non–Ischemic Cardiomyopathy Without Congestive Heart Failure. *Circ. Heart Fail.* **2014**, *7*, 448–456. [CrossRef] [PubMed]

12. Yu, L.; Ruifrok, W.P.T.; Meissner, M.; Bos, E.M.; van Goor, H.; Sanjabi, B.; van der Harst, P.; Pitt, B.; Goldstein, I.J.; Koerts, J.A.; et al. Genetic and Pharmacological Inhibition of Galectin-3 Prevents Cardiac Remodeling by Interfering With Myocardial Fibrogenesis. *Circ. Heart Fail.* **2013**, *6*, 107–117. [CrossRef]

13. Agoston-Coldea, L.; Lupu, S.; Petrovai, D.; Mocan, T.; Mousseaux, E. Correlations between echocardiographic parameters of right ventricular dysfunction and Galectin-3 in patients with chronic obstructive pulmonary disease and pulmonary hypertension. *Med. Ultrason.* **2015**, *17*, 486–495. [CrossRef]

14. de Boer, R.A.; Lok, D.J.A.; Jaarsma, T.; van der Meer, P.; Voors, A.A.; Hillege, H.L.; van Veldhuisen, D.J. Predictive value of plasma galectin-3 levels in heart failure with reduced and preserved ejection fraction. *Ann. Med.* **2011**, *43*, 60–68. [CrossRef] [PubMed]

15. Vergaro, G.; Del Franco, A.; Giannoni, A.; Prontera, C.; Ripoli, A.; Barison, A.; Masci, P.G.; Aquaro, G.D.; Cohen Solal, A.; Padeletti, L.; et al. Galectin-3 and myocardial fibrosis in nonischemic dilated cardiomyopathy. *Int. J. Cardiol.* **2015**, *184*, 96–100. [CrossRef]

16. Raafs, A.G.; Verdonschot, J.A.J.; Henkens, M.T.; Adriaans, B.P.; Wang, P.; Derks, K.; Abdul Hamid, M.A.; Knackstedt, C.; Empel, V.P.M.; Díez, J.; et al. The combination of carboxy-terminal propeptide of procollagen type I blood levels and late gadolinium enhancement at cardiac magnetic resonance provides additional prognostic information in idiopathic dilated cardiomyopathy— A multilevel assessment of myocardial fibrosis in dilated cardiomyopathy. *Eur. J. Heart Fail.* **2021**, *23*, 933–944. [CrossRef]

17. Pinto, Y.M.; Elliott, P.M.; Arbustini, E.; Adler, Y.; Anastasakis, A.; Böhm, M.; Duboc, D.; Gimeno, J.; de Groote, P.; Imazio, M.; et al. Proposal for a revised definition of dilated cardiomyopathy, hypokinetic non-dilated cardiomyopathy, and its implications for clinical practice: A position statement of the ESC working group on myocardial and pericardial diseases. *Eur. Heart J.* **2016**, *37*, 1850–1858. [CrossRef] [PubMed]

18. Kocaoglu, M.; Pednekar, A.; Tkach, J.A.; Taylor, M.D. Quantitative assessment of velocity and flow using compressed SENSE in children and young adults with adequate acquired temporal resolution. *J. Cardiovasc. Magn. Reson.* **2021**, *23*, 113. [CrossRef]

19. Arenja, N.; Andre, F.; Riffel, J.H.; Hegenbart, U.; Schönland, S.; Kristen, A.V.; Katus, H.A.; Buss, S.J. Prognostic value of novel imaging parameters derived from standard cardiovascular magnetic resonance in high risk patients with systemic light chain amyloidosis. *J. Cardiovasc. Magn. Reson.* **2019**, *21*, 53. [CrossRef]

20. Liang, Y.; Li, W.; Zeng, R.; Sun, J.; Wan, K.; Xu, Y.; Cao, Y.; Zhang, Q.; Han, Y.; Chen, Y. Left Ventricular Spherical Index Is an Independent Predictor for Clinical Outcomes in Patients with Nonischemic Dilated Cardiomyopathy. *JACC Cardiovasc. Imaging* **2019**, *12*, 1578–1580. [CrossRef]

21. Cerqueira, M.D.; Weissman, N.J.; Dilsizian, V.; Jacobs, A.K.; Kaul, S.; Laskey, W.K.; Pennell, D.J.; Rumberger, J.A.; Ryan, T.; Verani, M.S. Standardized Myocardial Segmentation and Nomenclature for Tomographic Imaging of the Heart. *Circulation* **2002**, *105*, 539–542. [CrossRef] [PubMed]

22. Bondarenko, O.; Beek, A.; Hofman, M.; Kühl, H.; Twisk, J.; van Dockum, W.; Visser, C.; van Rossum, A. Standardizing the Definition of Hyperenhancement in the Quantitative Assessment of Infarct Size and Myocardial Viability Using Delayed Contrast-Enhanced CMR. *J. Cardiovasc. Magn. Reson.* **2005**, *7*, 481–485. [CrossRef] [PubMed]

23. Cojan-Minzat, B.O.; Zlibut, A.; Agoston-Coldea, L. Non-ischemic dilated cardiomyopathy and cardiac fibrosis. *Heart Fail. Rev.* **2021**, *26*, 1081–1101. [CrossRef] [PubMed]

24. Halliday, B.P.; Baksi, A.J.; Gulati, A.; Ali, A.; Newsome, S.; Izgi, C.; Arzanauskaite, M.; Lota, A.; Tayal, U.; Vassiliou, V.S.; et al. Outcome in Dilated Cardiomyopathy Related to the Extent, Location, and Pattern of Late Gadolinium Enhancement. *JACC Cardiovasc. Imaging* **2019**, *12*, 1645–1655. [CrossRef] [PubMed]

25. Leong, D.P.; Chakrabarty, A.; Shipp, N.; Molaee, P.; Madsen, P.L.; Joerg, L.; Sullivan, T.; Worthley, S.G.; De Pasquale, C.G.; Sanders, P.; et al. Effects of myocardial fibrosis and ventricular dyssynchrony on response to therapy in new-presentation idiopathic dilated cardiomyopathy: Insights from cardiovascular magnetic resonance and echocardiography. *Eur. Heart J.* **2012**, *33*, 640–648. [CrossRef]

26. Nguyen, M.-N.; Ziemann, M.; Kiriazis, H.; Su, Y.; Thomas, Z.; Lu, Q.; Donner, D.G.; Zhao, W.-B.; Rafehi, H.; Sadoshima, J.; et al. Galectin-3 deficiency ameliorates fibrosis and remodeling in dilated cardiomyopathy mice with enhanced Mst1 signaling. *Am. J. Physiol. Circ. Physiol.* **2019**, *316*, H45–H60. [CrossRef]

27. López, B.; González, A.; Ravassa, S.; Beaumont, J.; Moreno, M.U.; San José, G.; Querejeta, R.; Díez, J. Circulating Biomarkers of Myocardial Fibrosis. *J. Am. Coll. Cardiol.* **2015**, *65*, 2449–2456. [CrossRef]

28. Aoki, T.; Fukumoto, Y.; Sugimura, K.; Oikawa, M.; Satoh, K.; Nakano, M.; Nakayama, M.; Shimokawa, H. Prognostic Impact of Myocardial Interstitial Fibrosis in Non-Ischemic Heart Failure. *Circ. J.* **2011**, *75*, 2605–2613. [CrossRef]

29. Vigliano, C.A.; Cabeza Meckert, P.M.; Diez, M.; Favaloro, L.E.; Cortés, C.; Fazzi, L.; Favaloro, R.R.; Laguens, R.P. Cardiomyocyte Hypertrophy, Oncosis, and Autophagic Vacuolization Predict Mortality in Idiopathic Dilated Cardiomyopathy With Advanced Heart Failure. *J. Am. Coll. Cardiol.* **2011**, *57*, 1523–1531. [CrossRef]

30. Ferreira, J.P.; Rossignol, P.; Pizard, A.; Machu, J.-L.; Collier, T.; Girerd, N.; Huby, A.-C.; Gonzalez, A.; Diez, J.; López, B.; et al. Potential spironolactone effects on collagen metabolism biomarkers in patients with uncontrolled blood pressure. *Heart* **2019**, *105*, 307–314. [CrossRef]

31. Xu, Y.; Li, W.; Wan, K.; Liang, Y.; Jiang, X.; Wang, J.; Mui, D.; Li, Y.; Tang, S.; Guo, J.; et al. Myocardial Tissue Reverse Remodeling After Guideline-Directed Medical Therapy in Idiopathic Dilated Cardiomyopathy. *Circ. Heart Fail.* **2021**, *14*, e007944. [CrossRef] [PubMed]

Article

Stress Perfusion Cardiac Magnetic Resonance in Long-Standing Non-Infarcted Chronic Coronary Syndrome with Preserved Systolic Function

Pierpaolo Palumbo [1,2,*], **Ester Cannizzaro** [1], **Annamaria Di Cesare** [3], **Federico Bruno** [2,4], **Francesco Arrigoni** [1], **Alessandra Splendiani** [4], **Antonio Barile** [4], **Carlo Masciocchi** [4] and **Ernesto Di Cesare** [5,*]

[1] Department of Diagnostic Imaging, Area of Cardiovascular and Interventional Imaging, Abruzzo Health Unit 1, Via Saragat, Località Campo di Pile, 67100 L'Aquila, Italy; estercannizzaro@hotmail.it (E.C.); arrigoni.francesco@gmail.com (F.A.)

[2] SIRM Foundation, Italian Society of Medical and Interventional Radiology (SIRM), 20122 Milan, Italy; federico.bruno.1988@gmail.com

[3] Ospedale "Infermi" di Rimini, Viale Luigi Settembrini, 2, 47923 Rimini, Italy; annamariadicesare.adc@gmail.com

[4] Department of Applied Clinical Sciences and Biotechnology, University of L'Aquila, Via Vetoio 1, 67100 L'Aquila, Italy; alessandra.splendiani@univaq.it (A.S.); antonio.barile@univaq.it (A.B.); carlo.masciocchi@univaq.it (C.M.)

[5] Department of Life, Health and Environmental Sciences, University of L'Aquila, Piazzale Salvatore Tommasi 1, 67100 L'Aquila, Italy

* Correspondence: palumbopierpaolo89@gmail.com (P.P.); ernesto.dicesare@univaq.it (E.D.C.)

Citation: Palumbo, P.; Cannizzaro, E.; Di Cesare, A.; Bruno, F.; Arrigoni, F.; Splendiani, A.; Barile, A.; Masciocchi, C.; Di Cesare, E. Stress Perfusion Cardiac Magnetic Resonance in Long-Standing Non-Infarcted Chronic Coronary Syndrome with Preserved Systolic Function. *Diagnostics* **2022**, *12*, 786. https://doi.org/10.3390/diagnostics12040786

Academic Editors: Minjie Lu and Arlene Sirajuddin

Received: 31 January 2022
Accepted: 22 March 2022
Published: 23 March 2022

Publisher's Note: MDPI stays neutral with regard to jurisdictional claims in published maps and institutional affiliations.

Abstract: (1) Background: The impact of imaging-derived ischemia is still under debate and the role of stress perfusion cardiac magnetic resonance (spCMR) in non-high-risk patient still needs to be clarified. The aim of this study was to evaluate the impact of spCMR in a case series of stable long-standing chronic coronary syndrome (CCS) patients with ischemia and no other risk factor. (2) Methods: This is a historical prospective study including 35 patients with history of long-standing CCS who underwent coronary CT angiography (CCTA) and additional adenosine spCMR. Clinical and imaging findings were included in the analysis. Primary outcomes were HF (heart failure) and all major cardiac events (MACE) including death from cardiovascular causes, myocardial infarction, or hospitalization for unstable angina, or resuscitated cardiac arrest. (3) Results: Mean follow-up was 3.7 years (IQR: from 1 to 6). Mean ejection fraction was $61 \pm 8\%$. Twelve patients (31%) referred primary outcomes. Probability of experiencing primary outcomes based on symptoms was 62% and increased to 67% and 91% when multivessel disease and ischemia, respectively, were considered. Higher ischemic burden was predictive of disease progression (OR: 1.59, 95%CI: 1.18–2.14; *p*-value = 0.002). spCMR model resulted non inferior to the model comprising all variables (4) Conclusions: In vivo spCMR-modeling including perfusion and strain anomalies could represent a powerful tool in long-standing CCS, even when conventional imaging predictors are missing.

Keywords: stress perfusion CMR; ischemia; CAD; CCS; long-standing CCS; CCTA; strain; CAD extension; heart failure

1. Introduction

Chronic coronary syndrome (CCS) includes a wide spectrum of clinical scenarios involving patients with known or suspected coronary artery disease (CAD) [1].

CAD is a chronic and often progressive disease, and CCS definition has recently been introduced to differentiate a clinical stable presentation from an acute presentation (acute coronary syndrome or ACS). Different outcomes, however, can occur due to the dynamic nature of CAD, and ACS can also destabilize a long-standing (i.e., more than one year after initial diagnosis or revascularization) apparently stable clinical scenario. Therefore, correct risk stratification and adequate clinical management of CCS is essential to reduce the risk of major cardiac events [1].

Advanced cardiac imaging plays a primary role in the assessment of heart disease, both in ischemic and non-ischemic cardiomyopathy [2–7].

Based on the Bayesian probability of CAD, anatomical strategy with coronary CT angiography (CCTA) finds a prevalent role in patients with a low likelihood to have CAD [1,8,9]. From the PROMISE study, the high ability of CCTA to identify a low-risk group corresponds to an event rate of 0.9% vs. 2.1% observed in patients managed with conventional stress testing (over a two-year period of observation) [10–12]. Conversely, CCTA finds only a marginal role in long-standing CCS for the lack of functional information related to ischemia [1].

On the other hand, functional tests imaging for ischemia detection finds a primary role in patients with an intermediate-to-high probability of CAD and in patients with long-standing CCS, both in symptomatic or asymptomatic patients, given the risk for complications also in an otherwise asymptomatic patient. However, the impact of non-invasive imaging strategies for ischemia detection to guide initial coronary revascularization and improve long-term outcomes is still under debate [13].

Recently, the ISCHEMIA trial caused major controversy reporting no substantial benefit of ischemia testing in CAD prognostication and patient stratification, especially in the early time-window of observation and in patient with a good systolic performance, proving no superiority of an initial invasive vs. conventional medical treatment when moderate-to-severe ischemia is detected [14,15].

From ISCHEMIA, different questions arise regarding which ischemia test should be performed and which subcategories can benefit from a more aggressive treatment [16]. In this regard, from the extended STICHES, revascularization strategies plus medical therapies seem more effective than medical therapy alone in treatment of patients with reduced left ventricular ejection fraction (LVEF), thus suggesting a real benefit to coronary revascularization in patients with both ischemia and reduced EF heart failure (HFrEF) [17,18].

During the latest years, stress perfusion cardiac magnetic resonance (spCMR) showed a relevant impact in CAD stratification in many trial and registry studies [16,19–25]. Differently from other imaging techniques, spCMR offers a holistic approach to the heart patient through the simultaneous evaluation of the triad systolic function–perfusion abnormalities–tissue characterization [26].

CMR showed high accuracy in ischemia detection and is currently considered the gold standard for cardiac volume and systolic function evaluation [27,28]. Moreover, late gadolinium enhancement (LGE) as an imaging marker of myocardial scarring is a well-known predictor of all-cause mortality from different studies including ischemic and non-ischemic cardiomyopathies [29–33].

Given these discrepancies, spCMR may impact more efficiently than other conventional ischemia tests (largely involved in ISCHEMIA), although its clinical utility should be defined especially if conventional outcome predictors are missing.

The purpose of our study was to assess the impact of spCMR findings in a case series of apparently clinically stable long-standing CAD patients with preserved EF and no previous infarction or signs of HF during a long-term follow-up.

2. Materials and Methods

This study was carried out after the approval of our university's Internal Review Board.

This is a retrospective assessment of prospectively followed-up patients (historical prospective/cohort study).

We screened our database to identify patients referred to our institution for a history of long-standing CCS who had been re-submitted to CCTA (to identify unprotected CAD) and additional adenosine spCMR in a short time interval (less than 6 months) for a comprehensive evaluation and were deemed able to complete a long-term follow-up.

Long-standing CCS was defined in accordance with the latest ESC guidelines (i.e., more than one year after initial diagnosis and medical treatment or revascularization) [1].

We identified and analyzed 87 patients. Those with a history of myocardial infarction (MI) were excluded. 23 participants reported previous ACS or showed ischemic-type myocardial scarring, and therefore were excluded. Another 29 patients were excluded for the following: (a) time interval between CCTA and stress CMR examination longer than 1 year; (b) clinical condition not specifically attributable to CAD for concomitant morbidities; and (c) moderate to severe systolic dysfunction.

In the end, 35 patients matched our inclusion criteria. Eligible participants were recalled for an on-site interview performed by two specialists. All information reported within the radiology information system (RIS) was also collected.

Cardiovascular symptoms (i.e., angina and/or ischemic equivalent as dyspnea) and CAD extension (i.e., a single or multivessel disease—2 or 3 major epicardial vessel) were collected.

A healthy control group was also recruited. A healthy control group was defined for: (i). preserved EF; (ii). absence of clinical history of CAD, myocardial injury and/or systemic disease; (iii). no cardiovascular symptoms or risk factor; (iv). absence of signs of structural heart disease or LGE (23 participants; 12 males, 44 ± 9 years). The healthy controls were recruited among people referred to our center for echocardiographic suspicion of cardiomyopathy but not confirmed with CMR.

2.1. Cardiac Magnetic Resonance Imaging Protocol

Stress CMR protocol included assessment of cardiac function, ischemia testing, and LGE imaging to exclude myocardial scarring. Resting perfusion was not included in our standard protocol.

For the assessment of LV volumes/systolic function steady-state free precession cine images (echo time/reception time 1.5/3.0 ms, flip angle 60°) were acquired on short-axis (slice thickness 8 mm, spacing 0 mm) and radial long-axis views (ten slices covering the entire circumference of the ventricle, planned on short-axis pilots at 18° angles to each other to visualize all 17 segments) and analyzed with dedicated software (Circle, cvi42, Calgary, AB, Canada; version 5.11.4).

Tissue tracking (TT) analysis was also performed to obtain strain data. TT analysis was performed on resting cine imagines. LVOT and mitral valve planes were excluded from the analysis.

Global 2 d longitudinal (GLS), circumferential (GCS), and radial (GRS) strain values were recorded.

Standard stress protocol included infusion of 140–210 mg/kg/min of adenosine (heart rate increase at peak stress >10% above baseline), for up to 6 min. First-pass perfusion data were acquired after injection of Gadobutrol 0.05 mmol/kg (Gadovist®, Bayer AG, Zurich, Switzerland) at 5 mL/s, followed by a 15-mL saline in 3 short axis slices using a breath-hold T1-weighted fast gradient echo sequence. Beta-blocker drugs were stopped five days prior to examination while nitrates, calcium-channel blockers and ACE inhibitors were interrupted two days before, as for caffeine or theine.

Ischemia was defined as a sub-endocardial hypointense area in the left ventricle wall during first-pass perfusion, evident in at least three frames beyond peak contrast enhancement. Significant cut-off considered was two or more neighboring segments, two adjacent slices, or a single transmural segment (approximately 6% of the myocardium). Extension of ischemia (ischemic burden) was defined as the sum of involved segments.

LGE sequences were analyzed to exclude from the analysis all patients with myocardial scarring indicative of previous MI.

2.2. Study Outcomes and Patient's Follow-Up

Follow-up time was considered as the time lapse from the last examination to the interview. Primary outcomes were HF and all major cardiac events (MACE) including death from cardiovascular causes, myocardial infarction, hospitalization for unstable angina, or resuscitated cardiac arrest. Secondary outcomes were HF.

2.3. Statistical Analysis

Descriptive variables are presented as average and correspondent confidential intervals or as percentages (frequencies). The Shapiro–Wilk (SW) test was used to evaluate data distribution. A *t*-test was used for normal variables comparison; a chi-squared test was used with nominal (dichotomic) variables. The healthy group was used to define normal strain values. A comparison of strain data between long-standing CCS patients and healthy participants was performed. The probability of having primary outcomes given variables was estimated as odds/1 + odds. These analyses were performed with SPSS (IBM Corp. Released 2016. IBM SPSS Statistics for Mac, Version 26.0. Armonk, NY, USA: IBM Corp.). The outcome has been modelled performing exact logic regression to account for the small sample size. Model fitting has been assessed using the probability score of each model. Model diagnostic performance has been addressed carrying out a ROC analysis. Exact logistic regression was performed via Stata (StataCorp. 2021. Stata Statistical Software: Release 17. College Station, TX, USA: StataCorp LLC.). This analysis was also re-tested with a surrogate test via NCSS 2022 Statistical Software (2022, NCSS, LLC., Kaysville, UT, USA), which confirmed the same results. A non-inferiority test (with 0.1 margin) for two AUCs built up using the models described above was performed via NCSS 2022 Statistical Software. An alpha error of 5% was used as a threshold of significance.

3. Results

3.1. Patient Characteristics

The study population consisted of 35 patients (29 M, 6 F, mean age of 69 ± 9 years) referred for long-standing CCS, all managed with MT at the time of the scan. Mean follow-up was 3.73 years (interquartile range: from 1 to 6).

No major nor minor complication occurred during spCMR.

The baseline characteristics of study participants are listed in Table 1.

Table 1. Baseline patient characteristics.

		PO (No)	PO (Yes)	*p*-Value
All *n* (%)	35 (100)	23 (66)	12 (34)	
Sex (male) *n* (%)	29 (83)	21 (60)	8 (23)	0.089
Sex (female) *n* (%)	6 (17)	2 (6)	4 (11)	
Age (years)	69 ± 9	67 ± 9	72 ± 7	0.122
EF (%)	61 ± 8	60 ± 7	63 ± 10	0.415
Diabetes *n* (%)	6 (17)	3 (9)	3 (9)	0.329
Hypertension *n* (%)	13 (37)	9 (26)	4 (11)	0.517
Smoking habits *n* (%)	8 (23)	6 (17)	2 (6)	0.429
Familiarity for CHD *n* (%)	10 (29)	8 (23)	2 (6)	0.236
Dyslipidemia *n* (%)	19 (54)	12 (34)	7 (20)	0.505
Symptoms *n* (%)	16 (46)	6 (17)	10 (29)	0.002 **
Multivessel CAD *n* (%)	21 (60)	11 (31)	10 (29)	0.045 *
Ischemia *n* (%)	12 (34)	3 (9)	9 (26)	0.0001 **
Ischemic burden (%)	9 ± 3	1 ± 2	7 ± 5	0.0001 **
CMR-Tissue Tracking CCS group				
GLS (%)	-16 ± 2	-17 ± 1	-14 ± 2	0.0001 **
GCS (%)	-17 ± 3	-17 ± 2	-17 ± 4	0.489
GRS (%)	28 ± 7	27 ± 6	29 ± 10	0.147
CMR-Tissue Tracking Healthy group				
GLS (%)	-18 ± 1			
GCS (%)	-20 ± 2			
GRS (%)	36 ± 6			

PO: primary outcomes; EF: ejection fraction; CHD: coronary heart disease; CAD: coronary artery disease; CMR: cardiac magnetic resonance; GLS: global longitudinal strain; GCS: global circumferential strain; GRS: global radial strain. significant difference *: level of significance: *p*-value < 0.05; ** significant difference: level of significance: *p*-value < 0.01.

Perfusion images and LGE were considered of good quality in all cases examined.

Mean ejection fraction was 61% ± 8%, meaning an overall preserved systolic function of the study participants.

Twelve patients (31%) referred primary outcomes, seven of which (20%) with heart failure syndrome, and one died (3%).

3.2. Chronic Coronary Syndrome Characteristics

Twenty-one patients (60%) showed a multivessel disease and were more likely symptomatic, referring typical chest pain or dyspnea (13 out of 16 symptomatic patients).

Ischemia was detected in 12 patients. No LGE was included in the analysis.

Multivessel disease was significantly associated with ischemia (10 out of 12 ischemic patients: *p*-value 0.045).

In patients with ischemia, mean ischemic burden (expressed as a percentage) was 9% ± 3%.

All strain values were lower when compared with healthy group (GLS: −16 ± 2 vs. −18 ± 1, *p*-value: 0.001; GCS: −17 ± 3 vs. −20 ± 2, *p*-value: 0.0001; GRS: 28 ± 7 vs. 36 ± 6, *p*-value: 0.0001).

However, the GLS only resulted significantly different for CCS patients categorized for the presence of ischemia (−14 ± 2% vs. −17 ± 1%, *p*-value 0.0001) (Figure 1).

No significant difference was detected between ischemic patients for other strain values.

Ischemic burden correlates with GLS (r: 0.699, *p*-value: 0.0001). Moreover, when GLS was categorized for reduced or preserved values, ischemic burden was predictive of GLS impairment (OR: 1.33, 95%CI: 1.08–1.64; *p*-value 0.008).

Among patients with multivessel diseases, patients with a three-vessel disease (TVD) were more likely associated also with all-strain anomalies (4 out of 5 patients with all-strain anomalies showed a TVD; *p*-value 0.007).

3.3. Association with Outcomes

Among all variables, multivessel diseases, symptoms, ischemia, and GLS involvement were associated with primary outcomes (Table 2). Probability to experience primary outcomes based on symptoms was 62% and increase to 67% when multivessel disease was also considered.

Table 2. Primary and Secondary Outcome According to Symptoms, Multivessel Disease, Ischemia and Strain.

	Symptoms	Multivessel Disease	Ischemia	GLS Impairment	Ischemia and GLS Impairment
No of patients	16 (46)	21 (60)	12 (34)	13 (37)	8 (23)
MACE *n* (%)	10 (63)	10 (48)	9 (75)	9 (69)	8 (100)
HF *n* (%)	7 (44)	6 (29)	6 (50)	5 (38)	5 (63)
no MACE or HF *n* (%)	6 (37)	11 (52)	3 (25)	4 (31)	0

MACE: major adverse cardiac events; HF: heart failure.

Probability to experience primary outcomes when ischemia only was detected was 75% and increased to 91% when multivessel disease and symptoms also were considered.

Higher burden of ischemia was predictive of disease progression (i.e., occurrence of primary or secondary outcomes) (OR: 1.59, 95% CI: 1.18–2.14; *p*-value 0.002).

A predictive model including all multivessel diseases, symptoms, ischemia, and strain anomalies (Model I) reached an AUC of 0.93 (95% CI: 0.83–1).

Model II (including only ischemia and strain anomalies) reached an AUC of 0.89 (95% CI: 0.74–0.97).

Lastly, Model III (including only symptoms and multivessel diseases) reached an AUC of 0.82 (95% CI: 0.67–0.97).

Using Model I as reference, Model II was non-inferior to model I (AUC difference: −0.04; One-Sided 95% lower limit: −0.09; Non-Inferiority *p*-value: 0.033). Conversely, the non-inferiority test failed for Model III (AUC difference: −0.1; One-Sided 95% lower limit: −0.22; Non-Inferiority *p*-value: 0.524) (Figure 2).

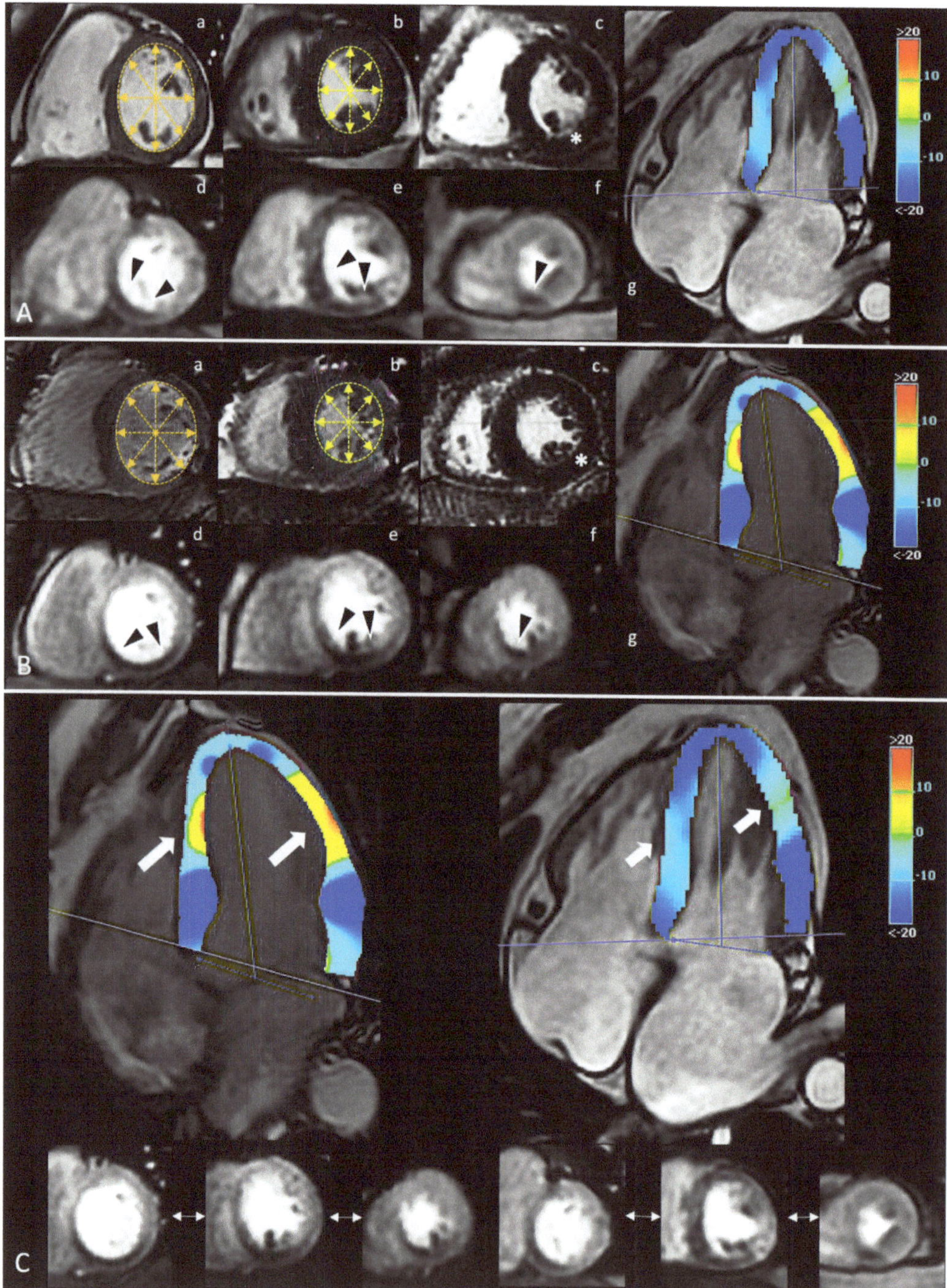

Figure 1. In panel (**A**,**B**), spCMR findings of two CCS patients. In (a,b), a single frame from SA cine sequence in diastolic and systolic phase, respectively, changes of inner double-arrows lines length highlight preserved systolic contraction. In (c), evidence of no enhancement in LGE sequences (white asterisk). In (d–f), a single frame from first-pass perfusion during adenosine infusion with evidence

of similar ischemic burden involved inferior segments from the base to the apex (black arrowheads). In (g), a single systolic frame from HLA cine sequence with superimposed colorimetric map of GLS. On the right, the legend of colorimetric map with correlation between values and colors. In panel (**C**), the comparison between the different GLS. Despite a similar ischemic burden, the two patients report different GLS abnormalities (white thick arrows), suggesting a different impact of ischemia on global deformability. GLS acts as accurate index of early global impairment beyond the focal injury. spCMR: stress perfusion cardiac magnetic resonance; CCS: chronic coronary syndrome; SA: short-axis; LGE: late gadolinium enhancement; HLA: horizontal long-axis; GLS: global longitudinal strain.

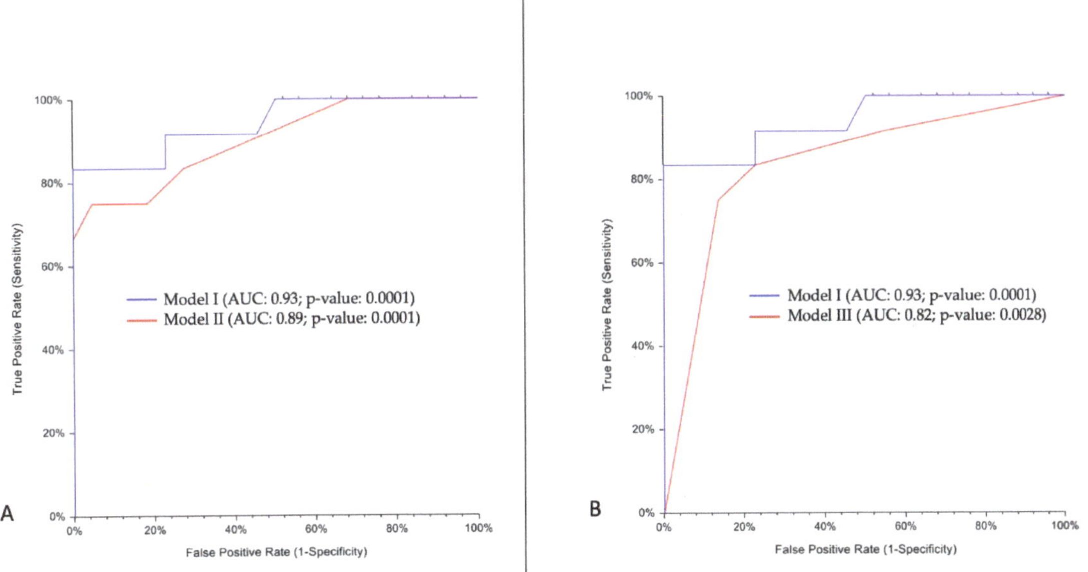

Figure 2. Non-inferiority test for two AUCs. Model I (symptoms, multivessel diseases, ischemia, and strain anomalies); Model II (ischemia and strain anomalies); Model III (symptoms and multivessel diseases). Panel (**A**) shows comparison between Model I (blue line) and II (red line) AUCs. In panel (**B**), comparison between Model I (blue line) and Model III (red line) AUCs. AUC: area under curve.

4. Discussion

This study is a historical prospective analysis of patients referred to our institution for a long-standing CCS, managed with MT and evidence of preserved systolic function and no previous MI, consecutively undergoing both CCTA and spCMR over a short period for a comprehensive evaluation.

In 100% of cases, it was possible to carry out a clinical follow-up, including both clinical evaluations obtained through an on-site interview and all information collected through RIS for any hospitalizations.

Our analysis highlights some important findings:

(i) stratification considering symptoms, anatomical extension of disease, and ischemia showed significant association with primary outcomes including all composite MACE;

(ii) extensive disease was more likely associated with ischemia, and when extensive disease and ischemia were considered, high probability of early alterations of myocardial deformability were also detected in patients with preserved EF; and

(iii) inducible ischemia and early alteration of myocardial deformability showed a non-inferior ability to predict clinical evolution of long-standing CCS compared to the model also including symptoms and anatomical extension of disease.

Long-standing CCS affects a highly heterogeneous population in need of complex therapies, and management of these "complex" patients is a tricky process, which historically has considered several factors other than ischemia [34].

The main decision-making point was in fact based on anatomic level of lesions, symptoms, and clinical conditions.

Other conventional imaging outcome predictors include LVEF, more recently echocardiographic GLS, and LGE, with LVEF considered as a major predictor of long-term survival in patients with CAD.

In latest years, cardiac imaging has shown high accuracy in guiding standard-of-care coronary revascularization, although the recent ISCHEMIA trial (or the controverse COURAGE or BARI-2D) failed to show a real advantage in using an early invasive revascularization in patients stratified for moderate-to-severe ischemia [14,15,35,36]. Moreover, in a sub study of ISCHEMIA by Reynolds et al., extent of ischemia is also a poor discriminator of risk for most clinical end-points [13–15].

It is therefore legitimate to question the role of imaging findings in a real-world scenario of these complex patients, especially when conventional outcome predictors are missing [37,38].

ISCHEMIA reveals that 70% of all ischemia-tests performed were perfusion imaging tests and that the predominant tool was nuclear perfusion imaging. To confirm this, spCMR continues to be an underutilized tool, accounting for <0.1% of all tests used in 2018, according to US statistics [39].

However, other focused trials and large registry studies have shown that spCMR had a high clinical impact in stratification of ischemic patients [16,19–23,40,41].

Regarding this strongest evidence, recent AHA guidelines for the evaluation and diagnosis of chest pain emphasize the primary role of spCMR in the identification of myocardial ischemia for a proper management of patients with acute chest pain and no known CAD [42].

In MR-INFORM RCT, spCMR-related ischemia proved similar to FFR in stratifying ischemic patients with no significant difference in MACE occurrence between spCMR and FFR-harm [43].

In SPINS registry, similarly to other studies, extensive ischemic burden was related to a higher risk of long-term, all-cause mortality, and revascularization was associated with a protective effect only in the restricted subset of patients with extensive spCMR-related ischemia [44–46].

Similarly, our results showed that quantification of ischemic burden improved the prognostication of CCS patients.

Moreover, as shown by Ge et al. through the same SPINS registry, spCMR did not suffer from the same limitation of CT (i.e., elevated BMI did not negatively impact its diagnostic quality) [47]. In our case series, spCMR confirmed high diagnostic quality in all examinations, irrespective of age.

spCMR has several advantages to other perfusion techniques, allowing an in vivo modeling of the heart including perfusion, tissue characterization and systolic function. High attention is paid also on CMR strain which has shown relevant impact in several cardiomyopathies [48–54].

CMR-derived cardiac model could therefore represent an effective tool in guide IHD management.

CCS and Outcome Association

During the latest years, CCTA strongly impact the management of heart disease both in routine and emergency settings [55–67]. Clinical trial also highlighted the impact of CAD definition from CCTA [68–70].

From PROMISE, CCTA detects CAD better than conventional stress testing (i.e., not including spCMR), and thus, with a better prediction of cardiac event, especially in non-obstructive CAD. In this regard, the analysis by Hoffmann et al. shows that the ability of

CCTA to identify a low-risk group correspond to an event rate of 0.9% over a two-year period vs. 2.1% observed in patients managed with normal stress test [10–12].

Similarly, from SCOT-HEART, management of CAD patients based on CT findings is revealed to be effective vs. the standard-of-care only, also guiding the clinicians in adopting medical treatment to prevent major events [59,71].

Presence, extension, and severity showing advantages in stratifying CAD patients also in CONFIRM study vs. clinical scoring, e.g., Framingham or Morise scores [72].

Therefore, CCTA is primarily adopted in CAD rule-out, with a negative predictive value close to 100%, enhanced by the effective dose optimization protocols and current technologies improved though the use of artificial intelligence, but is limited in detecting ischemia, the effective myocardial injurer, unless functional markers as with FFR-CT or CT-MPI are adopted [73–78].

In our study, CAD anatomy proved effective in the definition of outcomes, with the CAD extension significantly associated with composite MACE. However, the current inability of standard analysis to provide functional information capable to predict the clinical evolution of longstanding CCS patients affect the overall accuracy of Model III, which resulted inferior to the reference (i.e., the model including all variables).

As highlighted in FAME(s) and similar studies, the most important prognostic factor of a given coronary artery stenosis with respect to cardiac death or MI is indeed its ability to produce myocardial ischemia [79–84].

Ischemia is also the main predictor of HF evolution of CCS patients.

The pathophysiological process underlying the development of HF in ischemic patients can be variable and recognize different and specific therapeutic management [85,86].

In HFpEF models, recent evidence suggests that the onset of coronary microvascular dysfunctions (CMD) in non-infarcted areas contributes to the recurrence of ischemia which determines the progression of organ dysfunction as the onset of congestion symptoms even in absence of a real impairment of global systolic function [87–91].

Evidence suggests the potential occurrence of CMD in patients with CCS. In the CE-MARC 2 coronary physiology sub-study, a high incidence of CMD was found in patients with obstructive and non-obstructive CAD [92].

The identification of ischemic substrate in absence of obstructive CAD (more likely due to microvascular injury) is of primary importance considering that standardized approaches often fail in a correct assessment and management of CMD patients as is evident by the recent CorMicA trial [93–98].

Therefore, the identification of ischemia as a territory-specific assessment, irrespective of lesion-specific assessments, proves necessary for a proper treatment.

The following evidence, in concordance with our results, bring about some considerations:

1. spCMR findings result as good predictors of clinical evolution of CCS patients beyond symptoms and CAD extension. The probability of developing MACE was about 90% when ischemia was detected, with a high prevalence of HF syndrome during follow-up.
2. CMR-related strain confirms its ability to stage myocardial damage, which could translate into a critical ability to predict disease progression [99,100]. Among CCS patients with ischemia and no other conventional imaging predictor, GLS resulted highly impaired with a good correlation with the ischemic burden. This correlation proves GLS (an indicator of global function) as effective in describing the real impact of ischemia on cardiac function beyond the localized distribution of ischemic damage [101].
3. Despite GLS significantly differing between ischemic and non-ischemic CCS patients, GCS and GRS results were impaired when compared to a healthy population. Actually, in our series, GCS showed a stable early impairment compared to healthy volunteers. GCS impairment is indeed more likely related to a transmural injury/advanced disease, while GLS resulted most sensitive to a subendocardial/early injury [99]. On the other hand, the lack of a significant difference of GRS between ischemic and

non-ischemic longstanding CCS patients could be explained by a relatively preserved compensating mechanism offered by circumferential fibers, since radial strain is tethered with other longitudinal and circumferential fibers and no radially oriented fibers are disposed within the myocardium.

This study presents some obvious limits. (i) This study is based on a retrospective analysis of a small study sample with the consequent risk of a potential selection bias, although all patients meeting the inclusion criteria were included in the prospective follow-up, uncensored real world picture of patients with longstanding CCS; (ii) this is a single-center study, although the single-center reference allowed for the obtaining of all the clinical information also included in the RIS and the standardization of the approach to patients and clinical information; (iii) although the individual therapeutic schemes were included in the interview, we are not able to identify all cardiological therapeutic modifications based on CCTA and stress CMR findings; thus, therapeutic modifications that did not include an invasive approach were not considered for the overall clinical evolution; (iv) CMR-TT analysis was performed on cine images acquired only during rest condition (i.e., not with increased strain values), and therefore it is not possible to conclude about strain anomalies during stress condition.

5. Conclusions

In vivo spCMR-modeling including perfusion and strain anomalies could represent a powerful tool in long-standing CCS, even when conventional imaging predictors are missing. In addition to the definition of death and MI risk, spCMR-modeling could also predict the clinical evolution trends toward HF as in our series of long-standing CCS patients with preserved systolic function and no previous MI, thus identifying patients who deserve more aggressive treatment, although larger studies are needed to fully clarify this issue.

Author Contributions: Conceptualization, P.P.; methodology, P.P.; formal analysis, P.P.; investigation, P.P. and E.C.; writing—original draft preparation, P.P.; writing—review and editing, P.P.; visualization, P.P., F.B., F.A., A.B., A.S., C.M. and A.D.C.; supervision, E.D.C.; project administration, E.D.C. All authors have read and agreed to the published version of the manuscript.

Funding: This research received no external funding.

Institutional Review Board Statement: The study complies with the Declaration of Helsinki principles and the Institutional Review Board has granted its ethics approval (sequential number 28/2019).

Informed Consent Statement: Reasonable efforts were conducted to obtain a written informed consent from all patients for the publication of this article, as stated by the recommendation of our Internal Review Board.

Data Availability Statement: Data available to request.

Acknowledgments: Authors wish to thank Evan David Wellmeyer and Angela Martella for English revision.

Conflicts of Interest: The authors declare no conflict of interest.

References

1. Knuuti, J.; Wijns, W.; Saraste, A.; Capodanno, D.; Barbato, E.; Funck-Brentano, C.; Prescott, E.; Storey, R.F.; Deaton, C.; Cuisset, T.; et al. 2019 ESC Guidelines for the diagnosis and management of chronic coronary syndromes: The Task Force for the diagnosis and management of chronic coronary syndromes of the European Society of Cardiology (ESC). *Eur. Heart J.* **2020**, *41*, 407–477. [CrossRef] [PubMed]
2. Danad, I.; Szymonifka, J.; Twisk, J.W.; Nørgaard, B.; Zarins, C.K.; Knaapen, P.; Min, J.K. Diagnostic performance of cardiac imaging methods to diagnose ischaemia-causing coronary artery disease when directly compared with fractional flow reserve as a reference standard: A meta-analysis. *Eur. Heart J.* **2016**, *38*, 991–998 [CrossRef] [PubMed]
3. Esposito, A.; Gallone, G.; Palmisano, A.; Marchitelli, L.; Catapano, F.; Francone, M. The current landscape of imaging recommendations in cardiovascular clinical guidelines: Toward an imaging-guided precision medicine. *Radiol. Med.* **2020**, *125*, 1013–1023. [CrossRef] [PubMed]

4. Ciancarella, P.; Ciliberti, P.; Santangelo, T.P.; Secchi, F.; Stagnaro, N.; Secinaro, A. Noninvasive imaging of congenital cardiovascular defects. *Radiol. Med.* 2020, *125*, 1167–1185. [CrossRef] [PubMed]
5. La Grutta, L.; Toia, P.; Grassedonio, E.; Pasta, S.; Albano, D.; Agnello, F.; Maffei, E.; Cademartiri, F.; Bartolotta, T.V.; Galia, M.; et al. TAVI imaging: Over the echocardiography. *Radiol. Med.* 2020, *125*, 1148–1166. [CrossRef]
6. Takehara, Y. 4D Flow when and how? *Radiol. Med.* 2020, *125*, 838–850. [CrossRef]
7. Palmisano, A.; Darvizeh, F.; Cundari, G.; Rovere, G.; Ferrandino, G.; Nicoletti, V.; Cilia, F.; De Vizio, S.; Palumbo, R.; Esposito, A.; et al. Advanced cardiac imaging in athlete's heart: Unravelling the grey zone between physiologic adaptation and pathology. *Radiol. Med.* 2021, *126*, 1518–1531. [CrossRef]
8. Schicchi, N.; Fogante, M.; Palumbo, P.; Agliata, G.; Pirani, P.E.; Di Cesare, E.; Giovagnoni, A. The sub-millisievert era in CTCA: The technical basis of the new radiation dose approach. *Radiol. Med.* 2020, *125*, 1024–1039. [CrossRef]
9. Ledda, R.E.; Milanese, G.; Cademartiri, F.; Maffei, E.; Benedetti, G.; Goldoni, M.; Silva, M.; Sverzellati, N. Association of hepatic steatosis with epicardial fat volume and coronary artery disease in symptomatic patients. *Radiol. Med.* **2021**, *126*, 652–660. [CrossRef]
10. Hoffmann, U.; Truong, Q.A.; Schoenfeld, D.A.; Chou, E.T.; Woodard, P.K.; Nagurney, J.T.; Pope, J.H.; Hauser, T.H.; White, C.S.; Weiner, S.; et al. Coronary CT Angiography versus Standard Evaluation in Acute Chest Pain. *N. Engl. J. Med.* **2012**, *367*, 299–308. [CrossRef]
11. Douglas, P.S.; Hoffmann, U.; Patel, M.R.; Mark, D.B.; Al-Khalidi, H.R.; Cavanaugh, B.; Cole, J.; Dolor, R.; Fordyce, C.B.; Huang, M.; et al. Outcomes of Anatomical versus Functional Testing for Coronary Artery Disease. *N. Engl. J. Med.* **2015**, *372*, 1291–1300. [CrossRef] [PubMed]
12. Hoffmann, U.; Ferencik, M.; Udelson, J.E.; Picard, M.H.; Truong, Q.A.; Patel, M.R.; Huang, M.; Pencina, M.; Mark, D.B.; Heitner, J.F.; et al. Prognostic Value of Noninvasive Cardiovascular Testing in Patients With Stable Chest Pain. *Circulation* **2017**, *135*, 2320–2332. [CrossRef] [PubMed]
13. Newby, D.E.; Williams, M.C.; Dweck, M.R. Forget Ischemia: It's All About the Plaque. *Circulation* **2021**, *144*, 1039–1041. [CrossRef] [PubMed]
14. Reynolds, H.R.; Picard, M.H.; Spertus, J.A.; Peteiro, J.; Sendon, J.L.L.; Senior, R.; El-Hajjar, M.C.; Celutkiene, J.; Shapiro, M.D.; Pellikka, P.A.; et al. Natural History of Patients With Ischemia and No Obstructive Coronary Artery Disease. *Circulation* **2021**, *144*, 1008–1023. [CrossRef]
15. Reynolds, H.R.; Shaw, L.J.; Min, J.K.; Page, C.B.; Berman, D.S.; Chaitman, B.R.; Picard, M.H.; Kwong, R.Y.; O'Brien, S.M.; Huang, Z.; et al. Outcomes in the ISCHEMIA Trial Based on Coronary Artery Disease and Ischemia Severity. *Circulation* **2021**, *144*, 1024–1038. [CrossRef]
16. Pezel, T.; Silva, L.M.; Bau, A.A.; Teixiera, A.; Jerosch-Herold, M.; Coelho-Filho, O.R. What Is the Clinical Impact of Stress CMR After the ISCHEMIA Trial? *Front. Cardiovasc. Med.* 2021, *8*, 683434. [CrossRef]
17. Velazquez, E.J.; Lee, K.L.; Jones, R.H.; Al-Khalidi, H.R.; Hill, J.A.; Panza, J.A.; Michler, R.E.; Bonow, R.O.; Doenst, T.; Petrie, M.C.; et al. Coronary-Artery Bypass Surgery in Patients with Ischemic Cardiomyopathy. *N. Engl. J. Med.* **2016**, *374*, 1511–1520. [CrossRef]
18. Ge, Y.; Antiochos, P.; Steel, K.; Bingham, S.; Abdullah, S.; Chen, Y.-Y.; Mikolich, J.R.; Arai, A.E.; Bandettini, W.P.; Shanbhag, S.M.; et al. Prognostic Value of Stress CMR Perfusion Imaging in Patients With Reduced Left Ventricular Function. *JACC Cardiovasc. Imaging* 2020, *13*, 2132–2145. [CrossRef]
19. Patel, A.R.; Salerno, M.; Kwong, R.Y.; Singh, A.; Heydari, B.; Kramer, C.M. Stress Cardiac Magnetic Resonance Myocardial Perfusion Imaging. *J. Am. Coll. Cardiol.* 2021, *78*, 1655–1668. [CrossRef]
20. Pezel, T.; Garot, P.; Hovasse, T.; Unterseeh, T.; Champagne, S.; Kinnel, M.; Toupin, S.; Louvard, Y.; Morice, M.C.; Sanguineti, F.; et al. Vasodilatation stress cardiovascular magnetic resonance imaging: Feasibility, workflow and safety in a large prospective registry of more than 35,000 patients. *Arch. Cardiovasc. Dis.* 2021, *114*, 490–503. [CrossRef]
21. Pezel, T.; Unterseeh, T.; Garot, P.; Hovasse, T.; Kinnel, M.; Champagne, S.; Toupin, S.; Sanguineti, F.; Garot, J. Prognostic value of vasodilator stress perfusion cardiovascular magnetic resonance after inconclusive stress testing. *J. Cardiovasc. Magn. Reson.* **2021**, *23*, 89. [CrossRef] [PubMed]
22. Pezel, T.; Unterseeh, T.; Garot, P.; Hovasse, T.; Sanguineti, F.; Toupin, S.; Morisset, S.; Champagne, S.; Garot, J. Long-Term Prognostic Value of Stress Cardiovascular Magnetic Resonance–Related Coronary Revascularization to Predict Death: A Large Registry With >200,000 Patient-Years of Follow-Up. *Circ. Cardiovasc. Imaging* **2021**, *14*, e012789. [CrossRef] [PubMed]
23. Pavon, A.G.; Porretta, A.P.; Arangalage, D.; Domenichini, G.; Rutz, T.; Hugelshofer, S.; Pruvot, E.; Monney, P.; Pascale, P.; Schwitter, J. Feasibility of adenosine stress cardiovascular magnetic resonance perfusion imaging in patients with MR-conditional transvenous permanent pacemakers and defibrillators. *J. Cardiovasc. Magn. Reson.* **2022**, *24*, 1–11. [CrossRef]
24. Lipinski, M.J.; McVey, C.M.; Berger, J.; Kramer, C.M.; Salerno, M. Prognostic Value of Stress Cardiac Magnetic Resonance Imaging in Patients With Known or Suspected Coronary Artery Disease: A Systematic Review and Meta-Analysis. *J. Am. Coll. Cardiol.* **2013**, *62*, 826–838. [CrossRef] [PubMed]
25. Centonze, M.; Steidler, S.; Casagranda, G.; Alfonsi, U.; Spagnolli, F.; Rozzanigo, U.; Palumbo, D.; Faletti, R.; De Cobelli, F. Cardiac-CT and cardiac-MR cost-effectiveness: A literature review. *Radiol. Med.* 2020, *125*, 1200–1207. [CrossRef] [PubMed]
26. Buffa, V.; Di Renzi, P. CMR in the diagnosis of ischemic heart disease. *Radiol. Med.* 2020, *125*, 1114–1123. [CrossRef]

27. Desai, R.R.; Jha, S. Diagnostic Performance of Cardiac Stress Perfusion MRI in the Detection of Coronary Artery Disease Using Fractional Flow Reserve as the Reference Standard: A Meta-Analysis. *Am. J. Roentgenol.* **2013**, *201*, W245–W252. [CrossRef]
28. Leiner, T.; Bogaert, J.; Friedrich, M.G.; Mohiaddin, R.; Muthurangu, V.; Myerson, S.; Powell, A.J.; Raman, S.V.; Pennell, D.J. SCMR Position Paper (2020) on clinical indications for cardiovascular magnetic resonance. *J. Cardiovasc. Magn. Reson.* **2020**, *22*, 1–37. [CrossRef]
29. Klem, I.; Klein, M.; Khan, M.; Yang, E.Y.; Nabi, F.; Ivanov, A.; Bhatti, L.; Hayes, B.; Graviss, E.A.; Nguyen, D.T.; et al. Relationship of LVEF and Myocardial Scar to Long-Term Mortality Risk and Mode of Death in Patients With Nonischemic Cardiomyopathy. *Circulation* **2021**, *143*, 1343–1358. [CrossRef]
30. Hachamovitch, R. Impact of ischemia and scar on therapeutic benefit of myocardial revascularization. *Herz* **2013**, *38*, 344–349. [CrossRef]
31. Kwon, D.H.; Obuchowski, N.A.; Marwick, T.H.; Menon, V.; Griffin, B.; Flamm, S.D.; Hachamovitch, R. Jeopardized Myocardium Defined by Late Gadolinium Enhancement Magnetic Resonance Imaging Predicts Survival in Patients With Ischemic Cardiomyopathy: Impact of Revascularization. *J. Am. Heart Assoc.* **2018**, *7*, e009394. [CrossRef] [PubMed]
32. Craft, J.; Li, Y.; Bhatti, S.; Cao, J.J. How to do left atrial late gadolinium enhancement: A review. *Radiol. Med.* **2021**, *126*, 1159–1169. [CrossRef] [PubMed]
33. Palmisano, A.; Vignale, D.; Benedetti, G.; Del Maschio, A.; De Cobelli, F.; Esposito, A. Late iodine enhancement cardiac computed tomography for detection of myocardial scars: Impact of experience in the clinical practice. *Radiol. Med.* **2020**, *125*, 128–136. [CrossRef] [PubMed]
34. Kip, K.E.; Hollabaugh, K.; Marroquin, O.C.; Williams, D.O. The Problem With Composite End Points in Cardiovascular Studies: The Story of Major Adverse Cardiac Events and Percutaneous Coronary Intervention. *J. Am. Coll. Cardiol.* **2008**, *51*, 701–707. [CrossRef]
35. Boden, W.E.; O'Rourke, R.A.; Teo, K.K.; Hartigan, P.M.; Maron, D.J.; Kostuk, W.J.; Knudtson, M.; Dada, M.; Casperson, P.; Harris, C.L.; et al. Optimal Medical Therapy with or without PCI for Stable Coronary Disease. *N. Engl. J. Med.* **2007**, *356*, 1503–1516. [CrossRef]
36. BARI 2D Study Group; Frye, R.L.; August, P.; Brooks, M.M.; Hardison, R.M.; Kelsey, S.F.; MacGregor, J.M.; Orchard, T.J.; Chaitman, B.R.; Genuth, S.M.; et al. A Randomized Trial of Therapies for Type 2 Diabetes and Coronary Artery Disease. *N. Engl. J. Med.* **2009**, *360*, 2503–2515. [CrossRef]
37. Xie, J.X.; Winchester, D.E.; Phillips, L.M.; Hachamovitch, R.; Berman, D.S.; Blankstein, R.; Di Carli, M.F.; Miller, T.D.; Al-Mallah, M.H.; Shaw, L.J. The elusive role of myocardial perfusion imaging in stable ischemic heart disease: Is ISCHEMIA the answer? *J. Nucl. Cardiol.* **2017**, *24*, 1610–1618. [CrossRef]
38. Mani, P.; Hachamovitch, R. Can Stress Cardiac Magnetic Resonance Identify Potential Survival Benefit With Revascularization in Stable Ischemic Heart Disease? *JACC Cardiovasc. Imaging* **2020**, *13*, 1687–1689. [CrossRef]
39. Schwitter, J. The SPINS Trial: Building Evidence and a Consequence? *J. Am. Coll. Cardiol.* **2019**, *74*, 1756–1759. [CrossRef]
40. Hendel, R.C.; Friedrich, M.G.; Schulz-Menger, J.; Zemmrich, C.; Bengel, F.; Berman, D.S.; Camici, P.G.; Flamm, S.D.; Le Guludec, D.; Kim, R.; et al. CMR First-Pass Perfusion for Suspected Inducible Myocardial Ischemia. *JACC Cardiovasc. Imaging* **2016**, *9*, 1338–1348. [CrossRef]
41. Kwong, R.Y.; Ge, Y.; Steel, K.; Bingham, S.; Abdullah, S.; Fujikura, K.; Wang, W.; Pandya, A.; Chen, Y.-Y.; Mikolich, J.R.; et al. Cardiac Magnetic Resonance Stress Perfusion Imaging for Evaluation of Patients With Chest Pain. *J. Am. Coll. Cardiol.* **2019**, *74*, 1741–1755. [CrossRef] [PubMed]
42. Gulati, M.; Levy, P.D.; Mukherjee, D.; Amsterdam, E.; Bhatt, D.L.; Birtcher, K.K.; Blankstein, R.; Boyd, J.; Bullock-Palmer, R.P.; Conejo, T.; et al. 2021 AHA/ACC/ASE/CHEST/SAEM/SCCT/SCMR Guideline for the Evaluation and Diagnosis of Chest Pain: Executive Summary: A Report of the American College of Cardiology/American Heart Association Joint Committee on Clinical Practice Guidelines. *Circulation* **2021**, *144*, 368–454. [CrossRef] [PubMed]
43. Nagel, E.; Greenwood, J.P.; McCann, G.P.; Bettencourt, N.; Shah, A.M.; Hussain, S.T.; Perera, D.; Plein, S.; Bucciarelli-Ducci, C.; Paul, M.; et al. Magnetic Resonance Perfusion or Fractional Flow Reserve in Coronary Disease. *N. Engl. J. Med.* **2019**, *380*, 2418–2428. [CrossRef]
44. Farzaneh-Far, A.; Borges-Neto, S. Ischemic Burden, Treatment Allocation, and Outcomes in Stable Coronary Artery Disease. *Circ. Cardiovasc. Imaging* **2011**, *4*, 746–753. [CrossRef] [PubMed]
45. Hachamovitch, R. Does Ischemia Burden in Stable Coronary Artery Disease Effectively Identify Revascularization Candidates? *Circ. Cardiovasc. Imaging* **2015**, *8*, 8. [CrossRef]
46. Marcos-Garces, V.; Gavara, J.; Monmeneu, J.V.; Lopez-Lereu, M.P.; Bosch, M.J.; Merlos, P.; Perez, N.; Rios-Navarro, C.; De Dios, E.; Bonanad, C.; et al. Vasodilator Stress CMR and All-Cause Mortality in Stable Ischemic Heart Disease. *JACC Cardiovasc. Imaging* **2020**, *13*, 1674–1686. [CrossRef] [PubMed]
47. Ge, Y.; Steel, K.; Antiochos, P.; Bingham, S.; Abdullah, S.; Mikolich, J.R.; Arai, A.E.; Bandettini, W.P.; Shanbhag, S.M.; Patel, A.R.; et al. Stress CMR in patients with obesity: Insights from the Stress CMR Perfusion Imaging in the United States (SPINS) registry. *Eur. Heart J. Cardiovasc. Imaging* **2021**, *22*, 518–527. [CrossRef]
48. Galea, N.; Polizzi, G.; Gatti, M.; Cundari, G.; Figuera, M.; Faletti, R. Cardiovascular magnetic resonance (CMR) in restrictive cardiomyopathies. *Radiol. Med.* **2020**, *125*, 1072–1086. [CrossRef]

49. Liguori, C.; Farina, D.; Vaccher, F.; Ferrandino, G.; Bellini, D.; Carbone, I. Myocarditis: Imaging up to date. *Radiol. Med.* **2020**, *125*, 1124–1134. [CrossRef]

50. Palumbo, P.; Cannizzaro, E.; Di Cesare, A.; Bruno, F.; Schicchi, N.; Giovagnoni, A.; Splendiani, A.; Barile, A.; Masciocchi, C.; Di Cesare, E. Cardiac magnetic resonance in arrhythmogenic cardiomyopathies. *Radiol. Med.* **2020**, *125*, 1087–1101. [CrossRef]

51. Palumbo, P.; Masedu, F.; De Cataldo, C.; Cannizzaro, E.; Bruno, F.; Pradella, S.; Arrigoni, F.; Valenti, M.; Splendiani, A.; Barile, A.; et al. Real-world clinical validity of cardiac magnetic resonance tissue tracking in primitive hypertrophic cardiomyopathy. *Radiol. Med.* **2021**, *126*, 1532–1543. [CrossRef] [PubMed]

52. Pierpaolo, P.; Rolf, S.; Manuel, B.-P.; Davide, C.; Dresselaers, T.; Claus, P.; Bogaert, J. Left ventricular global myocardial strain assessment: Are CMR feature-tracking algorithms useful in the clinical setting? *Radiol. Med.* **2020**, *125*, 444–450. [CrossRef] [PubMed]

53. Pradella, S.; Grazzini, G.; De Amicis, C.; Letteriello, M.; Acquafresca, M.; Miele, V. Cardiac magnetic resonance in hypertrophic and dilated cardiomyopathies. *Radiol. Med.* **2020**, *125*, 1056–1071. [CrossRef] [PubMed]

54. Russo, V.; Lovato, L.; Ligabue, G. Cardiac MRI: Technical basis. *Radiol. Med.* **2020**, *125*, 1040–1055. [CrossRef]

55. Hadamitzky, M.; Freissmuth, B.; Meyer, T.; Hein, F.; Kastrati, A.; Martinoff, S.; Schömig, A.; Hausleiter, J. Prognostic Value of Coronary Computed Tomographic Angiography for Prediction of Cardiac Events in Patients With Suspected Coronary Artery Disease. *JACC Cardiovasc. Imaging* **2009**, *2*, 404–411. [CrossRef] [PubMed]

56. Hadamitzky, M.; Distler, R.; Meyer, T.; Hein, F.; Kastrati, A.; Martinoff, S.; Schömig, A.; Hausleiter, J. Prognostic Value of Coronary Computed Tomographic Angiography in Comparison With Calcium Scoring and Clinical Risk Scores. *Circ. Cardiovasc. Imaging* **2011**, *4*, 16–23. [CrossRef]

57. Andreini, D.; Pontone, G.; Mushtaq, S.; Bartorelli, A.L.; Bertella, E.; Antonioli, L.; Formenti, A.; Cortinovis, S.; Veglia, F.; Annoni, A.; et al. A Long-Term Prognostic Value of Coronary CT Angiography in Suspected Coronary Artery Disease. *JACC Cardiovasc. Imaging* **2012**, *5*, 690–701. [CrossRef]

58. Motoyama, S.; Ito, H.; Sarai, M.; Kondo, T.; Kawai, H.; Nagahara, Y.; Harigaya, H.; Kan, S.; Anno, H.; Takahashi, H.; et al. Plaque Characterization by Coronary Computed Tomography Angiography and the Likelihood of Acute Coronary Events in Mid-Term Follow-Up. *J. Am. Coll. Cardiol.* **2015**, *66*, 337–346. [CrossRef]

59. Nadjiri, J.; Hausleiter, J.; Jähnichen, C.; Will, A.; Hendrich, E.; Martinoff, S.; Hadamitzky, M. Incremental prognostic value of quantitative plaque assessment in coronary CT angiography during 5 years of follow up. *J. Cardiovasc. Comput. Tomogr.* **2016**, *10*, 97–104. [CrossRef]

60. Marano, R.; Rovere, G.; Savino, G.; Flammia, F.C.; Carafa, M.R.P.; Steri, L.; Merlino, B.; Natale, L. CCTA in the diagnosis of coronary artery disease. *Radiol. Med.* **2020**, *125*, 1102–1113. [CrossRef]

61. Valente, T.; Pignatiello, M.; Sica, G.; Bocchini, G.; Rea, G.; Cappabianca, S.; Scaglione, M. Hemopericardium in the acute clinical setting: Are we ready for a tailored management approach on the basis of MDCT findings? *Radiol. Med.* **2020**, *126*, 527–543. [CrossRef] [PubMed]

62. Şeker, M. Prevalence and morphologic features of dual left anterior descending artery subtypes in coronary CT angiography. *Radiol. Med.* **2019**, *125*, 247–256. [CrossRef] [PubMed]

63. Rovere, G.; Meduri, A.; Savino, G.; Flammia, F.C.; Piccolo, F.L.; Carafa, M.R.P.; Larici, A.R.; Natale, L.; Merlino, B.; Marano, R. Practical instructions for using drugs in CT and MR cardiac imaging. *Radiol. Med.* **2021**, *126*, 356–364. [CrossRef] [PubMed]

64. Palumbo, P.; Cannizzaro, E.; Bruno, F.; Schicchi, N.; Fogante, M.; Agostini, A.; De Donato, M.C.; De Cataldo, C.; Giovagnoni, A.; Barile, A.; et al. Coronary artery disease (CAD) extension-derived risk stratification for asymptomatic diabetic patients: Usefulness of low-dose coronary computed tomography angiography (CCTA) in detecting high-risk profile patients. *Radiol. Med.* **2020**, *125*, 1249–1259. [CrossRef]

65. Esposito, A.; Francone, M.; Andreini, D.; Buffa, V.; Cademartiri, F.; Carbone, I.; Clemente, A.; Guaricci, A.I.; Guglielmo, M.; Indolfi, C.; et al. SIRM—SIC appropriateness criteria for the use of Cardiac Computed Tomography. Part 1: Congenital heart diseases, primary prevention, risk assessment before surgery, suspected CAD in symptomatic patients, plaque and epicardial adipose tissue characterization, and functional assessment of stenosis. *Radiol. Med.* **2021**, *126*, 1236–1248. [CrossRef]

66. De Rubeis, G.; Marchitelli, L.; Spano, G.; Catapano, F.; Cilia, F.; Galea, N.; Carbone, I.; Catalano, C.; Francone, M. Radiological outpatient' visits to avoid inappropriate cardiac CT examinations: An 8-year experience report. *Radiol. Med.* **2021**, *126*, 214–220. [CrossRef]

67. Pontone, G.; Di Cesare, E.; Castelletti, S.; De Cobelli, F.; De Lazzari, M.; Esposito, A.; Focardi, M.; Di Renzi, P.; Indolfi, C.; Lanzillo, C.; et al. Appropriate use criteria for cardiovascular magnetic resonance imaging (CMR): SIC—SIRM position paper part 1 (ischemic and congenital heart diseases, cardio-oncology, cardiac masses and heart transplant). *Radiol. Med.* **2021**, *126*, 365–379. [CrossRef]

68. Motoyama, S.; Sarai, M.; Narula, J.; Ozaki, Y. Coronary CT angiography and high-risk plaque morphology. *Cardiovasc. Interv. Ther.* **2013**, *28*, 1–8. [CrossRef]

69. Pontone, G.; Andreini, D.; Bartorelli, A.L.; Bertella, E.; Cortinovis, S.; Mushtaq, S.; Foti, C.; Annoni, A.; Formenti, A.; Baggiano, A.; et al. A Long-Term Prognostic Value of CT Angiography and Exercise ECG in Patients with Suspected CAD. *JACC Cardiovasc. Imaging* **2013**, *6*, 641–650. [CrossRef]

70. Seitun, S.; Clemente, A.; Maffei, E.; Toia, P.; La Grutta, L.; Cademartiri, F. Prognostic value of cardiac CT. *Radiol. Med.* **2020**, *125*, 1135–1147. [CrossRef]

71. Nakanishi, R.; Osawa, K.; Kurata, A.; Miyoshi, T. Role of coronary computed tomography angiography (CTA) post the ISCHEMIA trial: Precision prevention based on coronary CTA-derived coronary atherosclerosis. *J. Cardiol.* **2021**. [CrossRef] [PubMed]

72. Van Rosendael, A.R.; Bax, A.M.; van den Hoogen, I.J.; Smit, J.M.; Al'Aref, S.J.; Achenbach, S.; Al-Mallah, M.H.; Andreini, D.; Berman, D.S.; Budoff, M.J.; et al. Associations between dyspnoea, coronary atherosclerosis, and cardiovascular outcomes: Results from the long-term follow-up CONFIRM registry. *Eur. Heart J. Cardiovasc. Imaging* **2020**, *23*, 266–274. [CrossRef] [PubMed]

73. Schicchi, N.; Mari, A.; Fogante, M.; Pirani, P.E.; Agliata, G.; Tosi, N.; Palumbo, P.; Cannizzaro, E.; Bruno, F.; Splendiani, A.; et al. In vivo radiation dosimetry and image quality of turbo-flash and retrospective dual-source CT coronary angiography. *Radiol. Med.* **2020**, *125*, 117–127. [CrossRef] [PubMed]

74. Van Assen, M.; Muscogiuri, G.; Caruso, D.; Lee, S.J.; Laghi, A.; De Cecco, C.N. Artificial intelligence in cardiac radiology. *Radiol. Med.* **2020**, *125*, 1186–1199. [CrossRef]

75. Scapicchio, C.; Gabelloni, M.; Barucci, A.; Cioni, D.; Saba, L.; Neri, E. A deep look into radiomics. *Radiol. Med.* **2021**, *126*, 1296–1311. [CrossRef]

76. Nardone, V.; Reginelli, A.; Grassi, R.; Boldrini, L.; Vacca, G.; D'Ippolito, E.; Annunziata, S.; Farchione, A.; Belfiore, M.P.; Desideri, I.; et al. Delta radiomics: A systematic review. *Radiol. Med.* **2021**, *126*, 1571–1583. [CrossRef]

77. Coppola, F.; Faggioni, L.; Regge, D.; Giovagnoni, A.; Golfieri, R.; Bibbolino, C.; Miele, V.; Neri, E.; Grassi, R. Artificial intelligence: Radiologists' expectations and opinions gleaned from a nationwide online survey. *Radiol. Med.* **2021**, *126*, 63–71. [CrossRef]

78. Cicero, G.; Ascenti, G.; Albrecht, M.H.; Blandino, A.; Cavallaro, M.; D'Angelo, T.; Carerj, M.L.; Vogl, T.J.; Mazziotti, S. Extra-abdominal dual-energy CT applications: A comprehensive overview. *Radiol. Med.* **2020**, *125*, 384–397. [CrossRef]

79. Tonino, P.A.L.; De Bruyne, B.; Pijls, N.H.J.; Siebert, U.; Ikeno, F.; van't Veer, M.; Klauss, V.; Manoharan, G.; Engstrøm, T.; Oldroyd, K.G.; et al. Fractional flow reserve versus angiography for guiding percutaneous coronary intervention. *N. Engl. J. Med.* **2009**, *360*, 213–224. [CrossRef]

80. Pijls, N.H.; Fearon, W.F.; Tonino, P.A.; Siebert, U.; Ikeno, F.; Bornschein, B.; Veer, M.V.; Klauss, V.; Manoharan, G.; Engstrøm, T.; et al. Fractional Flow Reserve Versus Angiography for Guiding Percutaneous Coronary Intervention in Patients With Multivessel Coronary Artery Disease: 2-Year Follow-Up of the FAME (Fractional Flow Reserve Versus Angiography for Multivessel Evaluation) Study. *J. Am. Coll. Cardiol.* **2010**, *56*, 177–184. [CrossRef] [PubMed]

81. De Bruyne, B.; Pijls, N.H.; Kalesan, B.; Barbato, E.; Tonino, P.A.; Piroth, Z.; Jagic, N.; Mobius-Winckler, S.; Rioufol, G.; Witt, N.; et al. Fractional Flow Reserve–Guided PCI versus Medical Therapy in Stable Coronary Disease. *N. Engl. J. Med.* **2012**, *367*, 991–1001. [CrossRef] [PubMed]

82. Patel, M.R.; Jeremias, A.; Maehara, A.; Matsumura, M.; Zhang, Z.; Schneider, J.; Tang, K.; Talwar, S.; Marques, K.; Shammas, N.W.; et al. 1-Year Outcomes of Blinded Physiological Assessment of Residual Ischemia After Successful PCI. *JACC Cardiovasc. Interv.* **2022**, *15*, 52–61. [CrossRef] [PubMed]

83. Zhang, D.; Lv, S.; Song, X.; Yuan, F.; Xu, F.; Zhang, M.; Yan, S.; Cao, X. Fractional flow reserve versus angiography for guiding percutaneous coronary intervention: A meta-analysis. *Heart* **2015**, *101*, 455–462. [CrossRef] [PubMed]

84. Xaplanteris, P.; Fournier, S.; Pijls, N.H.; Fearon, W.F.; Barbato, E.; Tonino, P.A.; Engstrøm, T.; Kääb, S.; Dambrink, J.-H.; Rioufol, G.; et al. Five-Year Outcomes with PCI Guided by Fractional Flow Reserve. *N. Engl. J. Med.* **2018**, *379*, 250–259. [CrossRef]

85. Elgendy, I.Y.; Mahtta, D.; Pepine, C.J. Medical Therapy for Heart Failure Caused by Ischemic Heart Disease. *Circ. Res.* **2019**, *124*, 1520–1535. [CrossRef]

86. Di Cesare, E.; Carerj, S.; Palmisano, A.; Carerj, M.L.; Catapano, F.; Vignale, D.; Di Cesare, A.; Milanese, G.; Sverzellati, N.; Francone, M.; et al. Multimodality imaging in chronic heart failure. *Radiol. Med.* **2021**, *126*, 231–242. [CrossRef]

87. Masi, S.; Rizzoni, D.; Taddei, S.; Widmer, R.J.; Montezano, A.C.; Lüscher, T.F.; Schiffrin, E.L.; Touyz, R.M.; Paneni, F.; Lerman, A.; et al. Assessment and pathophysiology of microvascular disease: Recent progress and clinical implications. *Eur. Heart J.* **2021**, *42*, 2590–2604. [CrossRef]

88. Crea, F.; Camici, P.G.; Merz, C.N.B. Coronary microvascular dysfunction: An update. *Eur. Heart J.* **2014**, *35*, 1101–1111. [CrossRef]

89. Taqueti, V.R.; Solomon, S.D.; Shah, A.M.; Desai, A.S.; Groarke, J.D.; Osborne, M.; Hainer, J.; Bibbo, C.F.; Dorbala, S.; Blankstein, R.; et al. Coronary microvascular dysfunction and future risk of heart failure with preserved ejection fraction. *Eur. Heart J.* **2017**, *39*, 840–849. [CrossRef]

90. Sechtem, U.; Brown, D.L.; Godo, S.; Lanza, G.A.; Shimokawa, H.; Sidik, N. Coronary microvascular dysfunction in stable ischaemic heart disease (non-obstructive coronary artery disease and obstructive coronary artery disease). *Cardiovasc. Res.* **2020**, *116*, 771–786. [CrossRef]

91. Padro, T.; Manfrini, O.; Bugiardini, R.; Canty, J.; Cenko, E.; De Luca, G.; Duncker, D.J.; Eringa, E.C.; Koller, A.; Tousoulis, D.; et al. ESC Working Group on Coronary Pathophysiology and Microcirculation position paper on 'coronary microvascular dysfunction in cardiovascular disease'. *Cardiovasc. Res.* **2020**, *116*, 741–755. [CrossRef] [PubMed]

92. Corcoran, D.; Young, R.; Adlam, D.; McConnachie, A.; Mangion, K.; Ripley, D.; Cairns, D.; Brown, J.; Bucciarelli-Ducci, C.; Baumbach, A.; et al. Coronary microvascular dysfunction in patients with stable coronary artery disease: The CE-MARC 2 coronary physiology sub-study. *Int. J. Cardiol.* **2018**, *266*, 7–14. [CrossRef] [PubMed]

93. Ford, T.; Stanley, B.; Good, R.; Rocchiccioli, P.; McEntegart, M.; Watkins, S.; Eteiba, H.; Shaukat, A.; Lindsay, M.; Robertson, K.; et al. Stratified Medical Therapy Using Invasive Coronary Function Testing in Angina. *J. Am. Coll. Cardiol.* **2018**, *72*, 2841–2855. [CrossRef] [PubMed]

94. Ford, T.; Berry, C. How to Diagnose and Manage Angina Without Obstructive Coronary Artery Disease: Lessons from the British Heart Foundation CorMicA Trial. *Interv. Cardiol. Rev. Res. Resour.* **2019**, *14*, 76–82. [CrossRef] [PubMed]
95. Ford, T.; Ong, P.; Sechtem, U.; Beltrame, J.; Camici, P.G.; Crea, F.; Kaski, J.-C.; Merz, C.N.B.; Pepine, C.J.; Shimokawa, H.; et al. Assessment of Vascular Dysfunction in Patients Without Obstructive Coronary Artery Disease. *JACC Cardiovasc. Interv.* **2020**, *13*, 1847–1864. [CrossRef]
96. Ford, T.J.; Stanley, B.; Sidik, N.; Good, R.; Rocchiccioli, P.; McEntegart, M.; Watkins, S.; Eteiba, H.; Shaukat, A.; Lindsay, M.; et al. 1-Year Outcomes of Angina Management Guided by Invasive Coronary Function Testing (CorMicA). *JACC Cardiovasc. Interv.* **2020**, *13*, 33–45. [CrossRef]
97. Ford, T.J.; Corcoran, D.; Berry, C. Stable coronary syndromes: Pathophysiology, diagnostic advances and therapeutic need. *Heart* **2018**, *104*, 284–292. [CrossRef]
98. Ford, T.J.; Corcoran, D.; Oldroyd, K.G.; McEntegart, M.; Rocchiccioli, P.; Watkins, S.; Brooksbank, K.; Padmanabhan, S.; Sattar, N.; Briggs, A.; et al. Rationale and design of the British Heart Foundation (BHF) Coronary Microvascular Angina (CorMicA) stratified medicine clinical trial. *Am. Heart J.* **2018**, *201*, 86–94. [CrossRef]
99. Geyer, H.; Caracciolo, G.; Abe, H.; Wilansky, S.; Carerj, S.; Gentile, F.; Nesser, H.-J.; Khandheria, B.; Narula, J.; Sengupta, P.P. Assessment of Myocardial Mechanics Using Speckle Tracking Echocardiography: Fundamentals and Clinical Applications. *J. Am. Soc. Echocardiogr.* **2010**, *23*, 351–369. [CrossRef]
100. Voigt, J.-U.; Cvijic, M. 2- and 3-Dimensional Myocardial Strain in Cardiac Health and Disease. *JACC Cardiovasc. Imaging* **2019**, *12*, 1849–1863. [CrossRef]
101. Holmes, A.A.; Romero, J.; Levsky, J.M.; Haramati, L.B.; Phuong, N.; Rezai-Gharai, L.; Cohen, S.; Restrepo, L.; Ruiz-Guerrero, L.; Fisher, J.D.; et al. Circumferential strain acquired by CMR early after acute myocardial infarction adds incremental predictive value to late gadolinium enhancement imaging to predict late myocardial remodeling and subsequent risk of sudden cardiac death. *J. Interv. Card. Electrophysiol.* **2017**, *50*, 211–218. [CrossRef] [PubMed]

Review

Cardiac Magnetic Resonance Imaging in Appraising Myocardial Strain and Biomechanics: A Current Overview

Alexandru Zlibut [1,2,*], Cosmin Cojocaru [2,3], Sebastian Onciul [2,3] and Lucia Agoston-Coldea [1,4]

[1] Department of Internal Medicine, Iuliu Hatieganu University of Medicine and Pharmacy, 400347 Cluj-Napoca, Romania
[2] Cardiology Department, Emergency Clinical Hospital of Bucharest, 014461 Bucharest, Romania
[3] Faculty of Medicine, "Carol Davila" University of Medicine and Pharmacy, 050474 Bucharest, Romania
[4] Department of Internal Medicine, Cluj County Emergency Hospital, 400347 Cluj-Napoca, Romania
* Correspondence: alex.zlibut@yahoo.com

Abstract: Subclinical alterations in myocardial structure and function occur early during the natural disease course. In contrast, clinically overt signs and symptoms occur during late phases, being associated with worse outcomes. Identification of such subclinical changes is critical for timely diagnosis and accurate management. Hence, implementing cost-effective imaging techniques with accuracy and reproducibility may improve long-term prognosis. A growing body of evidence supports using cardiac magnetic resonance (CMR) to quantify deformation parameters. Tissue-tagging (TT-CMR) and feature-tracking CMR (FT-CMR) can measure longitudinal, circumferential, and radial strains and recent research emphasize their diagnostic and prognostic roles in ischemic heart disease and primary myocardial illnesses. Additionally, these methods can accurately determine LV wringing and functional dynamic geometry parameters, such as LV torsion, twist/untwist, LV sphericity index, and long-axis strain, and several studies have proved their utility in prognostic prediction in various cardiovascular patients. More recently, few yet important studies have suggested the superiority of fast strain-encoded imaging CMR-derived myocardial strain in terms of accuracy and significantly reduced acquisition time, however, more studies need to be carried out to establish its clinical impact. Herein, the current review aims to provide an overview of currently available data regarding the role of CMR in evaluating myocardial strain and biomechanics.

Keywords: left ventricle active biomechanics; cardiac magnetic resonance imaging; left ventricle torsion; left ventricle twist and untwist; left ventricle strain

Citation: Zlibut, A.; Cojocaru, C.; Onciul, S.; Agoston-Coldea, L. Cardiac Magnetic Resonance Imaging in Appraising Myocardial Strain and Biomechanics: A Current Overview. *Diagnostics* **2023**, *13*, 553. https://doi.org/10.3390/diagnostics13030553

Academic Editors: Minjie Lu and Arlene Sirajuddin

Received: 23 January 2023
Revised: 30 January 2023
Accepted: 31 January 2023
Published: 2 February 2023

1. Background

Myocardial strain and biomechanics are the results of intrinsic normal functioning of the heart, expressing the dynamic interdependency between cardiac structure and its physiology. Usually, in the early stages, heart diseases are clinically silent, often resulting in a delayed diagnosis and poor prognosis. Recent technological advances have developed cardiovascular imaging modalities which are able to thoroughly characterize myocardial tissue and function. Nevertheless, studies evaluating their clinical utility in the diagnosis and prognosis of cardiovascular patients are still sparse. Their close description could provide valuable insights into myocardial functional performance [1]. Briefly, by non-invasively assessing myocardial deformation, one can provide supplementary information regarding disease diagnosis, risk stratification, and prognosis [2].

Cardiac magnetic resonance imaging (CMR) is the gold-standard imaging method used for characterizing heart function and tissue structure, thus providing important information about cardiomyocytes, interstitium, microvasculature, and metabolic abnormalities [3]. Furthermore, MRI has been shown to be useful in post-mortem morphological studies for the study of sudden cardiac death [4]. Increasing evidence is now proving the clinical utility of various CMR methods to determine left ventricle (LV) myocardial strain, torsion, twist, untwist, sphericity index, and long-axis strain determining myocardial strain and biomechanics using various methods [5–8]. Moreover, several studies have shown good agreement between CMR-based strain and speckle-tracking echocardiography (STE) [6], even though the two methods are still not interchangeable.

LV myocardial contraction and relaxation are complex phenomena that involve various yet synergistic contractions of all three myocardial layers, thus ensuring hemodynamic stability and optimal LV systolic function. At a glance, there are two ways in which myocardial strain can be outlined: the Lagrangian strain, which provides contractility changes using the own myocardium as a benchmark, and the Eulerian strain, which assesses changes of specific tissue zones with fixed baselines, while the material points differ over time [6]. The disposition of myocardial fibers in the LV is as follows: (1). longitudinal; (2). transversal, with a central distribution and systolic thickening; (3). circumferential, with a circular distribution as viewed from the transversal view of the myocardial [9,10]. The main purpose of these different orientations is to ensure efficient cardiac revolution and maintain global hemodynamics within its normal ranges. In the subendocardial layer, the fibers are longitudinally oriented, from the LV's base to its apex, while those from the subepicardial layer are inversely directed, from the LV's apex to its base, and, additionally, the fibers within the middle layer of the myocardium are circumferentially disposed of. All these particularities form a complex multi-layered helical layout, thus guaranteeing adequate longitudinal and circumferential myocardial strain and optimal LV wall shear stress [11].

The main advantage of CMR-determined myocardial strain is in patients with a poor acoustic window in which STE-based ones are not determinable. Additionally, the intra- and inter-observer biases are significantly reduced [6]. Another benefit could be in patients with arrhythmias for whom CMR might provide useful data, especially if single-heartbeat acquisition techniques are used [7]. Nonetheless, there are several disadvantages, such as prolonged evaluation time and extended dorsal decubitus, which in patients with heart failure is often not possible [6].

Until now, invasive heart catheterization has been considered the gold-standard method to evaluate cardiac function by using pressure-volume loops, which are valuable markers of myocardial contractility and stroke work, especially by determining cardiac output, end-systolic (ESPVR) and end-diastolic (EDPVR) pressure-volume relationships, or cardiac elastance [12]. Recently, CMR with or without inferior vena cava (IVC) temporary closure has been able to determine ESPVR and EDPVR-derived measurements with comparable accuracy [12,13].

The purpose of this review is to provide an overview of currently available data regarding the clinical role of CMR in evaluating myocardial strain and biomechanics.

2. Basics of Myocardial Deformation and Biomechanics

The concept of continuum mechanics in a completely isolated media, along with the properties of cardiac tissue governates the functioning of the cardiovascular system. Thus, myocardial deformation and spatial dynamic geometry are strongly related to these phenomena [14,15]. Moreover, the physical dependency of strain and biomechanics relies on intracellular, extracellular and molecular components of the myocardium. Passive biomechanical properties are ensured by titin, which is an intracellular protein with a high molecular weight that ensures the elastic properties of the myocardial fibers by linking the sarcomeres' Z lines with the M lines and, thus, preventing the overelongation of these fibers. In this sense, mutations in the titin's gene have been significantly associated with LV diastolic dysfunction and heart failure [16,17]. On the other hand, active processes, such as deformation, torsion, twist, untwist, and shear stress, are mainly determined by actin and myosin [18]. Other relevant cellular components which are linked to diseased myocardium and heart failure are collagen, which forms complex reticular structures, elastin with its microfibrils of fibulin, and fibrillin, fibronectin, proteoglycans, and glycosaminoglycans [19]. Nonetheless, usually, the myocardium is a soft, heterogenous, anisotropic tissue, which is subject to significant deformations [20], that is based on equation models from the mechanical physics of continuum mechanics. Moreover, several mathematical models, such as the Cauchy stress tensor, the deformation gradient and its Jacobian determiner, and the strain energy density function, have been used to characterize the relation between the LV wall shear stress and myocardial deformation [12,21].

Moreover, myocardial strain and dynamic spatial geometry rely on myocardial contraction forces, which can be assessed using a specific mathematical model that includes the cardiomyocytes' active tension and calcium ions concentration [22]. LV wall shear stress generates the required forces that strain the cardiomyocytes, and given the situation, they are responsible for the myocardial oxygen mismatch, being easily explained by the law of Laplace [23]. Nevertheless, when it comes to characterizing active biomechanics, there is significant variation depending on the region of interest. When the mid-myocardial layer and the LV's base are considered, it is recommended to apply different equations to assess the longitudinal and circumferential fibers, while for the LV's apex, one can use similar mathematical models for both types of myofibers [12].

The primary task of the LV is to ensure continuous blood flow through the vessels during the cardiac cycle. LV function is majorly conditioned by myocardial contraction, end-diastolic filling pressures, and its dynamic geometry, but also by the integrity and correct functioning of heart valves [24,25]. Initially, invasive catheterization was used to accurately describe the heart's biomechanical physiology, especially by using the curves of ESPVR and EDPVR. Additionally, it has been shown that volume overload increases LV wall shear stress and tension [25]. Furthermore, as postulated in Frank-Starling's law, increased diastolic filling leads to a higher LV stroke volume due to better functioning of the sarcomeres [26]. Moreover, the maximum elastance, which is the slope between the direct relation between the end-systolic blood volume and the aortic pressure, can accurately assess the contractile ability of the LV, and changes within its inotropy will automatically modify the LV stroke volume [27].

Accordingly, using imaging methods, such as STE and, lately, CMR, deformations of all three myocardial fibers can be globally and regionally assessed, resulting in parameters with important diagnostic and prognostic values: global longitudinal (GLS), circumferential (GCS) and radial (GRS) strains. Various studies have confirmed their paramount roles in diagnosis, risk stratification, and prognosis prediction in many cardiovascular diseases [2]. To identify subclinical LV dysfunction and to subdue the main limitations of standard LV systolic function measurements, international guidelines recommend the comprehensive evaluation of LV strain parameters by echocardiography. These parameters are useful in approaching myocardial ischemia and viability, infraclinical dysfunction in patients with dilated cardiomyopathy, hypertrophic cardiomyopathy, arrhythmogenic cardiomyopathy, cardiac amyloidosis, chemotherapy-induced cardiotoxicity, heart failure, valvular heart diseases, and also in improving the selection of patients who might benefit from cardiac resynchronization therapy [28]. Several studies regarding the importance of LV myocardial strain using STE in various cardiovascular diseases are presented in Table 1 [29–45].

Table 1. Speckle-tracking echocardiography studies in evaluating LV GLS.

Authors	Year	Ref	*n*	Illness	Endpoint	GLS	LVEF
Janwanishstaporn et al.	2022	[32]	289	HFimprEF	CVD, HFH	−12.7%	53%
Thellier et al.	2020	[33]	332	AS	ACM	−15%	55%
Goedemans et al.	2018	[34]	143	AMI	ACM, HFH	−14.4%	N/A
Iacoviello et al.	2013	[35]	308	HF	ACM, HFH, CVD, VT	−10.2%	33%
Ersboll et al.	2013	[36]	849	AMI	ACM, CVD, HFH	−14.6%	53.5%
Yingchoncharoen et al.	2012	[37]	79	AS	CVD	−15.2%	63.4%

Table 1. *Cont.*

Authors	Year	Ref	n	Illness	Endpoint	GLS	LVEF
Munk et al.	2012	[38]	576	AMI	ACM, CVD, HFH, AMI	−14.3%	49.2%
Kearney et al.	2012	[39]	146	AS	ACM, AMI, CVD, HFH, VT	−15%	59%
Dahl et al.	2012	[40]	125	HT	ACM, CVD, HFH	−15.5%	34.1%
Buss et al.	2012	[41]	206	AL	ACM, CVD	−13.1%	51.7%
Bertini et al.	2012	[42]	1060	IHD	CVD, HFH	−11.5%	34%
Woo et al.	2011	[43]	98	AMI	CVD, HFH	−15.8%	56%
Nahum et al.	2010	[44]	125	HF	ACM, CVD, HFH	−8%	31%
Antoni et al.	2010	[45]	659	AMI	ACM, AMI, HFH	−15.3%	46%
Stanton et al.	2009	[46]	546	Various	ACM	−16.6%	58%
Cho et al.	2009	[47]	201	HF	CVD, HFH	−10.5%	34.1%
Lancellotti et al.	2008	[48]	163	AS	CVD, HF	−15.7%	66%

Abbreviations: ACM, all-cause mortality; AMI, acute myocardial infarction; AS, aortic stenosis; CVD, cardiovascular death; GLS, global longitudinal strain; HF, heart failure; HFH, heart failure hospitalization; HFimpEF, heart failure with improved ejection fraction; IHD, ischemic heart disease; LVEF, left ventricle ejection fraction; LVEF, left ventricle; N, number of patients; VT, ventricular tachyarrhythmias.

3. CMR Methods for Assessing Myocardial Strain and Biomechanics

Although echocardiography is considered the gold-standard imaging technique in assessing LV strain and strain rates, recently, increasing evidence has shown that some CMR techniques are able to appraise myocardial deformation using either specific acquisition variants or post-processing software [6]. Growing evidence has shown their usefulness in patients with ischemic heart disease, various cardiomyopathies, pulmonary hypertension, and congenital heart disease [46].

From a technical point of view, the first and foremost magnetic resonance system that has allowed a usable approach to assessing myocardial strain, functioning geometry, and active biomechanics is tissue-tagging CMR (TT-CMR), despite its poor spatial resolution [47]. Subsequently, this shortcoming was overcome by complementary spatial modulation of magnetization, which improved the spatial resolution of the myocardium and the grids [5]. Briefly, in the pre-acquisition phase, tags and lines need to be positioned over the myocardium to track myocardial spatial deformation, angulations, torsion, twist and untwist, and, further, specific sequences are recorded during the LV systole [47]. Nonetheless, to provide an objective and clear upshot, post-acquisition analysis software has been created: FINDTAGS, which quantifies the pixels' motion during the cardiac cycle, and a more improved one called harmonic phase (HARP), which is fully automated, being the most used for TT-CMR [48]. Still, the main shortcomings of this method comprise prolonged acquisition time and questionable ability to evaluate thin myocardial layers. Likewise, it has been shown that phase-velocity mapping CMR, the method of choice in approaching trans-valvular flows, could become another option for myocardial deformation and biomechanics assessment. By evaluating the spatial differences between each myocardial pixel, phase-velocity mapping CMR can provide all three deformation parameters, through a single breath-hold, by measuring the dynamic differences between pixels [49]. Moreover, fast cine displacement encoding with stimulated echoes (DENSE) uses balanced standard steady-state free precession (b-SSFP) CMR to encode myocardial tissue displacements with intrinsic phase correction to evaluate myocardial deformation. Nonetheless, its main limitation is that it cannot completely assess during the full cardiac cycle [50].

Furthermore, fast Strain-Encoding (fast-SENC) is a valuable CMR technique that uses myocardial magnetization tags, but the main difference from TT-CMR is that the tags are parallelly overlaid on the myocardium, allowing the evaluation of longitudinal and circumferential strain, while radial deformation remains unfortunately unquantifiable [6]. Fast-SENC has increased accuracy and significantly lowered acquisition time due to single-heartbeat free breathing, thus providing increased spatial resolution, as compared to other CMR methods, having also increased ability in detecting a wide range of cardiovascular illnesses [51].

On the other hand, a post-processing software package that could be applied to standard b-SSFP-CMR would hypothetically be the most convenient option to assess LV strain and biomechanics. FT-CMR is an optical flow magnetic resonance method, being derived from the technique which evaluates the motion of fluids. With proper optimization and adjustments, FT-CMR images are comparable to those obtained using speckle-tracking echocardiography and being applicable to standard cine-CMR, it might become highly usable soon [7], but optimization studies need to be further conducted.

4. Clinical Utility of CMR in Assessing LV Myocardial Strain

4.1. LV Myocardial Strain by CMR in Normal Individuals

Increasing evidence supports the role of CMR in assessing LV myocardial strain in different categories of patients. FT-CMR can determine LV strain measurements in both 2-dimensional and 3-dimensional approaches, with the latter requiring more studies for appropriate validation [7]. It has been reported that the normal values for FT-CMR were $-21.3 \pm 4.8\%$ for GLS, $-26.1 \pm 3.8\%$ for GCS, and $39.8 \pm 8.3\%$ for GRS [52], while the global rather than regional strain parameters, performed better in terms of reproducibility [53,54]. With a view to validate LV strain analysis by CMR, substantiation research using STE has recently been conducted. In a research paper that compared FT-CMR and strain-encoding (SENC)-CMR with STE, GLS, and GCS determined by both CMR methods had good performances in terms of inter-modality agreement [55]. By comparing FT-CMR with fast-SENC in healthy individuals, it was shown that all three strains had lower values in males than in females, with age being a minor but slightly notable determinant for their variation. In addition, fast-SENC reported a significantly higher value for GLS ($-20.3 \pm 1.8\%$) than FT-CMR ($-16.9 \pm 1.8\%$), whereas those of GCS were similar ($-19.2 \pm 2.1\%$ vs. $-19.2 \pm 1.8\%$) [56]. Therefore, normal myocardial strain values determined by CMR vary widely depending on various clinical and technical parameters. In the study conducted by Pierpaolo et al., which compared the agreement between manually traced strain and FT-CMR, they have shown poor agreement between the two methods, especially for GLS and GRS [57]. Other variabilities in terms of normal strain values and CMR techniques are presented in Table 2 [58–61].

Recently, an interesting article sought to assay the capacity and accuracy of fast-SENC to evaluate LV volumes, function, and mass. Almost all the following measurements were precisely determined using fast-SENC, requiring under two minutes of the total study time and being way faster than standard cine-CMR. Nevertheless, LV end-diastolic mass was underestimated by 7% [62]. Furthermore, in another study that aimed to test the accuracy of fast-SENC-based LV myocardial strain, it has been shown that the intra- and inter-observer reproducibility of this CMR method was excellent in terms of LV myocardial functioning assessment [63]. Further studies that could expand the examination in acquiring LV myocardial strain as well might be further conducted, being a very rapid CMR technique.

Table 2. LV myocardial strain variations in normal individuals (miscellaneous).

Authors	Year	*n*	Method	Findings
Mangion et al.	2019	88 healthy individuals	FT-CMR with 3 T MR	GLS different significantly between genders: $-18.48 \pm 3.65\%$ (m) vs. $-21.91 \pm 3.01\%$ (f) GCS did not differ considerably Aging did not influence GLS or GCS

Table 2. *Cont.*

Authors	Year	*n*	Method	Findings
Andre et al.	2015	150 healthy individuals	FT-CMR with 1.5 T MR	All the following varied significantly: GLS endocardial: $-22.2 \pm 3.4\%$ (m) vs. $-24.6 \pm 2.9\%$ (f) GLS myocardial: $-20.4 \pm 3.1\%$ (m) vs. $-22.9 \pm 2.7\%$ (f) GRS: $37.9 \pm 8.2\%$ (m) vs. $34.8 \pm 8.9\%$ (f) GCS endocardial: $-26.5 \pm 4.2\%$ (m) vs. $-27.9 \pm 3.7\%$ (f) GCS myocardial: $-22.2 \pm 3.4\%$ (m) vs. $-24.6 \pm 2.9\%$ (f)
Aurich et al.	2016	47 healthy individuals	FT-CMR vs. FT-Echo vs. STE	STE: GLS: $-15.7 \pm 5.0\%$ GCS: $-14.6 \pm 4.5\%$ GRS: $21.6 \pm 13.3\%$ FT-Echo GLS: -13.1 ± 4.0, GCS: -13.6 ± 4.0, GRS: 20.3 ± 9.5, FT-CMR GLS: -15.0 ± 4.0, GCS: -16.9 ± 5.4 GRS: 35.0 ± 10.8 Best agreement was between FT-Echo and FT-CMR for GLS
Bucius et al.	2019	11 healthy individuals + 7 with heart failure	FT-CMR vs. TT-CMR vs. fast-SENC	FT-CMR GLS: -23.5% $(-22.0\text{--}-25.9)$ GCS: -26.1% $(-21.8\text{--}-27.8)$ TT-CMR GLS: -14.9% $(-11.8\text{--}-16.9)$ GCS: -17.8% $(-16.4\text{--}-19.5)$ Fast-SENC GLS: -19.4% $(17.1\text{--}20.7)$ GCS: -20.3% $(16.5\text{--}22.3)$

Abbreviations: f, female subjects; Fast-SENC, fast Strain-encoding cardiac magnetic resonance; FT-CMR, feature-tracking cardiac magnetic resonance; GCS, global circumferential strain; GLS, global longitudinal strain; GRS, global radial strain; m, male subjects; *n*, number of subjects; TT-CMR, tissue-tagging cardiac magnetic resonance.

4.2. LV Myocardial Strain by CMR in Various Cardiovascular Diseases

In a recently published systematic review that evaluated the impact of GLS by both echocardiography and CMR in patients with acute myocardial infarction, it was shown that the latter technique exhibited major advantages in matters of tissue characterization and resolution, regardless of the acoustic window. Nonetheless, larger cohort studies are needed to objectify the real incremental prognostic value that might be deployed by CMR in terms of LV strain characterization [64]. Moreover, a clinical-based study conducted on 232 patients with ST-elevated myocardial infarction searched to appraise the ability of LV myocardial strain determined by FT-CMR in predicting LV post-infarction remodeling. All three global deformation measurements were associated with adverse myocardial remodeling, although only GLS was an independent predictor for it after the adjustment for imaging covariates. Furthermore, a GLS of over -14% increased the risk of adverse remodeling 4 times, with an odds ratio of 4.16, $p = 0.005$, and provided significant incremental predicting value for it [65]. Similarly, the same findings in terms of GLS were also reported by Cha [66]. Likewise, the role of GCS as an independent predictor for late LV myocardial remodeling after myocardial infarction has been proved in the study of Holmes et al. [67].

Latterly, fast-SENC is gaining more and more ground even in patients with ischemic heart disease, due to its rapidity and reproducibility. In a recently published study, which sought to compare fast-SENC and FT-CMR with STE in patients with acute myocardial infarction, El-Saadi et al., have shown that in terms of GLS, fast-SENC provided higher values than FT-CMR, but without any statistical significance as compared to STE. Moreover, for GCS, the parameters determined by fast-SENC were almost equal to FT-CMR, while as concerns the regional strain in the infarct-related artery, fast-SENC had a significantly higher area under the curve in properly identifying the injured myocardial segments, in contrast with FT-CMR [68]. Furthermore, Fong et al., conducted a systematic review and meta-analysis in which they compared the utility of GLS in patients with both ischemic and non-ischemic dilated cardiomyopathy. They found GLS as a prognostic predictor for mortality in both groups of patients, however, its predictive ability was lower in those with LVEF of under 30% [69].

In patients with dilated cardiomyopathy, TT-CMR was able to accurately determine impaired LV strain parameters, even within the early stage of the disease [70]. Moreover, LV deformation measurements by FT-CMR were related to the severity of basal dysfunction, whereas GCS alone predicted the recovery of LV ejection fraction [71]. In another cohort of 210 patients with dilated cardiomyopathy, GLS by FT-CMR was also an independent predictor for cardiac death, heart transplant, and ventricular tachyarrhythmias, overcoming GCS, GRS, LVEF, and biomarkers of heart failure [72]. In the study of Korosoglu et al., conducted on 1169 patients with various cardiovascular diseases, the authors sought to evaluate the ability of a fast-SENC-derived strain to diagnose and stratify heart failure. They have shown that the percentage of myocardial segments with impaired strain was able to better identify patients with subclinical heart failure and to improve their risk stratification than standard LV functional parameters [73]. Additionally, in the FT-CMR-based study conducted on 740 patients with myocarditis, GLS was significantly associated with the occurrence of major adverse cardiovascular events, including ventricular tachyarrhythmias, heart failure hospitalization, and all-cause mortality, and proved to be an independent predictor for them [74].

Moreover, in patients suffering from hypertrophic cardiomyopathy, impaired GCS determined by FT-CMR along with LGE were found as independent predictors of ventricular tachyarrhythmias [75]. In addition, a recently published study that sought to evaluate the ability of LV deformation parameters to differentiate between hypertrophic cardiomyopathy and hypertensive heart disease showed that GLS by FT-CMR significantly discriminated between these two illnesses and was also strongly correlated to LGE, T1-mapping, and LV mass [76]. Similarly, LV deformation parameters determined by fast-SENC-CMR were also able to differentiate between athletes' hearts, hypertrophic cardiomyopathy, and hypertensive heart disease, respectively [77].

Lastly, the role of the LV strain by CMR to identify subclinical myocardial impairment has been recently appraised. In paediatric patients with end-stage renal disease, GLS, GCS, and GRS by TT-CMR, along with LV ejection fraction and mass, were significantly inflicted, whereas GCS and GRS were associated with poor outcomes [78]. Furthermore, more recently, it was shown that LV strain parameters by FT-CMR significantly improved in pediatric patients with end-stage renal disease 1 year after renal transplantation [79]. Similarly, GLS and GCS by FT-CMR were significantly impaired in patients with rheumatoid arthritis, even though standard LV systolic function parameters remained unmodified. In Table 3 [55,56,65–68,70–73,75,77,80–86] are summarized various CMR-based studies on LV myocardial strain.

Table 3. LV myocardial strain assessed by various CMR techniques.

Authors	Ref	Year	*n*	Method	Diagnose	Strain	Findings
El-Saadi et al.	[13]	2022	30	Fast-SENC vs. FT-CMR	AMI	GLS, GCS	Fast-SENC was superior to FT-CMR
Reindl et al.	[60]	2021	232	FT-CMR	AMI	GLS, GCS, GRS	GLS > −14%—independent predictor LV remodeling
Cha et al.	[61]	2019	82	FT-CMR	AMI	GLS	Independent predictor for LV remodeling
Holmes et al.	[62]	2017	141	FT-CMR	AMI	GCS	Independent predictor for LV remodeling
Singh et al.	[73]	2015	18	FT-CMR	AS	GLS, GCS	Higher values
Pozo Osinalde et al.	[64]	2021	N/A	FT-CMR	DCM	GCS	Predictor for LV systolic function recovery
Yu et al.	[63]	2017	48	TT-CMR	DCM	GLS, GCS	Impaired parameters
Moody et al.	[74]	2015	45	FT-CMR	DCM	GLS, GCS	Good agreement
Buss et al.	[65]	2015	210	FT-CMR	DCM	GLS	Independent predictor for outcome
Hor et al.	[78]	2010	233	FT-CMR	DMD	GLS	−13.3%
Pu et al.	[68]	2021	93	FT-CMR	HCM	GCS	Independent predictor for VT
Harrild et al.	[77]	2012	24	FT-CMR	HCM	GLS	Good agreement
Giusca et al.	[70]	2021	214	Fast-SENC	HCM, Athletes' hearts, AHT	GLS	Disease discrimination
Weise Valdes et al.	[57]	2021	181	fast-SENC	Healthy	GLS, GCS	−20.3%, −19.2%
Erley et al.	[79]	2019	50	fast-SENC	Healthy	GLS, GCS	Good agreement
Taylor et al.	[53]	2015	100	FT-CMR	Healthy	GLS, GCS, GRS	−21.3%, −26.1%, 39.8%
Korosoglu et al.	[66]	2021	1169	Fast-SENC	Heart failure	GLS, GCS	Independent prognostic predictors for outcome
Wu et al.	[75]	2014	30	FT-CMR	LBBB, HCM	GCS	Good agreement
Augustine et al.	[76]	2013	145	FT-CMR	normal	GLS, GCS, GRS	Good agreement

Abbreviations: AHT, arterial hypertension; AMI, acute myocardial infarction; AS, aortic stenosis; DCM, dilated cardiomyopathy; DMD, Duchenne muscular dystrophy; Fast-SENC, fast Strain-encoding cardiac magnetic resonance imaging; FT-CMR, feature-tracking cardiac magnetic resonance imaging; GCS, global circumferential strain; GLS, global longitudinal strain; GRS, global radial strain; HCM, hypertrophic cardiomyopathy; LBBB, left bundle branch block; LV, left ventricle;. N, number of patients; TT-CMR, tissue-tagging cardiac magnetic resonance imaging; VT, ventricular tachyarrhythmias.

5. Clinical Utility of CMR in Evaluating LV Biomechanics

5.1. LV Wringing Parameters

Thus far, the association between LV torsion (Figure 1), twist and untwist, and cardiac diseases has been supported by several experimental and clinical CMR studies. At a glance, myocardial fibers strain following base-to-apex and endo-to-epicardium patterns ensure a constant LV circumferential-to-longitudinal shear angle. In this regard, LV torsion is molded during systole when the base and the apex rotate in opposite directions, clockwise and counterclockwise, respectively. This phenomenon results from the normal physiology of the myocardial fibers [87], which is significantly altered in pathological states. In patients with ischemic heart disease, basal rotation was impaired at exertion leading to afflicted LV torsion, while the apical spin remained unchanged, presumably as a compensatory mechanism [88,89]. Conversely, in patients with dilated cardiomyopathy, inverted apical rotation was the main reason for abnormal LV torsion [90].

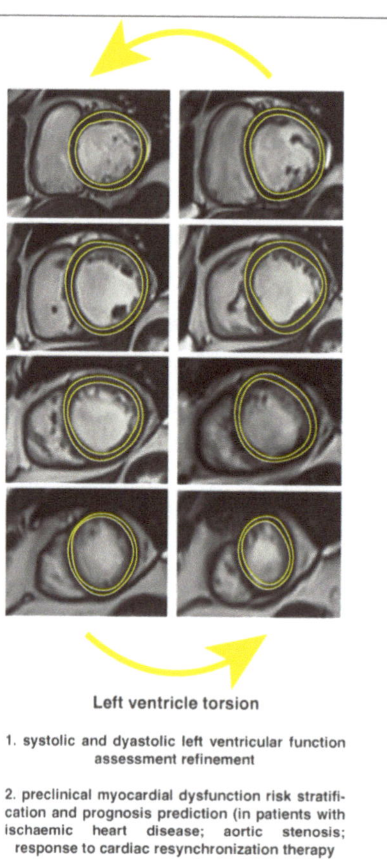

Figure 1. Left ventricle torsion by feature-tracking cardiac magnetic resonance imaging.

As previously stated by Rosen et al., when TT-CMR is used to determine LV torsion, the peak systolic LV twist is divided by the inter-slice distance to ensure standardization [91]. Regarding FT-CMR, Kowallick et al., suggested that the best accuracy and feasibility are guaranteed when apical and basal rotations are measured at 25% and 75% of the total LV's tip-to-base distance [92]. Additionally, if reference values are provided, various post-processing software for CMR can be used to assess LV torsion [93]. Furthermore, this parameter has promising results in risk stratification and prognosis prediction of cardiovascular patients, although studies are just at the beginning. In both apparently healthy elder subjects and diabetics, advanced age and hypertension were associated with higher LV torsion, probably as a result of a balancing mechanism. In addition, it was inversely correlated to the LV sphericity index [94,95]. In contrast, patients with myocardial infarction showed considerably lower LV torsion and the severity of its impairment was significantly associated with an increased risk of cardiovascular death, re-infarction, heart failure hospitalization, and stroke [96]. Withal the standardization of LV torsion to LV long-axis size and radius has led to the development of LV torsion shear angle as a more precise parameter for myocardial remodeling and diastolic function [97]. By normalizing its change rate to analogous variation in LV volume, the LV torsion shear angle was able to accurately identify LV diastolic dysfunction invasively defined as elevated LV end-diastolic pressure and prolonged time of LV relaxation in patients with heart failure and preserved LVEF [98].

LV torsion-to-shortening ratio, which is defined as the ratio between inner wall shortening and torsion at ejection, was developed to accurately characterize endocardial strain and wringing. Essentially, this parameter is a precise marker of subendocardial myocardial impairment, especially in subjects with LV hypertrophy [98]. In patients with aortic valve stenosis (AS), LV torsion-to-shortening ratio determined by TT-CMR was considerably higher when compared to controls, and, in addition, it significantly decreased at three months after aortic valve replacement procedures [99]. FT-CMR has been recently proved to be equally useful in determining impaired LV torsion-to-shortening ratio in patients with AS [100]. Correspondingly, impaired LV torsion-to-shortening ratio and LV torsion have been found even in patients with hypertrophy cardiomyopathy mutation and without clinically overt disease, presumably due to subendocardial malfunction [101].

Over and above, a co-dependency between LV wringing and myocardial scarring has been latterly reported. Intriguingly, reduced LV torsion, along with other abnormal LV systolic parameters, was strongly related to the magnitude of myocardial fibrosis evidenced by Masson's staining [102]. Due to clinical availability, specific CMR techniques use late gadolinium enhancement (LGE) and native and post-contrast T1-mapping techniques to quantify irreversible replacement and diffuse interstitial fibrosis, respectively [103]. In dilated cardiomyopathy, the presence of LGE was associated with increased basal rotation and decreased apical rotation, which led to defective LV torsion. Additionally, the load of myocardial fibrosis was even higher in those with inverted apical rotation [104]. The presence of LV mid-wall fibrosis, a scarring pattern that is particular for dilated cardiomyopathy, was also closely related to impaired LV torsion and rotation [84]. In contrast, Csecs et al., have failed to prove a significant correlation between the presence and extent of myocardial fibrosis and LV torsion and twist parameters in a well-defined cohort of 239 patients with nonischemic dilated cardiomyopathy, thus suggesting that merely the LV dilation and dysfunction themselves are responsible for impaired LV wringing [105,106]. Therefore, further research is required to correctly ascertain these findings.

As for LV twist, certain evidence concerning the impact of cardiac dysfunction on LV twisting is beginning to emerge to expand the clinical utility of CMR [106]. FT-CMR has been recently shown to have high feasibility and reproducibility in the evaluation of ventricular twist and untwist [54,92]. Therefore, afflicted LV twist was associated with LV enlargement and systolic dysfunction [107]. A recently published systematic review has endorsed the utility of CMR to accurately determine LV untwist [108]. Moreover, Paetsch et al., first demonstrated that in a low-dose dobutamine stress-CMR, LV untwist accurately distinguished patients with ischemic heart disease from controls [109].

5.2. LV Functional Dynamic Geometry Measurements

Compelling evidence renders the utility of CMR-derived LV sphericity index in various cardiovascular diseases. Some reports have shown that LV sphericity is inversely associated with LVEF, LV torsion, and mass-to-volume ratio, as well as with both global and regional LV trabeculation indexes [95,110]. Likewise, it was able to correctly identify dilated cardiomyopathy, since it is closely related to increased LV end-systolic volume and decreased LVEF [111]. Correspondingly, by being directly related to sera levels of N-terminal prohormone of brain natriuretic peptide, it may be used for the risk stratification of patients with heart failure [110,112].

Furthermore, the LV sphericity index (Figure 2) is emerging as a novel tool to predict the cardiovascular outcome. In patients with dilated cardiomyopathy, the LV sphericity index significantly predicted major adverse cardiovascular events, including heart failure hospitalization, ventricular tachyarrhythmias, and cardiac death, independent of decreased LVEF and LGE [113,114]. Additionally, in the study of Nakamori et al., the LV sphericity index was an effective marker of appropriate implantable cardioverter defibrillator therapy, thus rightly forecasting ventricular tachyarrhythmias in patients with heart failure and reduced LVEF [115]. Nonetheless, the LV sphericity index might also be useful to predict the occurrence of cardiovascular disease in healthy subjects. In the MESA cohort, the

LV sphericity index was found as a strong predictor for the occurrence of ischemic heart disease, heart failure, and atrial fibrillation in initially healthy subjects after 10 years of follow-up [116]. Conclusively, the LV sphericity index is a simple and reproducible parameter, and larger cohort studies should be further conducted to correctly establish its clinical utility.

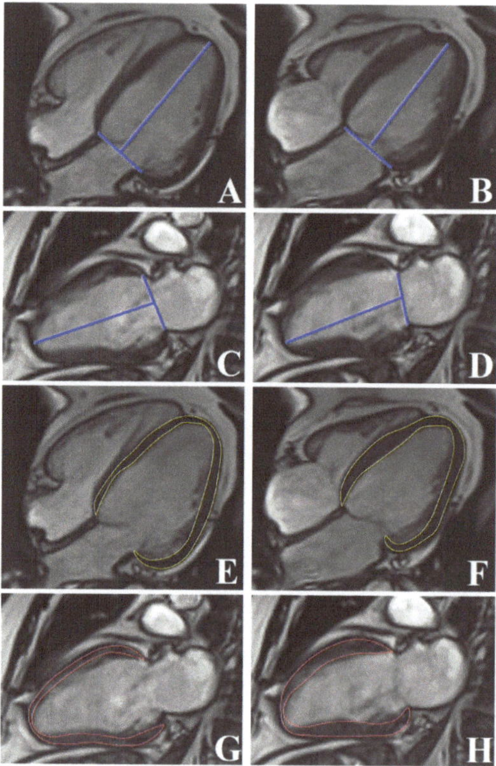

Figure 2. Left ventricular geometry and strain by cine-cardiac magnetic resonance imaging: Left ventricle long-axis strain (**A–D**) and sphericity index (**E–H**).

LV long-axis strain (Figure 2) is a novel indicator of LV systolic function, which can be easily determined by FT-CMR, having high reproducibility and considerable predictive ability [117]. Recently, Leng et al. have shown that standard cine-CMR can also deploy effective and reproducible LV systolic parameters, including LV long-axis strain [118]. Cine-CMR-derived LV long-axis strain has proven non-inferior to FT-CMR-derived one in identifying patients with various cardiomyopathies, being also more time-efficient [119]. Moreover, it was significantly impaired in diabetic patients without clinically overt cardiac disease, even after adjustment for clinical and biological covariates [92]. In the MESA cohort population, impaired LV long-axis strain significantly predicted congestive heart failure, cardiovascular death, stroke, and myocardial infarction, even in subjects without clinically overt cardiovascular illnesses [120]. Likewise, the utility of LV long-axis strain for the prediction of cardiac outcome has also been shown in cardiac amyloidosis, aortic stenosis, and dilated cardiomyopathy [113,117,121]. In addition, it may also improve risk stratification in patients with non-ischemic dilated cardiomyopathy [122]. As for patients with myocardial infarction, impaired LV long-axis strain independently predicted major adverse cardiovascular events at the one-year follow-up [12].

5.3. Cardiac Pressure-Volume Loops by CMR

The basic principle of cardiac active biomechanics can be summed up by the relationship between the pressure and volume gradients that develop throughout every cardiac cycle. The close connections between these two physical phenomena have deployed specific pressure-volume curves, which can be used to accurately assess cardiac function. Moreover, specific surrogates of cardiac biodynamics which can precisely estimate myocardial contractility, ventricular-arterial coupling, end-systolic (ESPVR), and end-diastolic (EDPVR) pressure-volume relations can be derived from such measurements. However, the main disadvantage of these measurements is that, until now, they could have been accurately determined only by invasive conductance catheterization [12,123].

Recent studies have begun to deploy hybrid methods that may assess these parameters by combining cine- and velocity-encoded CMR with transient closure of IVC with venous catheters, thus mimicking the preload reduction in cardiac volumes. It was shown that CMR can evaluate the topmost right ventricular pressure during isovolumic normal heartbeats. This may be used to determine ESPVR (Figure 3), yielding it as a potential reliable option to accurately estimate myocardial contractility and ventricular-arterial coupling [124]. Subsequently, Kuehne et al., revealed that venous catheters can be positioned into the pulmonary artery under real-time CMR guidance and, by combining with CMR-determined ventricular volumes and mass, right ventricular pressure-volume loops and ESPVR can be determined. In murine models, they matched these measurements with those determined invasively and found excellent inter-agreements. Moreover, in human subjects, they tested the method on patients with pulmonary hypertension and healthy controls. They found that in the diseased group, cardiac index and ventricular-arterial coupling were significantly afflicted, while ESPVR was increased [125].

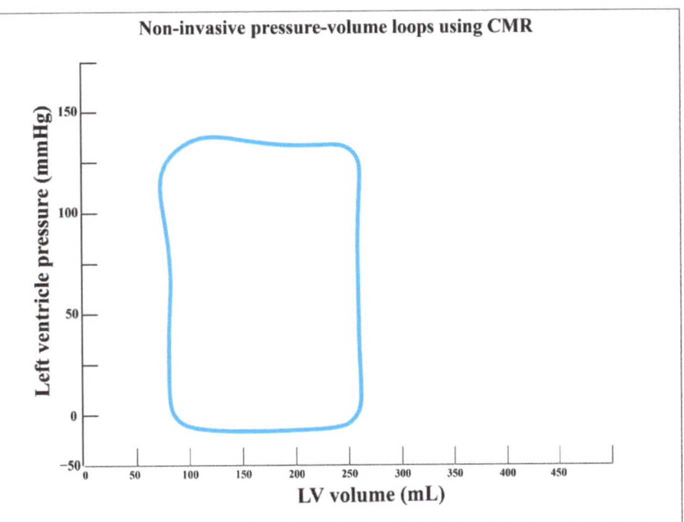

Figure 3. Non-invasive assessment of pressure-volume loops using cardiac magnetic resonance imaging.

Likewise, using a similar principle, few pilot studies rendered the utility of CMR to determine EDPVR. In the study of Schmitt et al., pressure-volume curves were initially determined invasively by conductance catheterization. Afterward, using cine- and velocity-encoded CMR along with cardiac pre-load decrease by temporary inferior vena cava occlusion, the authors deployed a hybrid method to estimate EDPVR as a marker of LV stiffness. Not only did they succeed to demonstrate excellent agreement between the two methods, but these measurements were dynamically influenced by pharmacological stress, thus improving diastolic function parameters in a similar manner to those determined strictly by conductance catheterization. Nevertheless, these promising findings require larger studies to validate their clinical utility [126]. Additionally, a murine study proposed a novel method that uses real-time CMR with shorter acquisition timing for cardiac pressure-volume assessment which can be used to continuously determine ventricular volumes, ESPVR, and preload recruitable stroke work as well as eliminate several existing shortcomings. Nonetheless, these methods require larger cohort validation [127]. By the same token, Giao et al., demonstrated the use of real-time CMR in the estimation of ESPVR during inferior vena cava obstruction. They showed that this method provides relevant data regarding LV geometry and regional function and, thus, emphasized the importance of LV shape and segmental biomechanics in maintaining cardiac performance [128].

Moreover, the clinical efficacy of these was first evaluated in murine models by Faragli et al., who sought to assess the relationship between determined LV strain parameters and hemodynamical parameters such as cardiac index, cardiac power output, and ESPVR determined by FT-CMR in various stress conditions. Despite several technical and analytical drawbacks that relatively lowered the statistical power, LV global longitudinal and circular strain were closely related to all LV hemodynamic measurements, regardless of the inotropic state, while LV global longitudinal performed best in assessing LV contractility, similar to LVEF. Therefore, FT-CMR might become a promising technique for evaluating LV hemodynamics; however, future studies are required for the optimization of this method [129].

Last but not least, by creating a time-variance elastance mode, Seeman et al., were the first to develop a completely non-invasive method that uses solely CMR and brachial pressure to assess LV pressure-volume loops, thus overcoming the need for IVC occlusion. Firstly, they tested and validated this method in murine models and further confirmed it in human subjects by comparing patients with heart failure with healthy controls [13].

6. Future Perspectives

Even though the role of CMR in determining myocardial strain and biomechanical parameters is gaining serious ground, there are still many uncertainties that need to be unraveled. FT-CMR has been proven to be a useful CMR method in assessing myocardial strain, but there is still insufficient evidence in terms of various cardiovascular diseases. Further studies should be conducted in patients with valvular heart disease, such as aortic stenosis or mitral regurgitation, in order to test the predictive ability of myocardial strain parameters in prognosis prediction. Moreover, there are not any available data regarding the role of FT-CMR-derived myocardial strain in patients with cardiac amyloidosis or other infiltrative heart diseases, which might provide information of tremendous importance in risk stratification and prognosis prediction. Further studies could also aim to test the ability of FT-CMR in creating multi-parametric predictive models based on LV wringing parameters in various cardiomyopathies or myocarditis. Additionally, there is little evidence of the role of FT-CMR-derived strain and biomechanics in patients with acute myocarditis.

As for fast-SENC, it is a promising valuable CMR imaging technique that might enter day-to-day practice in the future, but more studies still need to be conducted. Although few studies have shown the superior ability of fast-SENC-derived myocardial strain parameters in risk stratification and prognosis prediction of patients with acute myocardial infarction, there are no studies conducted in patients with various primary myocardial diseases, thus, this could represent a valuable research direction. Moreover, the ability of fast-SENC to deploy LV wringing and functional dynamic geometry parameters represents another uncharted territory.

As for routine clinical applicability, things are still in their infancy. FT-CMR might become a promising option in daily medical practice because it uses standard cine-CMR acquisitions. Additionally, semi-automated or automated software might aid the evaluation. Furthermore, fast-SENC is another promising technique that might significantly reduce the acquisition time for CMR examinations. Nonetheless, due to great disparities in results, which depend on the method of acquisition and data processing, further studies still need to be conducted.

7. Conclusions

The role of CMR in assessing LV myocardial strain and biomechanics is beginning to take shape. Recent technological advancements in the field of CMR, such as fast-SENC and FT-CMR, are able to ensure increased accuracy in evaluating myocardial strain, LV wringing, and active geometry parameters and, along with these developments, increasing evidence endorses their future clinical ability. Even though things are just at the beginning, few yet important studies have shown the tremendous potential which lies behind LV strain and biomechanics.

Author Contributions: Conceptualization, A.Z. and C.C.; methodology, A.Z.; software, S.O. and L.A.-C.; validation, L.A.-C.; investigation, A.Z. and C.C.; writing—original draft preparation, A.Z.; writing—review and editing, A.Z. and C.C.; visualization, S.O.; supervision, L.A.-C.; project administration, A.Z. All authors have read and agreed to the published version of the manuscript.

Funding: This research did not receive any external funding.

Acknowledgments: This work was supported by internal institutional doctoral fellowship from the Iuliu Hatieganu University of Medicine and Pharmacy, Cluj-Napoca, Romania.

Conflicts of Interest: The authors declare no competing interest.

References

1. Amzulescu, M.S.; de Craene, M.; Langet, H.; Pasquet, A.; Vancraeynest, D.; Pouleur, A.C.; Vanoverschelde, J.L.; Gerber, B.L. Myocardial strain imaging: Review of general principles, validation, and sources of discrepancies. *Eur. Heart J. Cardiovasc. Imaging* **2019**, *20*, 605–619. [CrossRef] [PubMed]
2. Narang, A.; Addetia, K. An introduction to left ventricular strain. *Curr. Opin. Cardiol.* **2018**, *33*, 455–463. [CrossRef] [PubMed]
3. Lee, E.; Ibrahim, E.-S.H.; Parwani, P.; Bhave, N.; Stojanovska, J. Practical Guide to Evaluating Myocardial Disease by Cardiac MRI. *Am. J. Roentgenol.* **2020**, *214*, 546–556. [CrossRef] [PubMed]
4. Bertozzi, G.; Cafarelli, F.P.; Ferrara, M.; Di Fazio, N.; Guglielmi, G.; Cipolloni, L.; Manetti, F.; La Russa, R.; Fineschi, V. Sudden Cardiac Death and Ex-Situ Post-Mortem Cardiac Magnetic Resonance Imaging: A Morphological Study Based on Diagnostic Correlation Methodology. *Diagnostics* **2022**, *12*, 218. [CrossRef]
5. Rutz, A.K.; Ryf, S.; Plein, S.; Boesiger, P.; Kozerke, S. Accelerated whole-heart 3D CSPAMM for myocardial motion quantification. *Magn. Reson. Med.* **2008**, *59*, 755–763. [CrossRef]
6. Scatteia, A.; Baritussio, A.; Bucciarelli-Ducci, C. Strain imaging using cardiac magnetic resonance. *Heart Fail. Rev.* **2017**, *22*, 465–476. [CrossRef]
7. Pedrizzetti, G.; Claus, P.; Kilner, P.J.; Nagel, E. Principles of cardiovascular magnetic resonance feature tracking and echocardiographic speckle tracking for informed clinical use. *J. Cardiovasc. Magn. Reson.* **2016**, *18*, 51. [CrossRef]
8. Xu, J.; Yang, W.; Zhao, S.; Lu, M. State-of-the-art myocardial strain by CMR feature tracking: Clinical applications and future perspectives. *Eur. Radiol.* **2022**, *32*, 5424–5435. [CrossRef]
9. Streeter, D.D.; Spotnitz, H.M.; Patel, D.P.; Ross, J.; Sonnenblick, E.H. Fiber Orientation in the Canine Left Ventricle during Diastole and Systole. *Circ. Res.* **1969**, *24*, 339–347. [CrossRef]
10. Hoshino, T.; Fujiwara, H.; Kawai, C.; Hamashima, Y. Myocardial fiber diameter and regional distribution in the ventricular wall of normal adult hearts, hypertensive hearts and hearts with hypertrophic cardiomyopathy. *Circulation* **1983**, *67*, 1109–1116. [CrossRef]
11. Jeung, M.-Y.; Germain, P.; Croisille, P.; el Ghannudi, S.; Roy, C.; Gangi, A. Myocardial Tagging with MR Imaging: Overview of Normal and Pathologic Findings. *Radiographics* **2012**, *32*, 1381–1398. [CrossRef]
12. Voorhees, A.P.; Han, H.-C. Biomechanics of Cardiac Function. In *Comprehensive Physiology*; John Wiley & Sons, Inc.: Hoboken, NJ, USA, 2015; pp. 1623–1644.
13. Seemann, F.; Arvidsson, P.; Nordlund, D.; Kopic, S.; Carlsson, M.; Arheden, H.; Heiberg, E. Noninvasive Quantification of Pressure-Volume Loops from Brachial Pressure and Cardiovascular Magnetic Resonance. *Circ. Cardiovasc. Imaging* **2019**, *12*, e008493. [CrossRef]
14. Avazmohammadi, R.; Soares, J.S.; Li, D.S.; Raut, S.S.; Gorman, R.C.; Sacks, M.S. A Contemporary Look at Biomechanical Models of Myocardium. *Ann. Rev. Biomed. Eng.* **2019**, *21*, 417–442. [CrossRef]
15. Rothermel, T.M.; Win, Z.; Alford, P.W. Large-Deformation Strain Energy Density Function for Vascular Smooth Muscle Cells. *J. Biomech.* **2020**, *111*, 110005. [CrossRef]

16. Buckberg, G.; Hoffman, J.I.E.; Nanda, N.C.; Coghlan, C.; Saleh, S.; Athanasuleas, C. Ventricular Torsion and Untwisting: Further Insights into Mechanics and Timing Interdependence: A Viewpoint. *Echocardiography* **2011**, *28*, 782–804. [CrossRef]
17. Hein, S.; Gaasch, W.H.; Schaper, J. Giant Molecule Titin and Myocardial Stiffness. *Circulation* **2002**, *106*, 1302–1304. [CrossRef]
18. De Tombe, P.P.; ter Keurs, H.E.D.J. The velocity of cardiac sarcomere shortening: Mechanisms and implications. *J. Muscle Res. Cell Motil.* **2012**, *33*, 431–437. [CrossRef]
19. Kadler, K.E.; Hill, A.; Canty-Laird, E.G. Collagen fibrillogenesis: Fibronectin, integrins, and minor collagens as organizers and nucleators. *Curr. Opin. Cell Biol.* **2008**, *20*, 495–501. [CrossRef]
20. Demer, L.L.; Yin, F.C. Passive biaxial mechanical properties of isolated canine myocardium. *J. Physiol.* **1983**, *339*, 615–630. [CrossRef]
21. Holzapfel, G.A.; Ogden, R.W. Constitutive modelling of passive myocardium: A structurally based framework for material characterization. *Philos. Trans. R. Soc. A Math. Phys. Eng. Sci.* **2009**, *367*, 3445–3475. [CrossRef]
22. Hunter, P.J.; McCulloch, A.D.; ter Keurs, H.E.D.J. Modelling the mechanical properties of cardiac muscle. *Prog. Biophys. Mol. Biol.* **1998**, *69*, 289–331. [CrossRef] [PubMed]
23. Casha, A.R.; Camilleri, L.; Manché, A.; Gatt, R.; Gauci, M.; Camilleri-Podesta, M.-T.; Grima, J.N.; Scarci, M.; Chetcuti, S. Physiological rules for the heart, lungs and other pressure-based organs. *J. Thorac. Dis.* **2017**, *9*, 3793–3801. [CrossRef] [PubMed]
24. Maurer, M.S.; Kronzon, I.; Burkhoff, D. Ventricular Pump Function in Heart Failure with Normal Ejection Fraction: Insights from Pressure-Volume Measurements. *Prog. Cardiovasc. Dis.* **2006**, *49*, 182–195. [CrossRef] [PubMed]
25. Cingolani, O.H.; Kass, D.A. Pressure-volume relation analysis of mouse ventricular function. *Am. J. Physiol. -Heart Circ. Physiol.* **2011**, *301*, H2198–H2206. [CrossRef] [PubMed]
26. Sequeira, V.; van der Velden, J. Historical perspective on heart function: The Frank–Starling Law. *Biophys. Rev.* **2015**, *7*, 421–447. [CrossRef]
27. Metra, M.; Bettari, L.; Carubelli, V.; Bugatti, S.; Cas, A.D.; Del Magro, F.; Lazzarini, V.; Lombardi, C.; Cas, L.D. Use of Inotropic Agents in Patients with Advanced Heart Failure. *Drugs* **2011**, *71*, 515–525. [CrossRef]
28. Smiseth, O.A.; Torp, H.; Opdahl, A.; Haugaa, K.H.; Urheim, S. Myocardial strain imaging: How useful is it in clinical decision making? *Eur. Heart J.* **2016**, *37*, 1196–1207. [CrossRef]
29. Janwanishstaporn, S.; Cho, J.Y.; Feng, S.; Brann, A.; Seo, J.-S.; Narezkina, A.; Greenberg, B. Prognostic Value of Global Longitudinal Strain in Patients with Heart Failure with Improved Ejection Fraction. *JACC Heart Fail.* **2022**, *10*, 27–37. [CrossRef]
30. Thellier, N.; Altes, A.; Appert, L.; Binda, C.; Leman, B.; Marsou, W.; Debry, N.; Joly, C.; Ennezat, P.-V.; Tribouilloy, C.; et al. Prognostic Importance of Left Ventricular Global Longitudinal Strain in Patients with Severe Aortic Stenosis and Preserved Ejection Fraction. *J. Am. Soc. Echocardiogr.* **2020**, *33*, 1454–1464. [CrossRef]
31. Goedemans, L.; Abou, R.; Hoogslag, G.E.; Ajmone Marsan, N.; Delgado, V.; Bax, J.J. Left ventricular global longitudinal strain and long-term prognosis in patients with chronic obstructive pulmonary disease after acute myocardial infarction. *Eur. Heart J. Cardiovasc. Imaging* **2019**, *20*, 56–65. [CrossRef]
32. Iacoviello, M.; Puzzovivo, A.; Guida, P.; Forleo, C.; Monitillo, F.; Catanzaro, R.; Lattarulo, M.S.; Antoncecchi, V.; Favale, S. Independent Role of Left Ventricular Global Longitudinal Strain in Predicting Prognosis of Chronic Heart Failure Patients. *Echocardiography* **2013**, *30*, 803–811. [CrossRef]
33. Ersbøll, M.; Valeur, N.; Mogensen, U.M.; Andersen, M.J.; Møller, J.E.; Velazquez, E.J.; Hassager, C.; Søgaard, P.; Køber, L. Prediction of All-Cause Mortality and Heart Failure Admissions from Global Left Ventricular Longitudinal Strain in Patients With Acute Myocardial Infarction and Preserved Left Ventricular Ejection Fraction. *J. Am. Coll. Cardiol.* **2013**, *61*, 2365–2373. [CrossRef]
34. Yingchoncharoen, T.; Gibby, C.; Rodriguez, L.L.; Grimm, R.A.; Marwick, T.H. Association of Myocardial Deformation with Outcome in Asymptomatic Aortic Stenosis With Normal Ejection Fraction. *Circ. Cardiovasc. Imaging* **2012**, *5*, 719–725. [CrossRef]
35. Munk, K.; Andersen, N.H.; Terkelsen, C.J.; Bibby, B.M.; Johnsen, S.P.; Bøtker, H.E.; Nielsen, T.T.; Poulsen, S.H. Global Left Ventricular Longitudinal Systolic Strain for Early Risk Assessment in Patients with Acute Myocardial Infarction Treated with Primary Percutaneous Intervention. *J. Am. Soc. Echocardiogr.* **2012**, *25*, 644–651. [CrossRef]
36. Kearney, L.G.; Lu, K.; Ord, M.; Patel, S.; Profitis, K.; Matalanis, G.; Burrell, L.M.; Srivastava, P. Global longitudinal strain is a strong independent predictor of all-cause mortality in patients with aortic stenosis. *Eur. Heart J. Cardiovasc. Imaging* **2012**, *13*, 827–833. [CrossRef]
37. Dahl, J.S.; Videbæk, L.; Poulsen, M.K.; Rudbæk, T.R.; Pellikka, P.A.; Møller, J.E. Global Strain in Severe Aortic Valve Stenosis. *Circ. Cardiovasc. Imaging* **2012**, *5*, 613–620. [CrossRef]
38. Buss, S.J.; Emami, M.; Mereles, D.; Korosoglou, G.; Kristen, A.V.; Voss, A.; Schellberg, D.; Zugck, C.; Galuschky, C.; Giannitsis, E.; et al. Longitudinal Left Ventricular Function for Prediction of Survival in Systemic Light-Chain Amyloidosis. *J. Am. Coll. Cardiol.* **2012**, *60*, 1067–1076. [CrossRef]
39. Bertini, M.; Ng, A.C.T.; Antoni, M.L.; Nucifora, G.; Ewe, S.H.; Auger, D.; Marsan, N.A.; Schalij, M.J.; Bax, J.J.; Delgado, V. Global Longitudinal Strain Predicts Long-Term Survival in Patients with Chronic Ischemic Cardiomyopathy. *Circ. Cardiovasc. Imaging* **2012**, *5*, 383–391. [CrossRef]
40. Woo, J.S.; Kim, W.-S.; Yu, T.-K.; Ha, S.J.; Kim, S.Y.; Bae, J.-H.; Kim, K.S. Prognostic Value of Serial Global Longitudinal Strain Measured by Two-Dimensional Speckle Tracking Echocardiography in Patients with ST-Segment Elevation Myocardial Infarction. *Am. J. Cardiol.* **2011**, *108*, 340–347. [CrossRef]

41. Nahum, J.; Bensaid, A.; Dussault, C.; Macron, L.; Clémence, D.; Bouhemad, B.; Monin, J.-L.; Rande, J.-L.D.; Gueret, P.; Lim, P. Impact of Longitudinal Myocardial Deformation on the Prognosis of Chronic Heart Failure Patients. *Circ. Cardiovasc. Imaging* **2010**, *3*, 249–256. [CrossRef]

42. Antoni, M.L.; Mollema, S.A.; Delgado, V.; Atary, J.Z.; Borleffs, C.J.W.; Boersma, E.; Holman, E.R.; van der Wall, E.E.; Schalij, M.J.; Bax, J.J. Prognostic importance of strain and strain rate after acute myocardial infarction. *Eur. Heart J.* **2010**, *31*, 1640–1647. [CrossRef] [PubMed]

43. Stanton, T.; Leano, R.; Marwick, T.H. Prediction of All-Cause Mortality from Global Longitudinal Speckle Strain. *Circ. Cardiovasc. Imaging* **2009**, *2*, 356–364. [CrossRef] [PubMed]

44. Cho, G.-Y.; Marwick, T.H.; Kim, H.-S.; Kim, M.-K.; Hong, K.-S.; Oh, D.-J. Global 2-Dimensional Strain as a New Prognosticator in Patients with Heart Failure. *J. Am. Coll. Cardiol.* **2009**, *54*, 618–624. [CrossRef] [PubMed]

45. Lancellotti, P.; Cosyns, B.; Zacharakis, D.; Attena, E.; Van Camp, G.; Gach, O.; Radermecker, M.; Piérard, L.A. Importance of Left Ventricular Longitudinal Function and Functional Reserve in Patients with Degenerative Mitral Regurgitation: Assessment by Two-Dimensional Speckle Tracking. *J. Am. Soc. Echocardiogr.* **2008**, *21*, 1331–1336. [CrossRef]

46. Rajiah, P.S.; Kalisz, K.; Broncano, J.; Goerne, H.; Collins, J.D.; François, C.J.; Ibrahim, E.-S.; Agarwal, P.P. Myocardial Strain Evaluation with Cardiovascular MRI: Physics, Principles, and Clinical Applications. *RadioGraphics* **2022**, *42*, 968–990. [CrossRef]

47. Moore, C.C.; Reeder, S.B.; McVeigh, E.R. Tagged MR imaging in a deforming phantom: Photographic validation. *Radiology* **1994**, *190*, 765–769. [CrossRef]

48. Pan, L.; Prince, J.L.; Lima, J.A.C.; Osman, N.F. Fast Tracking of Cardiac Motion Using 3D-HARP. *IEEE Trans. Biomed. Eng.* **2005**, *52*, 1425–1435. [CrossRef]

49. Föll, D.; Jung, B.; Germann, E.; Hennig, J.; Bode, C.; Markl, M. Magnetic Resonance Tissue Phase Mapping: Analysis of Age-Related and Pathologically Altered Left Ventricular Radial and Long-Axis Dyssynchrony. *J. Magn. Reson. Imaging* **2011**, *34*, 518–525. [CrossRef]

50. Wen, H.; Bennett, E.; Epstein, N.; Plehn, J. Magnetic resonance imaging assessment of myocardial elastic modulus and viscosity using displacement imaging and phase-contrast velocity mapping. *Magn. Reson. Med.* **2005**, *54*, 538–548. [CrossRef]

51. Korosoglou, G.; Giusca, S.; Hofmann, N.P.; Patel, A.R.; Lapinskas, T.; Pieske, B.; Steen, H.; Katus, H.A.; Kelle, S. Strain-encoded magnetic resonance: A method for the assessment of myocardial deformation. *ESC Heart Fail.* **2019**, *6*, 584–602. [CrossRef]

52. Taylor, R.J.; Moody, W.E.; Umar, F.; Edwards, N.C.; Taylor, T.J.; Stegemann, B.; Townend, J.; Hor, K.N.; Steeds, R.; Mazur, W.; et al. Myocardial strain measurement with feature-tracking cardiovascular magnetic resonance: Normal values. *Eur. Heart J. Cardiovasc. Imaging* **2015**, *16*, 871–881. [CrossRef]

53. Morton, G.; Schuster, A.; Jogiya, R.; Kutty, S.; Beerbaum, P.; Nagel, E. Inter-study reproducibility of cardiovascular magnetic resonance myocardial feature tracking. *J. Cardiovasc. Magn. Reson.* **2012**, *14*, 43. [CrossRef]

54. Lamata, P.; Hussain, S.T.; Kutty, S.; Steinmetz, M.; Sohns, J.M.; Fasshauer, M.; Staab, W.; Unterberg-Buchwald, C.; Lotz, J.; Schuster, A. Cardiovascular magnetic resonance myocardial feature tracking for the measurement of myocardial twist and untwist at rest and during dobutamine stress in healthy volunteers. *J. Cardiovasc. Magn. Reson.* **2014**, *16*, P14. [CrossRef]

55. Erley, J.; Genovese, D.; Tapaskar, N.; Alvi, N.; Rashedi, N.; Besser, S.A.; Kawaji, K.; Goyal, N.; Kelle, S.; Lang, R.M.; et al. Echocardiography and cardiovascular magnetic resonance based evaluation of myocardial strain and relationship with late gadolinium enhancement. *J. Cardiovasc. Magn. Reson.* **2019**, *21*, 46. [CrossRef]

56. Weise Valdés, E.; Barth, P.; Piran, M.; Laser, K.T.; Burchert, W.; Körperich, H. Left-Ventricular Reference Myocardial Strain Assessed by Cardiovascular Magnetic Resonance Feature Tracking and fSENC—Impact of Temporal Resolution and Cardiac Muscle Mass. *Front. Cardiovasc. Med.* **2021**, *8*, 764496. [CrossRef]

57. Pierpaolo, P.; Rolf, S.; Manuel, B.-P.; Davide, C.; Dresselaers, T.; Claus, P.; Bogaert, J. Left ventricular global myocardial strain assessment: Are CMR feature-tracking algorithms useful in the clinical setting? *Radiol. Med.* **2020**, *125*, 444–450. [CrossRef]

58. Mangion, K.; Burke, N.M.M.; McComb, C.; Carrick, D.; Woodward, R.; Berry, C. Feature-tracking myocardial strain in healthy adults- a magnetic resonance study at 3.0 tesla. *Sci. Rep.* **2019**, *9*, 3239. [CrossRef]

59. Andre, F.; Steen, H.; Matheis, P.; Westkott, M.; Breuninger, K.; Sander, Y.; Kammerer, R.; Galuschky, C.; Giannitsis, E.; Korosoglou, G.; et al. Age- and gender-related normal left ventricular deformation assessed by cardiovascular magnetic resonance feature tracking. *J. Cardiovasc. Magn. Reson.* **2015**, *17*, 25. [CrossRef]

60. Aurich, M.; Keller, M.; Greiner, S.; Steen, H.; Siepen, F.A.D.; Riffel, J.; Katus, H.A.; Buss, S.J.; Mereles, D. Left ventricular mechanics assessed by two-dimensional echocardiography and cardiac magnetic resonance imaging: Comparison of high-resolution speckle tracking and feature tracking. *Eur. Heart J. Cardiovasc. Imaging* **2016**, *17*, 1370–1378. [CrossRef]

61. Bucius, P.; Erley, J.; Tanacli, R.; Zieschang, V.; Giusca, S.; Korosoglou, G.; Steen, H.; Stehning, C.; Pieske, B.; Pieske-Kraigher, E.; et al. Comparison of feature tracking, fast-SENC, and myocardial tagging for global and segmental left ventricular strain. *ESC Heart Fail.* **2020**, *7*, 523–532. [CrossRef]

62. Lapinskas, T.; Zieschang, V.; Erley, J.; Stoiber, L.; Schnackenburg, B.; Stehning, C.; Gebker, R.; Patel, A.R.; Kawaji, K.; Steen, H.; et al. Strain-encoded cardiac magnetic resonance imaging: A new approach for fast estimation of left ventricular function. *BMC Cardiovasc. Disord.* **2019**, *19*, 52. [CrossRef] [PubMed]

63. Giusca, S.; Korosoglou, G.; Zieschang, V.; Stoiber, L.; Schnackenburg, B.; Stehning, C.; Gebker, R.; Pieske, B.; Schuster, A.; Backhaus, S.; et al. Reproducibility study on myocardial strain assessment using fast-SENC cardiac magnetic resonance imaging. *Sci. Rep.* **2018**, *8*, 14100. [CrossRef] [PubMed]

64. Mangion, K.; McComb, C.; Auger, D.A.; Epstein, F.H.; Berry, C. Magnetic Resonance Imaging of Myocardial Strain After Acute ST-Segment–Elevation Myocardial Infarction. *Circ. Cardiovasc. Imaging* **2017**, *10*, e006498. [CrossRef] [PubMed]

65. Reindl, M.; Tiller, C.; Holzknecht, M.; Lechner, I.; Eisner, D.; Riepl, L.; Pamminger, M.; Henninger, B.; Mayr, A.; Schwaiger, J.P.; et al. Global longitudinal strain by feature tracking for optimized prediction of adverse remodeling after ST-elevation myocardial infarction. *Clin. Res. Cardiol.* **2021**, *110*, 61–71. [CrossRef] [PubMed]

66. Cha, M.J.; Lee, J.H.; Jung, H.N.; Kim, Y.; Choe, Y.H.; Kim, S.M. Cardiac magnetic resonance-tissue tracking for the early prediction of adverse left ventricular remodeling after ST-segment elevation myocardial infarction. *Int. J. Cardiovasc. Imaging* **2019**, *35*, 2095–2102. [CrossRef]

67. Holmes, A.A.; Romero, J.; Levsky, J.M.; Haramati, L.B.; Phuong, N.; Rezai-Gharai, L.; Cohen, S.; Restrepo, L.; Ruiz-Guerrero, L.; Fisher, J.D.; et al. Circumferential strain acquired by CMR early after acute myocardial infarction adds incremental predictive value to late gadolinium enhancement imaging to predict late myocardial remodeling and subsequent risk of sudden cardiac death. *J. Interv. Card. Electrophysiol.* **2017**, *50*, 211–218. [CrossRef]

68. El-Saadi, W.; Engvall, J.E.; Alfredsson, J.; Karlsson, J.-E.; Martins, M.; Sederholm, S.; Zaman, S.F.; Ebbers, T.; Kihlberg, J. A head-to-head comparison of myocardial strain by fast-strain encoding and feature tracking imaging in acute myocardial infarction. *Front. Cardiovasc. Med.* **2022**, *9*, 949440. [CrossRef]

69. Fong, L.C.W.; Lee, N.H.C.; Poon, J.W.L.; Chin, C.W.L.; He, B.; Luo, L.; Chen, C.; Wan, E.Y.F.; Pennell, D.J.; Mohiaddin, R.; et al. Prognostic value of cardiac magnetic resonance derived global longitudinal strain analysis in patients with ischaemic and non-ischaemic dilated cardiomyopathy: A systematic review and meta-analysis. *Int. J. Cardiovasc. Imaging* **2022**, *38*, 2707–2721. [CrossRef]

70. Yu, Y.; Yu, S.; Tang, X.; Ren, H.; Li, S.; Zou, Q.; Xiong, F.; Zheng, T.; Gong, L. Evaluation of left ventricular strain in patients with dilated cardiomyopathy. *J. Int. Med. Res.* **2017**, *45*, 2092–2100. [CrossRef]

71. Pozo Osinalde, E.; Urmeneta Ulloa, J.; Rodriguez Hernandez, J.L.; De Isla, L.P.; Fernandez, H.M.; Islas, F.; Marcos-Alberca, P.; Mahia, P.; A Cobos, M.; Hernandez, P.; et al. Correlation between cardiac magnetic resonance feature tracking derived left ventricular strain and morphological characteristics of non-ischemic dilated cardiomyopathy at baseline and follow-up. *Eur. Heart J.* **2021**, *42*, ehab724. [CrossRef]

72. Buss, S.J.; Breuninger, K.; Lehrke, S.; Voss, A.; Galuschky, C.; Lossnitzer, D.; Andre, F.; Ehlermann, P.; Franke, J.; Taeger, T.; et al. Assessment of myocardial deformation with cardiac magnetic resonance strain imaging improves risk stratification in patients with dilated cardiomyopathy. *Eur. Heart J. Cardiovasc. Imaging* **2015**, *16*, 307–315. [CrossRef]

73. Korosoglou, G.; Giusca, S.; Montenbruck, M.; Patel, A.R.; Lapinskas, T.; Götze, C.; Zieschang, V.; Al-Tabatabaee, S.; Pieske, B.; Florian, A.; et al. Fast Strain-Encoded Cardiac Magnetic Resonance for Diagnostic Classification and Risk Stratification of Heart Failure Patients. *JACC Cardiovasc. Imaging* **2021**, *14*, 1177–1188. [CrossRef]

74. Fischer, K.; Obrist, S.J.; Erne, S.A.; Stark, A.W.; Marggraf, M.; Kaneko, K.; Guensch, D.P.; Huber, A.T.; Greulich, S.; Aghayev, A.; et al. Feature Tracking Myocardial Strain Incrementally Improves Prognostication in Myocarditis Beyond Traditional CMR Imaging Features. *JACC Cardiovasc. Imaging* **2020**, *13*, 1891–1901. [CrossRef]

75. Pu, C.; Fei, J.; Lv, S.; Wu, Y.; He, C.; Guo, D.; Mabombo, P.U.; Chooah, O.; Hu, H. Global Circumferential Strain by Cardiac Magnetic Resonance Tissue Tracking Associated with Ventricular Arrhythmias in Hypertrophic Cardiomyopathy Patients. *Front. Cardiovasc. Med.* **2021**, *8*, 670361. [CrossRef]

76. Neisius, U.; Myerson, L.; Fahmy, A.S.; Nakamori, S.; El-Rewaidy, H.; Joshi, G.; Duan, C.; Manning, W.J.; Nezafat, R. Cardiovascular magnetic resonance feature tracking strain analysis for discrimination between hypertensive heart disease and hypertrophic cardiomyopathy. *PLoS ONE* **2019**, *14*, e0221061. [CrossRef]

77. Giusca, S.; Steen, H.; Montenbruck, M.; Patel, A.R.; Pieske, B.; Erley, J.; Kelle, S.; Korosoglou, G. Multi-parametric assessment of left ventricular hypertrophy using late gadolinium enhancement, T1 mapping and strain-encoded cardiovascular magnetic resonance. *J. Cardiovasc. Magn. Reson.* **2021**, *23*, 92. [CrossRef]

78. Sobh, D.M.; Batouty, N.M.; Tawfik, A.M.; Gadelhak, B.; Elmokadem, A.H.; Hammad, A.; Eid, R.; Hamdy, N. Left Ventricular Strain Analysis by Tissue Tracking–Cardiac Magnetic Resonance for early detection of Cardiac Dysfunction in children with End-Stage Renal Disease. *J. Magn. Reson. Imaging* **2021**, *54*, 1476–1485. [CrossRef]

79. Gong, I.Y.; Al-Amro, B.; Prasad, G.V.R.; Connelly, P.W.; Wald, R.M.; Wald, R.; Deva, D.P.; Leong-Poi, H.; Nash, M.M.; Yuan, W.; et al. Cardiovascular magnetic resonance left ventricular strain in end-stage renal disease patients after kidney transplantation. *J. Cardiovasc. Magn. Reson.* **2018**, *20*, 83. [CrossRef]

80. Singh, A.; Steadman, C.D.; Khan, J.N.; Horsfield, M.A.; Bekele, S.; Nazir, S.A.; Kanagala, P.; Masca, N.G.; Clarysse, P.; McCann, G.P. Intertechnique agreement and interstudy reproducibility of strain and diastolic strain rate at 1.5 and 3 tesla: A comparison of feature-tracking and tagging in patients with aortic stenosis. *J. Magn. Reson. Imaging* **2015**, *41*, 1129–1137. [CrossRef]

81. Moody, W.E.; Taylor, R.J.; Edwards, N.C.; Chue, C.D.; Umar, F.; Taylor, T.J.; Ferro, C.J.; Young, A.A.; Townend, J.N.; Leyva, F.; et al. Comparison of magnetic resonance feature tracking for systolic and diastolic strain and strain rate calculation with spatial modulation of magnetization imaging analysis. *J. Magn. Reson. Imaging* **2015**, *41*, 1000–1012. [CrossRef]

82. Hor, K.N.; Gottliebson, W.M.; Carson, C.; Wash, E.; Cnota, J.; Fleck, R.; Wansapura, J.; Klimeczek, P.; Al-Khalidi, H.R.; Chung, E.S.; et al. Comparison of Magnetic Resonance Feature Tracking for Strain Calculation with Harmonic Phase Imaging Analysis. *JACC Cardiovasc. Imaging* **2010**, *3*, 144–151. [CrossRef] [PubMed]

83. Harrild, D.M.; Han, Y.; Geva, T.; Zhou, J.; Marcus, E.; Powell, A.J. Comparison of cardiac MRI tissue tracking and myocardial tagging for assessment of regional ventricular strain. *Int. J. Cardiovasc. Imaging* **2012**, *28*, 2009–2018. [CrossRef] [PubMed]

84. Taylor, R.J.; Umar, F.; Lin, E.L.S.; Ahmed, A.; Moody, W.E.; Mazur, W.; Stegemann, B.; Townend, J.N.; Steeds, R.P.; Leyva, F. Mechanical effects of left ventricular midwall fibrosis in non-ischemic cardiomyopathy. *J. Cardiovasc. Magn. Reson.* **2015**, *18*, 1–8. [CrossRef] [PubMed]

85. Wu, L.; Germans, T.; Güçlü, A.; Heymans, M.W.; Allaart, C.P.; van Rossum, A.C. Feature tracking compared with tissue tagging measurements of segmental strain by cardiovascular magnetic resonance. *J. Cardiovasc. Magn. Reson.* **2014**, *16*, 10. [CrossRef]

86. Augustine, D.; Lewandowski, A.J.; Lazdam, M.; Rai, A.; Francis, J.; Myerson, S.; Noble, A.; Becher, H.; Neubauer, S.; Petersen, S.E.; et al. Global and regional left ventricular myocardial deformation measures by magnetic resonance feature tracking in healthy volunteers: Comparison with tagging and relevance of gender. *J. Cardiovasc. Magn. Reson.* **2013**, *15*, 8. [CrossRef]

87. Rüssel, I.K.; Götte, M.J.W.; Bronzwaer, J.G.; Knaapen, P.; Paulus, W.J.; van Rossum, A.C. Left Ventricular Torsion. *JACC Cardiovasc. Imaging* **2009**, *2*, 648–655. [CrossRef]

88. Peteiro, J.; Bouzas-Mosquera, A.; Barge-Caballero, G.; Martinez, D.; Yañez, J.C.; Lopez-Perez, M.; Gargallo, P.; Castro-Beiras, A. Left Ventricular Torsion During Exercise in Patients with and Without Ischemic Response to Exercise Echocardiography. *Rev. Española Cardiol. (Engl. Ed.)* **2014**, *67*, 706–716. [CrossRef]

89. Peteiro, J.; Bouzas-Mosquera, A.; Broullon, J.; Sanchez-Fernandez, G.; Barbeito, C.; Perez-Cebey, L.; Martinez, D.; Rodriguez, J.M.V. Left ventricular torsion and circumferential strain responses to exercise in patients with ischemic coronary artery disease. *Int. J. Cardiovasc. Imaging* **2017**, *33*, 57–67. [CrossRef]

90. Popescu, B.A.; Calin, A.; Beladan, C.C.; Muraru, D.; Rosca, M.; Deleanu, D.; Lancellotti, P.; Antonini-Canterin, F.; Nicolosi, G.L.; Ginghina, C. Left ventricular torsional dynamics in aortic stenosis: Relationship between left ventricular untwisting and filling pressures. A two-dimensional speckle tracking study. *Eur. J. Echocardiogr.* **2010**, *11*, 406–413. [CrossRef]

91. Rosen, B.D.; Gerber, B.L.; Edvardsen, T.; Castillo, E.; Amado, L.C.; Nasir, K.; Kraitchman, D.L.; Osman, N.F.; Bluemke, D.A.; Lima, J.A.C. Late systolic onset of regional LV relaxation demonstrated in three-dimensional space by MRI tissue tagging. *Am. J. Physiol.-Heart Circ. Physiol.* **2004**, *287*, H1740–H1746. [CrossRef]

92. Kowallick, J.T.; Lamata, P.; Hussain, S.T.; Kutty, S.; Steinmetz, M.; Sohns, J.S.; Fasshauer, M.; Staab, W.; Unterberg-Buchwald, C.; Bigalke, B.; et al. Quantification of Left Ventricular Torsion and Diastolic Recoil Using Cardiovascular Magnetic Resonance Myocardial Feature Tracking. *PLoS ONE* **2014**, *9*, e109164. [CrossRef]

93. Lehmonen, L.; Jalanko, M.; Tarkiainen, M.; Kaasalainen, T.; Kuusisto, J.; Lauerma, K.; Savolainen, S. Rotation and torsion of the left ventricle with cardiovascular magnetic resonance tagging: Comparison of two analysis methods. *BMC Med. Imaging* **2020**, *20*, 73. [CrossRef]

94. Yoneyama, K.; Gjesdal, O.; Choi, E.-Y.; Wu, C.O.; Hundley, W.G.; Gomes, A.S.; Liu, C.-Y.; McClelland, R.L.; Bluemke, D.; Lima, J.A. Age, Sex, and Hypertension-Related Remodeling Influences Left Ventricular Torsion Assessed by Tagged Cardiac Magnetic Resonance in Asymptomatic Individuals. *Circulation* **2012**, *126*, 2481–2490. [CrossRef]

95. Yoneyama, K.; Venkatesh, B.A.; Wu, C.O.; Mewton, N.; Gjesdal, O.; Kishi, S.; McClelland, R.L.; Bluemke, D.A.; Lima, J.A.C. Diabetes mellitus and insulin resistance associate with left ventricular shape and torsion by cardiovascular magnetic resonance imaging in asymptomatic individuals from the multi-ethnic study of atherosclerosis. *J. Cardiovasc. Magn. Reson.* **2018**, *20*, 53. [CrossRef]

96. Wei, L.; Ge, H.; Pu, J. Prognostic implications of left ventricular torsion by feature-tracking cardiac magnetic resonance in patients with ST-elevation myocardial infarction. *Eur. Heart J. Cardiovasc. Imaging* **2021**, *22*, jeab090. [CrossRef]

97. Sharifov, O.F.; Schiros, C.G.; Aban, I.; Perry, G.J.; Dell'Italia, L.J.; Lloyd, S.G.; Denney, T.S., Jr.; Gupta, H. Left Ventricular Torsion Shear Angle Volume Approach for Noninvasive Evaluation of Diastolic Dysfunction in Preserved Ejection Fraction. *J. Am. Heart Assoc.* **2018**, *7*, jeab090. [CrossRef]

98. Young, A.A.; Cowan, B.R. Evaluation of left ventricular torsion by cardiovascular magnetic resonance. *J. Cardiovasc. Magn. Reson.* **2012**, *14*, 49. [CrossRef]

99. van der Toorn, A.; Barenbrug, P.; Snoep, G.; Van der Veen, F.H.; Delhaas, T.; Prinzen, F.W.; Maessen, J.; Arts, T. Transmural gradients of cardiac myofiber shortening in aortic valve stenosis patients using MRI tagging. *Am. J. Physiol.-Heart Circ. Physiol.* **2002**, *283*, H1609–H1615. [CrossRef]

100. Delhaas, T.; Kotte, J.; van der Toorn, A.; Snoep, G.; Prinzen, F.W.; Arts, T. Increase in left ventricular torsion-to-shortening ratio in children with valvular aortic stenosis. *Magn. Reson. Med.* **2004**, *51*, 135–139. [CrossRef]

101. Rüssel, I.K.; Brouwer, W.P.; Germans, T.; Knaapen, P.; Marcus, T.J.; Van Der Velden, J.; Götte, M.J.; Van Rossum, A.C. Increased left ventricular torsion in hypertrophic cardiomyopathy mutation carriers with normal wall thickness. *J. Cardiovasc. Magn. Reson.* **2011**, *13*, 3. [CrossRef]

102. Cameli, M.; Mondillo, S.; Righini, F.M.; Lisi, M.; Dokollari, A.; Lindqvist, P.; Maccherini, M.; Henein, M. Left Ventricular Deformation and Myocardial Fibrosis in Patients with Advanced Heart Failure Requiring Transplantation. *J. Card. Fail.* **2016**, *22*, 901–907. [CrossRef] [PubMed]

103. Mewton, N.; Liu, C.Y.; Croisille, P.; Bluemke, D.; Lima, J.A.C. Assessment of Myocardial Fibrosis with Cardiovascular Magnetic Resonance. *J. Am. Coll. Cardiol.* **2011**, *57*, 891–903. [CrossRef] [PubMed]

104. Karaahmet, T.; Gürel, E.; Tigen, K.; Güler, A.; Dündar, C.; Fotbolcu, H.; Basaran, Y. The effect of myocardial fibrosis on left ventricular torsion and twist in patients with non-ischemic dilated cardiomyopathy. *Cardiol. J.* **2013**, *20*, 276–286. [CrossRef] [PubMed]

105. Csecs, I.; Pashakhanloo, F.; Paskavitz, A.; Jang, J.; Al-Otaibi, T.; Neisius, U.; Manning, W.J.; Nezafat, R. Association Between Left Ventricular Mechanical Deformation and Myocardial Fibrosis in Nonischemic Cardiomyopathy. *J. Am. Heart Assoc.* **2020**, *9*, e016797. [CrossRef]

106. Badano, L.P.; Muraru, D. Twist Mechanics of the Left Ventricle. *Circ. Cardiovasc. Imaging* **2019**, *12*, e009085. [CrossRef]

107. Menting, M.E.; Eindhoven, J.A.; van den Bosch, A.E.; Cuypers, J.A.A.E.; Ruys, T.P.E.; Van Dalen, B.M.; McGhie, J.S.; Witsenburg, M.; Helbing, W.A.; Geleijnse, M.L.; et al. Abnormal left ventricular rotation and twist in adult patients with corrected tetralogy of Fallot. *Eur. Heart J. Cardiovasc. Imaging* **2014**, *15*, 566–574. [CrossRef]

108. Bojer, A.S.; Soerensen, M.H.; Gaede, P.; Myerson, S.; Madsen, P.L. Left Ventricular Diastolic Function Studied with Magnetic Resonance Imaging: A Systematic Review of Techniques and Relation to Established Measures of Diastolic Function. *Diagnostics* **2021**, *11*, 1282. [CrossRef]

109. Paetsch, I.; Föll, D.; Kaluza, A.; Luechinger, R.; Stuber, M.; Bornstedt, A.; Wahl, A.; Fleck, E.; Nagel, E. Magnetic resonance stress tagging in ischemic heart disease. *Am. J. Physiol.-Heart Circ. Physiol.* **2005**, *288*, H2708–H2714. [CrossRef]

110. Marchal, P.; Lairez, O.; Cognet, T.; Chabbert, V.; Barrier, P.; Berry, M.; Méjean, S.; Roncalli, J.; Rousseau, H.; Carrié, D.; et al. Relationship between left ventricular sphericity and trabeculation indexes in patients with dilated cardiomyopathy: A cardiac magnetic resonance study. *Eur. Heart J. Cardiovasc. Imaging* **2013**, *14*, 914–920. [CrossRef]

111. ben Halima, A.; Zidi, A. The cardiac magnetic resonance sphericity index in the dilated cardiomyopathy: New diagnostic and prognostic marker. *Arch. Cardiovasc. Dis. Suppl.* **2018**, *10*, 42. [CrossRef]

112. Krittayaphong, R.; Boonyasirinant, T.; Saiviroonporn, P.; Thanapiboonpol, P.; Nakyen, S.; Udompunturak, S. Correlation Between NT-Pro BNP Levels and Left Ventricular Wall Stress, Sphericity Index and Extent of Myocardial Damage: A Magnetic Resonance Imaging Study. *J. Card. Fail.* **2008**, *14*, 687–694. [CrossRef]

113. Cojan-Minzat, B.O.; Zlibut, A.; Muresan, I.D.; Cionca, C.; Horvat, D.; Kiss, E.; Revnic, R.; Florea, M.; Ciortea, R.; Agoston-Coldea, L. Left Ventricular Geometry and Replacement Fibrosis Detected by cMRI Are Associated with Major Adverse Cardiovascular Events in Nonischemic Dilated Cardiomyopathy. *J. Clin. Med.* **2020**, *9*, 1997. [CrossRef]

114. Yazaki, M.; Nabeta, T.; Inomata, T.; Maemura, K.; Oki, T.; Fujita, T.; Ikeda, Y.; Ishii, S.; Naruke, T.; Ako, J. Clinical significance of left atrial geometry in patients with dilated cardiomyopathy: A cardiovascular magnetic resonance study. *Eur. Heart J.* **2020**, *41*, ehaa946. [CrossRef]

115. Nakamori, S.; Ismail, H.; Ngo, L.H.; Manning, W.J.; Nezafat, R. Left ventricular geometry predicts ventricular tachyarrhythmia in patients with left ventricular systolic dysfunction: A comprehensive cardiovascular magnetic resonance study. *J. Cardiovasc. Magn. Reson.* **2017**, *19*, 79. [CrossRef]

116. Ambale-Venkatesh, B.; Yoneyama, K.; Sharma, R.K.; Ohyama, Y.; O Wu, C.; Burke, G.L.; Shea, S.; Gomes, A.S.; A Young, A.; Bluemke, D.; et al. Left ventricular shape predicts different types of cardiovascular events in the general population. *Heart* **2017**, *103*, 499–507. [CrossRef]

117. Arenja, N.; Andre, F.; Riffel, J.H.; Siepen, F.A.D.; Hegenbart, U.; Schönland, S.; Kristen, A.V.; Katus, H.A.; Buss, S.J. Prognostic value of novel imaging parameters derived from standard cardiovascular magnetic resonance in high risk patients with systemic light chain amyloidosis. *J. Cardiovasc. Magn. Reson.* **2019**, *21*, 53. [CrossRef]

118. Leng, S.; Tan, R.-S.; Zhao, X.; Allen, J.C.; Koh, A.S.; Zhong, L. Fast long-axis strain: A simple, automatic approach for assessing left ventricular longitudinal function with cine cardiovascular magnetic resonance. *Eur. Radiol.* **2020**, *30*, 3672–3683. [CrossRef]

119. Riffel, J.H.; Andre, F.; Maertens, M.; Rost, F.; Keller, M.G.P.; Giusca, S.; Seitz, S.; Kristen, A.V.; Müller, M.; Giannitsis, E.; et al. Fast assessment of long axis strain with standard cardiovascular magnetic resonance: A validation study of a novel parameter with reference values. *J. Cardiovasc. Magn. Reson.* **2015**, *17*, 69. [CrossRef]

120. Gjesdal, O.; Yoneyama, K.; Mewton, N.; Wu, C.; Gomes, A.S.; Hundley, G.; Prince, M.; Shea, S.; Liu, K.; Bluemke, D.A.; et al. Reduced long axis strain is associated with heart failure and cardiovascular events in the multi-ethnic study of Atherosclerosis. *J. Magn. Reson. Imaging* **2016**, *44*, 178–185. [CrossRef]

121. Agoston-Coldea, L.; Bheecarry, K.; Cionca, C.; Petra, C.; Strimbu, L.; Ober, C.; Lupu, S.; Fodor, D.; Mocan, T. Incremental Predictive Value of Longitudinal Axis Strain and Late Gadolinium Enhancement Using Standard CMR Imaging in Patients with Aortic Stenosis. *J. Clin. Med.* **2019**, *8*, 165. [CrossRef]

122. Riffel, J.H.; Keller, M.G.P.; Rost, F.; Arenja, N.; Andre, F.; aus dem Siepen, F.; Fritz, T.; Ehlermann, P.; Taeger, T.; Frankenstein, L.; et al. Left ventricular long axis strain: A new prognosticator in non-ischemic dilated cardiomyopathy? *J. Cardiovasc. Magn. Reson.* **2016**, *18*, 36. [CrossRef] [PubMed]

123. Bastos, M.B.; Burkhoff, D.; Maly, J.; Daemen, J.; Uil, C.A.D.; Ameloot, K.; Lenzen, M.; Mahfoud, F.; Zijlstra, F.; Schreuder, J.J.; et al. Invasive left ventricle pressure–volume analysis: Overview and practical clinical implications. *Eur. Heart J.* **2020**, *41*, 1286–1297. [CrossRef] [PubMed]

124. Brimioulle, S.; Wauthy, P.; Ewalenko, P.; Rondelet, B.; Vermeulen, F.; Kerbaul, F.; Naeije, R. Single-beat estimation of right ventricular end-systolic pressure-volume relationship. *Am. J. Physiol.-Heart Circ. Physiol.* **2003**, *284*, H1625–H1630. [CrossRef] [PubMed]

125. Kuehne, T.; Yilmaz, S.; Steendijk, P.; Moore, P.; Groenink, M.; Saaed, M.; Weber, O.; Higgins, C.B.; Ewert, P.; Fleck, E.; et al. Magnetic Resonance Imaging Analysis of Right Ventricular Pressure-Volume Loops. *Circulation* **2004**, *110*, 2010–2016. [CrossRef] [PubMed]
126. Schmitt, B.; Steendijk, P.; Lunze, K.; Ovroutski, S.; Falkenberg, J.; Rahmanzadeh, P.; Maarouf, N.; Ewert, P.; Berger, F.; Kuehne, T. Integrated Assessment of Diastolic and Systolic Ventricular Function Using Diagnostic Cardiac Magnetic Resonance Catheterization. *JACC Cardiovasc. Imaging* **2009**, *2*, 1271–1281. [CrossRef]
127. Witschey, W.R.T.; Contijoch, F.; McGarvey, J.R.; Ferrari, V.A.; Hansen, M.; Lee, M.E.; Takebayashi, S.; Aoki, C.; Chirinos, J.A.; Yushkevich, P.A.; et al. Real-Time Magnetic Resonance Imaging Technique for Determining Left Ventricle Pressure-Volume Loops. *Ann. Thorac. Surg.* **2014**, *97*, 1597–1603. [CrossRef]
128. Giao, D.M.; Wang, Y.; Rojas, R.; Takaba, K.; Badathala, A.; Spaulding, K.A.; Soon, G.; Zhang, Y.; Wang, V.Y.; Haraldsson, H.; et al. Left ventricular geometry during unloading and the end-systolic pressure volume relationship: Measurement with a modified real-time MRI-based method in normal sheep. *PLoS ONE* **2020**, *15*, e0234896. [CrossRef]
129. Faragli, A.; Tanacli, R.; Kolp, C.; Abawi, D.; Lapinskas, T.; Stehning, C.; Schnackenburg, B.; Muzio, F.P.L.; Fassina, L.; Pieske, B.; et al. Cardiovascular magnetic resonance-derived left ventricular mechanics—Strain, cardiac power and end-systolic elastance under various inotropic states in swine. *J. Cardiovasc. Magn. Reson.* **2020**, *22*, 79. [CrossRef]

Review

Nephrogenic Systemic Fibrosis in Patients with Chronic Kidney Disease after the Use of Gadolinium-Based Contrast Agents: A Review for the Cardiovascular Imager

Sebastian Gallo-Bernal [1,2,*,†], Nasly Patino-Jaramillo [3,†], Camilo A. Calixto [2,4], Sergio A. Higuera [3], Julian F. Forero [5], Juliano Lara Fernandes [6], Carlos Góngora [2,7], Michael S. Gee [1,2], Brian Ghoshhajra [2,7] and Hector M. Medina [3]

[1] Department of Radiology, Massachusetts General Hospital, Boston, MA 02114, USA; msgee@mgh.harvard.edu
[2] Department of Radiology, Harvard Medical School, Boston, MA 02115, USA; camilo.calixtonunez@childrens.harvard.edu
[3] Division of Cardiology, Fundacion Cardioinfantil-LaCardio, Bogota 110131, Colombia; naslygisellp@gmail.com (N.P.-J.); seanhile@hotmail.com (S.A.H.); hmedina@lacardio.org (H.M.M.)
[4] Department of Radiology Boston Children's Hospital, Boston, MA 02115, USA
[5] Division of Radiology, Fundacion Cardioinfantil-LaCardio, Bogota 110131, Colombia; jforero@lacardio.org
[6] Jose Michel Kalaf Research Institute, Radiologia Clinica de Campinas, São Paulo 13092-123, Brazil; jlaraf@terra.com.br
[7] Cardiovascular Imaging Research Center (CIRC), Division of Cardiology, Massachusetts General Hospital, Boston, MA 02114, USA; drcgongora@gmail.com (C.G.); bghoshhajra@mgh.harvard.edu (B.G.)
* Correspondence: sgallobernal@mgh.harvard.edu
† These authors contributed equally to this manuscript.

Citation: Gallo-Bernal, S.; Patino-Jaramillo, N.; Calixto, C.A.; Higuera, S.A.; Forero, J.F.; Lara Fernandes, J.; Góngora, C.; Gee, M.S.; Ghoshhajra, B.; Medina, H.M. Nephrogenic Systemic Fibrosis in Patients with Chronic Kidney Disease after the Use of Gadolinium-Based Contrast Agents: A Review for the Cardiovascular Imager. *Diagnostics* **2022**, *12*, 1816. https://doi.org/10.3390/diagnostics12081816

Academic Editors: Minjie Lu and Arlene Sirajuddin

Received: 10 June 2022
Accepted: 22 July 2022
Published: 28 July 2022

Abstract: Gadolinium-enhanced cardiac magnetic resonance has revolutionized cardiac imaging in the last two decades and has emerged as an essential and powerful tool for the characterization and treatment guidance of a wide range of cardiovascular diseases. However, due to the high prevalence of chronic renal dysfunction in patients with cardiovascular conditions, the risk of nephrogenic systemic fibrosis (NSF) after gadolinium exposure has been a permanent concern. Even though the newer macrocyclic agents have proven to be much safer in patients with chronic kidney disease and end-stage renal failure, clinicians must fully understand the clinical characteristics and risk factors of this devastating pathology and maintain a high degree of suspicion to prevent and recognize it. This review aimed to summarize the existing evidence regarding the physiopathology, clinical manifestations, diagnosis, and prevention of NSF related to the use of gadolinium-based contrast agents.

Keywords: nephrogenic systemic fibrosis; gadolinium-based contrast agents; chronic kidney disease; cardiac magnetic resonance

1. Introduction to Gadolinium-Based Contrast Agents

Magnetic resonance imaging (MRI) contrast agents serve to improve diagnostic images' sensitivity and specificity by altering the tissues' intrinsic properties. Contrast agents carry strong paramagnetic properties that can be exploited to provide enhanced contrast between healthy and diseased tissues. By shortening the T1 and T2 relaxation times of the contiguous hydrogen nuclei, paramagnetic elements such as gadolinium (Gd) enhance the soft tissue contrast and help characterize a wide array of pathologies [1,2].

Even though elemental Gd can be toxic for humans [3], this element can be safely administered when combined with organic chelates designed to reduce the release of free Gd ions. These Gd organic chelate compounds are the basic structure of gadolinium-based contrast agents (GBCAs).

The pharmacokinetics of GBCAs helps differentiate between normal and diseased myocardium. Once administered, GBCAs diffuse rapidly out of capillaries into tissues but cannot cross intact cell membranes and equilibrate with the extracellular space. As a result, both healthy and sick myocardium accumulate GBCAs in their interstitial fluid. However, a combination of an increased volume of distribution and slower washout kinetics in sick tissues with an expanded extracellular fluid promote prolonged retention of the GBCAs, which can be detected in the late washout phase [4].

The relative accumulation of gadolinium in areas of expanded extracellular space can be seen in multiple pathologic scenarios such as fibrosis, myocardial disarray, and pathological extracellular protein infiltration [4,5]. This characteristic of the GBCAs is the key to late gadolinium enhancement (LGE), which has revolutionized cardiac magnetic resonance (CMR) in the last two decades, allowing characterization of several types of cardiomyopathies based on scar distribution.

2. Linear vs. Macrocyclic GBCAs

GBCAs can be classified as linear (e.g., Gd ion bridging diethylenepenta-acetic acid) or macrocyclic (e.g., a rigid, cage-like tetra-azaclododecone compound rings Gd 3+ ions), depending on the structure that encapsulates the free Gd [5,6]. Some of the main physicochemical features of linear GBCAs (L-GBCAs) and macrocyclic GBCAs (M-GBCAs) are illustrated in Table 1 [7–14].

Intrinsic characteristics of GBCAs such as relaxivity and thermodynamic stability are described in Table 1. Relaxivity refers to the contrast's ability to increase the surrounding water proton relaxation rate. Higher relaxivities indicate a more potent agent that requires a lower dose in order to obtain clinically useful images [12]. The hydration state of the GBCAs is the most relevant determinant of an agent's relaxivity. By improving this parameter, it is possible to increase the clinical utility of the GBCAs [12].

On the other hand, the thermodynamic stability describes how much Gd is released at equilibrium under certain circumstances [12]. There is an inverse relationship between the complex hydration state and its thermodynamic stability [2]. As a result, a higher relaxivity decreases the thermodynamic stability of the complex, facilitating transmetalation (a process by which endogenous metals—e.g., Fe, Cu, Zn—replace Gd in the complex, freeing it from the chelate molecule) and rendering Gd more accessible to endogenous anions [12]. Transmetalation is responsible for the dissociation of GBCAs and Gd's release in vivo.

Once Gd escapes from its organic cage, competitive binding of this metal with endogenous anions such as $CO_3{}^{2-}$ and $PO_4{}^{3-}$ promotes the formation of insoluble compounds which are free to extravasate from the bloodstream and deposit in target tissues [12,16]. In clinical practice, the release of Gd (especially by L-GBACs) increases the risk of nephrogenic systemic fibrosis (NSF) [12,17,18]. As Gd is almost exclusively cleared by the kidney, patients with chronic kidney disease (CKD) have a significantly higher risk of NSF due to the higher half-life of this metal in this population and increased risk of Gd dissociation, transmetalation, and tissue deposition.

M-GBCAs have higher thermodynamic stabilities without a significant decrease in their potency to create clinically useful images. This phenomenon can be explained by the fact they avoid the freeing of Gd ions by a lower de-chelation rate of the macrocyclic ring due to its improved molecular stability compared with linear agents [6,13].

Since their development, clinicians have shown an increased interest in M-GBCAs, such as gadobutrol and gadoterate meglumine, due to their low theoretical risk for developing NSF in patients with and without CKD and ESRD [3,19]. Their physicochemical characteristics and the experience collected during the past decades point toward a relative safety superiority of these contrast agents, given their security profile and low incidence of adverse events.

Table 1. Physicochemical characteristics of gadolinium-based contrast agents [7–14].

Molecule, Trade Name, and (Vendor)	Type	Osmolality (mOsm/kg H₂O) *	Viscosity (mPa•) *	Thermodynamic Complex Stability (log Keq)	Relaxivities r1/r2 in Plasma (mmol/L) at 1.5 T *	Concentration (mol/L)	Approved Intravenous Dose (mmol/kg)	Comments
Gadobutrol, Gadavist (Bayer Healthcare)	Macrocyclic Non-ionic	1603	4.96	21.8	5.2/6.1	1	0.1–0.3	Highest viscosity. It is marketed as Gadovist outside the United States.
Gadoterate meglumine, Dotarem (Guerbet)	Macrocyclic Ionic	1350	2.4	25.6	3.6/4.3	0.5	0.1–0.2	In 2019, a generic version of Dotarem was introduced (Clariscan, GE Healthcare). Some animal studies have shown slightly higher levels of gadolinium deposition with Clariscan compared to Dotarem [15].
Gadoteridol, ProHance (Bracco)	Macrocyclic Non-ionic	630	1.3	23.8	4.1/5	0.5	0.1–0.2	Lowest viscosity and osmolality. Below average viscosity.
Gadopentetate dimeglumine, Magnevist (Bayer Healthcare)	Linear Ionic	1960	1.9	22.1	4.1/4.6	0.5	0.1–0.3	Oldest approved agent. Below average relaxivity. High risk of NSF.
Gadodiamide, Omniscan (GE Healthcare)	Linear Non-ionic	783	1.4	16.9	4.3/5.2	0.5	0.1–0.3	Low thermodynamic stability; very high risk of NSF. Use suspended in the European Union.
Gadobenate dimeglumine, MultiHance (Bracco)	Linear Ionic	1970	5.3	22.6	6.3/8.7	0.5	0.05–0.1	Highest relaxivity of extracellular GBCAs. EMA restricted to hepatobiliary imaging.
Gadoxetate disodium, Eovist/Primovist (Bayer Healthcare)	Linear Ionic	688	1.19	23.5	6.9/8.7	0.25	0.025	Designed for liver imaging. Renal and biliary excretion. Very high relaxivity. EMA restricted to hepatobiliary imaging.

* At 37 °C; NSF: nephrogenic systemic fibrosis; GBCA: gadolinium-based contrast agent; EMA: European Medicines Agency.

3. Clinical Use of GBCAs in Cardiovascular Magnetic Resonance

Gadobutrol 1.0 mmol/mL was approved for neuroimaging in January 2000 in Germany and in June 2000 in the United States. Subsequently, in the United States, it gained approval by the FDA for angiography in November 2003 [20] and CMR in 2005 [14]. GBCAs have several applications in CMR, including the characterization of a wide range of cardiomyopathies [21–23]. A meta-analysis of 164 studies found that the most common cardiovascular applications of GBCAs during CMR were myocardial infarction and functional testing followed by cardiomyopathies characterization. Other applications include the study of myocarditis, valvular diseases, cardiac masses, stable coronary disease, pulmonary hypertension, and right-sided heart failure [24]. Utility of CMR using gadolinium in a patient with borderline renal function is depicted in the clinical case shown next.

Clinical Case: Use of GBCAs in a Patient with Decreased Left Ventricular Ejection Fraction and Increased Thickness

A previously healthy 73-year-old woman was admitted to the emergency room due to a 7-month history of worsening dyspnea on exertion, orthopnea, paroxysmal nocturnal dyspnea, and lower extremity without prior medical history. Physical examination showed jugular venous distention, diffuse crackles on both lung fields, and bilateral grade III pitting edema. Initial laboratories revealed anemia (Hb—10.3 g/dL; normal 13–17 gr/dL) and elevated BNP (1256 pg/mL; normal <125 pg/mL) as well as elevated lactate dehydrogenase (682 IU/L; normal 105–333 IU/L) and creatinine (1.8 mg/dL; GFR 27 mL/min/1.73 m^2). No proteinuria was detected.

A transthoracic echocardiogram revealed thickened interatrial and interventricular septum, severe end-diastolic thickening of the left ventricle with a "granular sparkling" appearance, and a severely reduced left ventricular systolic function (LVEF) of 15%. Impaired relaxation and elevated filling pressures were consistent with severe diastolic dysfunction. A huge atrial thrombus protruding from the left appendage was noticed.

A CMR was then ordered using gadobutrol at 1 mL/kg—the latest serum creatinine was 1.5 mg/dL/GFR 34.2 mL/min/1.73 m^2—that showed global diffuse circumferential left ventricular LGE with papillary muscle involvement (Figure 1) and extension to both atria. Cardiac amyloidosis diagnosis was considered followed by serum protein electrophoresis with no abnormal bands; however, serum immunofixation revealed a monoclonal spike determined to be IgGλ (lambda). Serum-free light chains showed an elevation in free lambda with an abnormal κ/λ ratio. A bone marrow biopsy demonstrated 1.36% of monotypic plasma cells staining for lambda light chains.

The patient was ultimately diagnosed with cardiac amyloidosis, but the family declined a percutaneous biopsy for further characterization. In the patient with CKD, the utilization of gadolinium on CMR helped determine the final diagnosis. Additionally, the were no signs of NSF at 18-month follow-up.

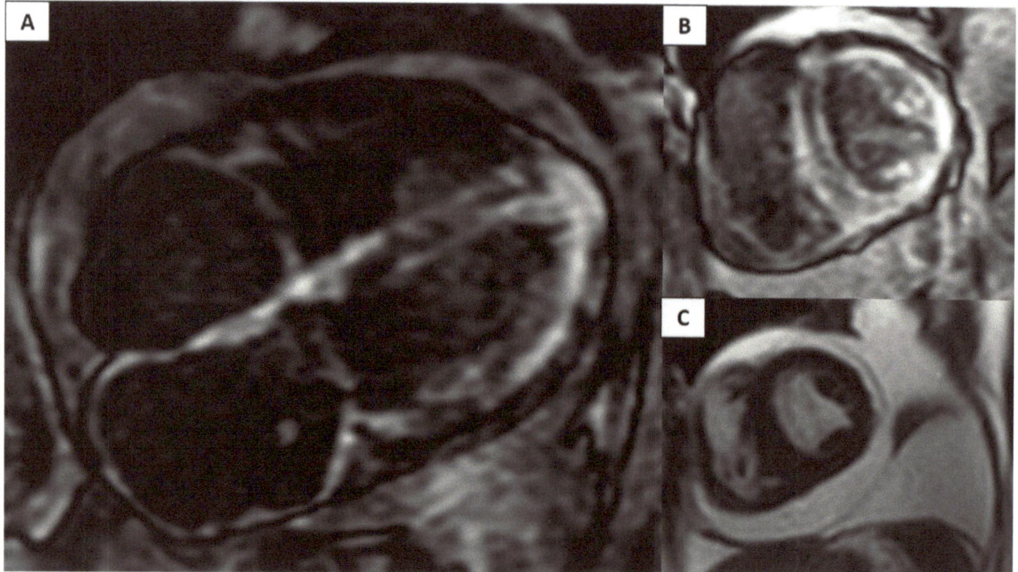

Figure 1. CMR in a patient with amyloidosis. (**A**) Apical-4 chamber showing diffuse LGE in the left ventricle and both atria. (**B**) Short-axis LGE with similar findings. (**C**) Short-axis cine with increased tele-diastolic thickness of the left ventricle, with pericardial and pleural effusions.

4. Overview of Nephrogenic Systemic Fibrosis

NSF is a multisystemic fibrotic disease that affects the skin, muscle, and other organs (including lung, esophagus, and kidney) described in patients with severe renal impairment exposed to a GBCA [3]. The pathophysiology and molecular mechanism of NSF are still a matter of debate. It is believed that the intravenous administration of some GBCAs causes a limited chelate instability that plays an essential role in the release of free Gd (Figure 2) [3,25].

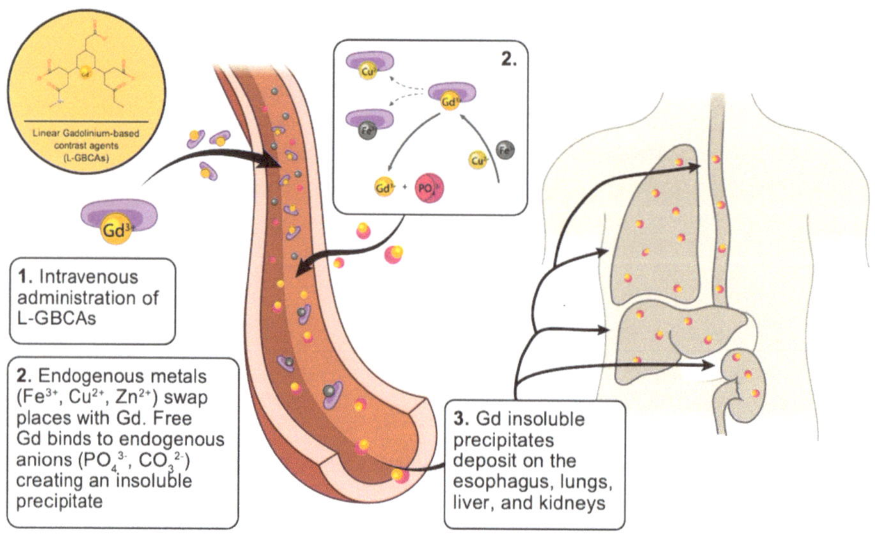

Figure 2. NSF pathophysiology: transmetalation and deposition of gadolinium-based contrast agents.

After de-chelation (by transmetalation), free Gd binds with endogenous anions, creating an insoluble precipitate that penetrates the interstitial tissue of the lung, esophagus, liver, and kidneys. In vitro studies have shown that Gd-anion complexes are highly immunogenic, binding to toll-like receptors (TLRs) on professional antigen-presenting cells (such as macrophages and dendritic cells) and leading to the release of pro-inflammatory and pro-fibrotic cytokines (Figure 3) [1].

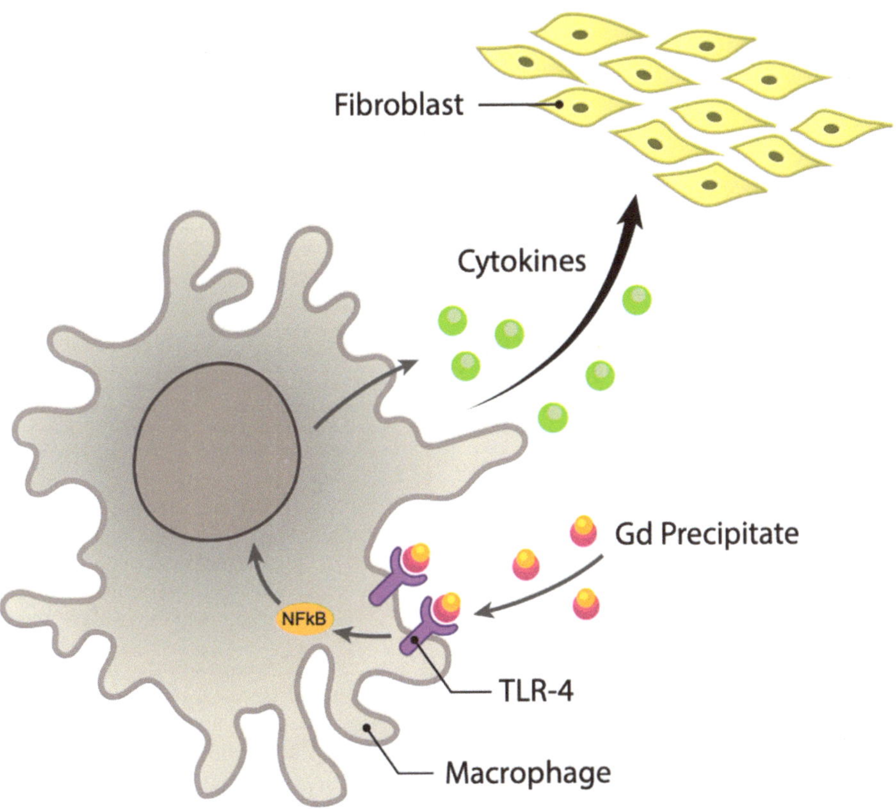

Figure 3. NSF: inflammatory response and subsequent systemic fibrotic reaction.

Most GBCAs have preferential renal elimination, and their clearance is highly dependent on the glomerular filtration rate. A small percentage of the administered dose is eliminated via the hepatobiliary route. In CKD patients, the prolonged GBCAs' half-life leads to a significant release of Gd, and therefore, a higher burden of Gd precipitates. Moreover, some of the anions that are thought to play a critical role in transmetalation (such as phosphates) are often elevated in CKD patients, facilitating this pathologic process by increasing the substrate availability for compound formation [26,27].

5. General Risk Factors for NSF

The main risk factor for NSF is the presence of severe acute or chronic renal insufficiency (estimated glomerular filtration rate (eGFR): 30 mL/min/1.73 m^2) or acute renal insufficiency of any severity due to hepato-renal syndrome or in the perioperative period after liver transplantation. In 2010, the European Medicines Agency (EMA) released a statement in which NSF was considered a potential side effect of GBCAs based on the number of published reports [3]. As a result, the EMA classified the different GBCAs depending on their individual risk of triggering NSF (Table 2) based on the existing evidence and the number of reported cases [3,28]. The American College of Radiology (ACR) also proposed a similar classification system based on similar criteria (Table 3) [25]. Of note, gadobutrol and gadoterate meglumine—the most frequently used M-GBCAs—are known to confer a significantly lower risk of NSF when compared to L-GBCAs.

Table 2. EMA classification for NSF risk among GBCAs [§].

High Risk	Intermediate Risk	Low Risk
Gadodiamide	Gadobenate dimeglumine	Gadobutrol
Gadoversetamide	Gadoxetate disodium	Gadoteridol
Gadopentetate dimeglumine	Gadofosvest	Gadoterate meglumine

[§]: Adapted with permission from: European Medicines Agency: Assessment report for Gadolinium-containing contrast agents; Reference: [10]. 2022, European Medicines Agency.

Table 3. ACR Manual Classification of Gadolinium-Based Agents Relative to NSF [†].

ACR Group I *	ACR Group II **	ACR Group III ***
Gadodiamide	Gadobenate dimeglumine	
Gadoversetamide	Gadobutrol	Gadoxetate disodium
Gadopentetate dimeglumine	Gadoteric acid	
	Gadoteridol	

[†]: Adapted with permission from: ACR Committee on Drugs and Contrast Media. ACR Manual on Contrast Media; Reference: [25]. 2022, American College of Radiology. *: Agents associate with the greatest number of NSF cases. **: Agents associated with few, if any, unconfounded NSF cases. ***: Agents for which data remain limited regarding NSF risk, but for which few, if any, unconfounded cases of NSF have been reported.

Other factors that may increase the risk of NSF include multiple contrast exposures, higher cumulative doses (specially gadodiamide) [29], acidosis, hypercalcemia, hyperphosphatemia, high-dose erythropoietin therapy, hepatorenal syndrome, immunosuppression, vasculopathy, and infection [25]. Additionally, it was suggested that NSF incidence in patients on renal replacement therapy (RRT) was 19% higher in those who received the highest approved dose of any GBCAs [30].

6. Clinical Approach for the Diagnosis of NSF

The first description of what would be known as NSF was published in the year 2000 and consisted of a series of 15 CKD patients from different cities in the United States of America presenting with a scleroderma-like disease [31]. Five years later, additional cases of patients presenting with renal and lung compromise were described, and the term NSF was subsequently coined [32]. In 2006, the first association between NSF and GBCAs was proposed. A case series showed that from a total of nine patients with end-stage renal disease (ESRD) who underwent MR angiography with gadopentetate dimeglumine (a L-GBCA), five developed NSF [33].

Since then, multiple cases with different presentations and histopathological findings have been published, and a scoring system to identify possible NSF cases was subsequently created. Girardi et al. proposed a system based on major and minor clinical and histopathological criteria based on the Yale International NSF Registry [34]. Major criteria include patterned plaques, joint contractures, or pronounced induration (*peau d' orange*), while minor criteria consist of linear banding, superficial plaques, dermal papules, and scleral plaques. NSF physical examination findings are summarized in Table 4 [34,35]. Histopathological criteria consist of increased dermal cellularity, CD34+ cells with tram-tracking, collagen bundles, septal involvement, and osseous metaplasia [34]. A high degree of suspicion should be maintained to detect and link these symptoms with a former GBCA exposure.

Table 4. Signs and symptoms of NSF [7,27,29,34,36,37].

Clinical History	Signs	Symptoms
L-GBCA exposure (2–8 weeks—10 years after gadolinium uptake) **Family history of NSF Renal:** AKI, history of chronic kidney disease, kidney transplantation, or hemodialysis	**Eye:** Whitish-yellow plaques with vascular ectasia **Skin changes:** Hyperpigmentation, symmetrical lesions, rash-patterned plaques (red to violaceous lesions), superficial papules (beefy lesions in upper extremities), macules, nodules, skin thickening (cobble stoning or peau d'orange appearance) **Renal:** Volume overload, uremia **Extremities:** Limited range motion, joint contractures (finger, elbows, toes, and ankles), symmetric edema (inferior limbs)	**Eye:** Vision impairment, conjunctival injection, and white-yellow scleral plaques **Skin:** Pruritus, burning pain, new skin lesion, induration, and swelling **Extremities:** Edema, pain, and decreased mobility of the joints **Urinary findings:** Anuria, oliguria

NSF usually manifests within 2–10 weeks after the initial exposure; however, clinical manifestations may only become apparent a couple of years after GBCA exposure. When inquiring about a former exposure to GBCAs, it should be categorized chronologically as acute (0–60 min), late (1 h–7 days), or very late (>7 days) in order to assess the linkage between the contrast administration and the clinical manifestations [3,38].

If NSF is suspected due to the formerly mentioned clinical and histopathological characteristics, a comprehensive review of risk factors and Gd chronological exposure should be performed. Calculation of the eGFR is vital for NSF diagnosis [3], as some degree of kidney dysfunction should be present in order to fulfill diagnostic criteria. In this way, for an accurate diagnosis, clinicians should focus their attention on the presence of a previous history of kidney transplantation, prior episodes of anuria or oliguria, a significant elevation in serum creatinine, a progressive decrease in the eGFR, or the presence of acute kidney injury (AKI) at the moment of contrast administration [3,29,34]. Such risk factors point toward the possibility of an underlying and undetected CKD, which may have prompted the onset of NSF.

Physicians should inquire for family and personal history of diseases with similar characteristics such as lipo-dermatosclerosis, chronic venous insufficiency, scleroderma, scleroderma diabeticorum, morphea, chronic graft-versus-host disease, amyloidosis, congenital fascial dystrophy, and porphyria cutanea tarda [34]. Even though these cases are rare and may have a sub-clinical course, most NSF patients have similar characteristics (especially in the initial phases) and should be excluded.

NSF severity is graded from 0–4 as follows: 0, asymptomatic; 1, mild physical, dermatologic, or neuropathic symptoms without any kind of disability; 2, moderate physical or neuropathic symptoms limiting physical performance; 3, severe symptoms limiting daily physical activities; and 4, severely disabling symptoms causing dependence on daily activities [38].

7. Histopathologic Examination

Histopathologic examination is essential in the definitive diagnosis of NSF. The presence of dermal hypercellularity, CD34+ cells, procollagen type I, thick and thin collagen bundles, and osseous metaplasia significantly point toward NSF diagnosis [34,35]. Additionally, some grade of fibrosis of skeletal muscle, diaphragm, heart, liver, and lung may be present and could facilitate the diagnosis (although their presence is not specific to NSF) [38].

8. GBCAs' Differential Risk of NSF

As it was previously stated, when compared to L-GBCAs, M-GBCAs confer a significantly lower risk of NSF. Table 5 summarizes some of the studies assessing the safety and tolerability of M-GBCAs and L-GBCAs, especially regarding the incidence of NSF.

As an example of this differential risk between the types of GBCAs, a recent study compared a cohort of 421 patients with a 3.1% NSF incidence exposed to L-GBCAs (gadodiamide, gadopentetate dimeglumine, and gadobenate dimeglumine) versus 0% incidence in those who were exposed to a M-GBCAs (gadoteridol) [30]. Based on various retrospective reports, gadodiamide has the largest number of reported NSF cases (n = 182), followed by gadopentetate dimeglumine (n = 26), gadoversetamide (n = 5), gadoterate meglumine (n = 7), and gadobutrol (n = 3) [16,47,48]. Nevertheless, NSF cases secondary to gadobutrol are still controversial, and a clear causal association has not been established [29,43].

Gadobutrol's safety and tolerability during contrast-enhanced MRI/angiography were evaluated in the GARDIAN study, a multicenter, international registry that included 23,708 patients [41]. The investigators concluded that gadobutrol was safe in patients with preserved kidney function and those with moderate (0.6%) or severe (0.6%) CKD. The frequency of adverse drug reactions (ADRs) was 0.7%. The most frequently reported ADRs were nausea (0.3%), followed by emesis (0.1%) and dizziness (0.1%). There were no NSF cases in the GARDIAN study after a mean follow-up of 2.8 years [41].

In a prospective, international, and multicenter study, Michaely et al. assessed the safety of gadobutrol-enhanced MRI in patients with moderate (n = 586; eGFR < 45 mL/min/1.73 m^2) and severe (n = 284; eGFR < 30 mL/min/1.73 m^2) CKD. A total of 927 patients was enrolled between 2008 to 2016. This study included patients with a history of organ transplantation (7.7%), hemodialysis (9.9%), diabetes (31.9%), and hypertension (58.5%) [43]. The investigators concluded that gadobutrol was safe in their patients with moderate and severe renal impairment, with no NSF cases reported after a two-year follow-up period [43].

The SECURE study assessed the safety and tolerability of gadoterate meglumine in a cohort of 35,499 patients. The total population included 514 patients that had some degree of renal impairment (eGFR less than 60 mL/min/1.73 m^2), including 417 patients with eGFR between 30–60 mL/min/1.73 m^2, 58 subjects with eGFR less than 30 mL/min/1.73 m^2, and 7 with eGFR less than 15 mL/min/1.73 m^2 on RRT. In this study, no NSF cases were observed after a 3-month follow-up period. The most frequent ADRs were urticaria (0.03%), nausea (0.02%), and emesis (0.01%) [42].

In the NSsaFe study, gadoterate meglumine was administered in 540 patients with moderate (69.4%), severe (16%), or end-stage renal impairment (12%). After a maximum follow-up of 2 years, there were no NSF reports, demonstrating gadoterate's safety in this specific group of patients [46].

A recent meta-analysis assessed the safety of ACR group-II GBCAs in patients with stage 4 or 5 CKD (eGFR, <30 mL/min/1.73 m^2) on RRT and concluded that the risk of NSF was less than 0.07% [19]. Interestingly, in the total population of 4931 patients included in this meta-analysis, not a single NSF case was reported secondary to group II GBCA exposure.

Table 5. Safety and tolerability studies of M-GBCAs and L-GBCAs.

Authors, Study Name	Year	Study Type	Total Number of Patients	GBCA	Number of Patients with Renal Impairment	NSF Cases at Maximum Follow-Up	Estimated NSF Incidence
Aneet et al. [39]	2007	Retrospective cohort	467 (87 with gadolinium exposure)	Gadopentetate diglumine (L) and gadodiamide (L)	87 patients with end-stage renal disease (patients in dialysis)	3	4.3 cases per 1000 patients-year (overall NSF incidence)
Wang Y et al. [40]	2011	Retrospective cohort	52,954 (after the 2007 Restrictive GBCA guidelines were implemented)	Gadopentetate diglumine (L) and gadobenate diglumine (L)	6454 patients with GFR between 30–59 mL/min/m^2; 36 patients with GFR lower than 30 mL/min/m^2	0	-
Prince MR et al. GARDIAN study [41]	2016	Prospective cohort	23,708	Gadobutrol (M)	100 patients with moderate renal impairment (GFR: 30–59 mL/min/1.73 m^2) and 31 patients with severe renal impairment (<30 mL/min/1.73 m^2)	0	-
Soyer P et al. SECURE study [42]	2017	Prospective cohort	35,499	Gadorate meglumine (M)	417 patients with moderate renal impairment (GFR: 30–59 mL/min/1.73 m^2); 58 patients with severe renal impairment (GFR: 15–39 mL/min/1.73 m^2); 7 patients with end-stage renal impairment (GFR: <15 mL/min/1.73 m^2) or dialysis.	0	-
Michaely HJ et al. GRIP study [43]	2017	Prospective cohort	908	Gadobutrol (M)	586 with moderate (GFR: 30–59 mL/min/1.73 m^2) and 284 with severe renal impairment (<30 mL/min/1.73 m^2)	0	-
Tsushima Y et al. [44]	2018	Prospective cohort	3337	Gadobutrol (M)	356 patients with GFR between 45–59 mL/min/m^2; 71 patients with GFR between 30–44 mL/min/m^2; 4 patients with GFR between 15–29 mL/min/m^2; 1 patient with GFR < 15 mL/min/m^2	0	-
Young LK [45]	2019	Retrospective cohort	22,897	Gadorate meglumine (M)	2570 patients with moderate renal impairment (GFR: 30–59 mL/min/1.73 m^2); 464 patients with severe renal impairment (GFR: 15–39 mL/min/1.73 m^2); 123 patients with end-stage renal impairment (GFR: <15 mL/min/1.73 m^2) or dialysis.	0	-
McWilliams RG et al. NSsaFe study [46]	2020	Prospective cohort	540	Gadorate meglumine (M)	226 patients with moderate renal impairment (GFR: 30–59 mL/min/1.73 m^2); 59 patients with severe renal impairment (GFR: 15–39 mL/min/1.73 m^2); 58 patients with end-stage renal impairment (GFR: <15 mL/min/1.73 m^2) or dialysis.	0	-

M: macrocyclic gadolinium-based contrast agent. L: linear gadolinium-based contrast agent.

9. Prevention and Treatment of NSF

There is not a specific prophylaxis regimen to prevent the onset of NSF. The current approach is based on minimizing the impact of predisposing risk factors and performing hemodialysis sessions right after GBCA exposure in patients with a history of ESRD on RRT [35]. Hemodialysis or peritoneal dialysis should take place the same day and within 2 or 3 h after contrast administration. Hemodialysis could be more efficient than peritoneal dialysis for gadolinium clearance; however, there is insufficient evidence supporting a clinical superiority of either technique for the prevention of NSF [49] and limited evidence in the use of peritoneal dialysis to effectively remove GBCAs [50]. Maintaining adequate hydration and minimizing the concomitant exposure to nephrotoxic agents (NSAIDs, diuretics, and certain antibiotics) are also recommended, as well as not exceeding the recommended dose of administration.

Although kidney transplant improves renal function, this may not help to treat NSF [29]. Dermatologic symptoms can be treated with thalidomide, calcipotriene, and clobetasol (high-potency topical corticosteroids). Extracorporeal photopheresis improves the articulations' range of motion and skin tightening as well. Finally, pentoxifylline demonstrated efficacy in ameliorating NSF symptoms [51].

10. Take-Home Messages and Clinical Applications

10.1. Estimation of the Glomerular Filtration Rate

- In the outpatient setting, eGFR should be estimated only in those patients with risk factors for CKD. Those patients with no risk factors or confirmed CKD should not undergo additional testing [25].
- Current evidence supports the usage of the Modification of the Diet in Renal Disease (MDRD) or the Chronic Kidney Disease Epidemiology Collaboration (CKD-EPI) in order to estimate the patient's eGFR and base clinical decisions regarding GBCA administration [25,28,36].
- A recent creatinine value should be used (<72 h) for eGFR estimation. However, there is no evidence regarding the most appropriate timing for eGFR estimation [49].

10.2. Patients at Risk for Chronic Kidney Disease

- Outpatients who may be receiving GBCAs should be screened for risk factors or conditions associated with CKD [25]. This assessment should include inquiring about a history of confirmed CKD or any kidney condition (dialysis, kidney transplant, glomerulopathies, single kidney, kidney surgery, or kidney neoplasm), hypertension (requiring medical therapy), cardiovascular disease (including heart failure or coronary disease), and diabetes mellitus on metformin. For those patients identified by screening with one or more risk factors, eGFR estimation with serum creatinine should be performed [25].
- For all inpatients, eGFR should be calculated within two days before the administration of a GBCA. Additionally, the possibility of an undetected AKI should always be considered [25,49].

10.3. Contrast Selection

- In patients with normal kidney function (eGFR > 60 mL/min/1.73 m^2) and no additional risk factors, the incidence of NSF after a GBCA infusion is negligible. As a result, any type of GBCAs can be safely used [43].
- In patients with stage 3 CKD (eGFR 30–50 mL/min/1.73 m^2) and no additional risk factors, NSF's risk is minimal. As a result, no additional actions are necessary.
- In patients with CKD stages 4 and 5 (eGFR < 30 mL/min/1.73 m^2) or patients on RRT, ACR group-I GBCAs are contraindicated (Table 3) [25]. Only ACR group-II GBCAs should be used in this circumstance.

- Acute kidney injury: the presence of AKI significantly increases the risk of NSF [36,39,47]. In addition, the incidence of AKI is significantly higher in patients with confirmed or suspected cardiovascular disease. As a result, additional precautions should be taken into account. In AKI, there is a lag between the serum creatinine values and the actual eGFR. As a result, the sole estimation of eGFR based on creatine values could be problematic. In this setting, the ACR group-I GBCA agents should be avoided in patients with confirmed or suspected AKI [25].

10.4. Dialysis: Specific Recommendation

- In those patients with terminal CKD already on RRT (hemodialysis or peritoneal dialysis), dialysis should continue after receiving a GBCA. GBCA infusion should be performed as closely before hemodialysis as is possible [25]. These patients should receive dialysis the same day of the procedure, ideally 2 to 3 h after the contrast infusion to minimize the possibility of transmetalation and NSF [25,49].
- There is insufficient evidence to support changing patients from peritoneal dialysis to hemodialysis or altering dialysis prescription after the infusion of a GBCA. Peritoneal dialysis may be less effective than hemodialysis in clearing circulating GBCA; however, there is no evidence regarding the superiority of a specific type of RTT in order to decrease the risk of NSF [37,49].

10.5. Patients Who Require Multiple Studies

- NSF occurs most commonly in patients who received high doses of GBCA, either as a single dose or cumulatively after multiple administrations [25]. In some circumstances, patients may require multiple doses of a GBCA within a short time frame; thus, these patients are at a higher risk of developing NSF.
- In patients with preserved or moderately reduced kidney function (eGFR > 30 mL/min/ 1.73 m^2), there is no contraindication if the examinations are determined to be necessary [25]. However, taking into account the elimination time of the GBCAs, it is advisable to wait at least 4 h between studies [37,49]. The usage of an ACR type-II GBCAs is advisable in this circumstance.
- In patients with residual kidney function who do not receive RRT, there should be at least 7 days between each study.
- Hemodialysis efficiently clears 70% of GBCA plasmatic concentrations after one session and more than 95% after three sessions [8,20,52]. As a result, the GBCAs' half-life in patients on hemodialysis is similar to an individual with normal kidney function.

11. Limitations

- Even though current studies may not suggest NSF cases with the use of group II GBCAs, there is still epidemiological limitations to consider NSF risk as zero. There is a small number of patients with CKD stage 5 involved which underestimates the NSF incidence rate [48]. CKD patients should be assessed with a complete medical history and risk factors to determine GBCA use.
- Long-term Gd+3 brain deposition should be taken with great caution in CKD patients. This association is more frequent with the use of L-GBCAs than group II GBCA injections [48,53].
- We did not mention some other alternatives for CKD patients with eGFR < 30 mL/min/ 1.73 m^2. For example, ferumoxytol is a vascular contrast agent for MR angiography with superparamagnetic properties useful to venous and arterial enhancement in stage 4 and 5 CKD patients [54].

12. Essentials

- Gadolinium-based contrast agents serve to improve diagnostic images' sensitivity and specificity and characterize a wide array of cardiovascular pathologies.

- NSF is a devastating, multisystemic fibrotic disease that affects the skin, muscle, and other organs (including lung, esophagus, and kidney) described in patients with severe renal impairment exposed to a gadolinium-based contrast agent.
- There is not a specific treatment or prophylaxis regimen to treat or prevent the onset of NSF.
- Even though the newer macrocyclic agents have proven to be much safer in patients with chronic kidney disease and end-stage renal failure, clinicians must fully understand the clinical characteristics and risk factors of this devastating pathology and maintain a high degree of suspicion to prevent and recognize it. Cardiac MRI with late gadolinium enhancement (LGE) has significantly impacted the management, decision making, and diagnosis of various cardiomyopathy or interstitial heart disease. However, the over-concerned about nephrogenic systemic fibrosis may make cardiac MRI with LGE be avoided inappropriately. The risk and benefits of this imaging study should be balanced.

Author Contributions: Conceptualization, S.G.-B., N.P.-J. and H.M.M.; methodology, N.P.-J., C.A.C., S.A.H. and H.M.M.; investigation, S.G.-B. and N.P.-J.: resources, J.F.F., J.L.F. and C.G.; writing—original draft preparation, S.G.-B., N.P.-J., C.A.C. and H.M.M.; writing—review and editing, all authors.; visualization, C.A.C.; supervision, M.S.G., B.G. and H.M.M.; project administration, S.G.-B. All authors have read and agreed to the published version of the manuscript.

Funding: This research received no external funding.

Conflicts of Interest: The authors declare no conflict of interest.

References

1. Wermuth, P.J.; Jimenez, S.A. Gadolinium compounds signaling through TLR 4 and TLR 7 in normal human macrophages: Establishment of a proinflammatory phenotype and implications for the pathogenesis of nephrogenic systemic fibrosis. *J. Immunol.* **2012**, *189*, 318–327. [CrossRef] [PubMed]
2. Caravan, P. Strategies for increasing the sensitivity of gadolinium based MRI contrast agents. *Chem. Soc. Rev.* **2006**, *35*, 512. [CrossRef] [PubMed]
3. Thomsen, H.S.; Morcos, S.K.; Almén, T.; Bellin, M.-F.; Bertolotto, M.; Bongartz, G.; Clement, O.; Leander, P.; Heinz-Peer, G.; Reimer, P.; et al. Nephrogenic systemic fibrosis and gadolinium-based contrast media: Updated ESUR contrast medium safety committee guidelines. *Eur. Radiol.* **2013**, *23*, 307–318. [CrossRef] [PubMed]
4. Moon, J.C.C.; Reed, E.; Sheppard, M.N.; Elkington, A.G.; Ho, S.; Burke, M.; Petrou, M.; Pennell, D.J. The histologic basis of late gadolinium enhancement cardiovascular magnetic resonance in hypertrophic cardiomyopathy. *J. Am. Coll. Cardiol.* **2004**, *43*, 2260–2264. [CrossRef]
5. Vogler, H.; Platzek, J.; Schuhmann-Giampieri, G.; Frenzel, T.; Weinmann, H.-J.; Radüchel, B.; Press, W.-R. Pre-clinical evaluation of gadobutrol: A new, neutral, extracellular contrast agent for magnetic resonance imaging. *Eur. J. Radiol.* **1995**, *21*, 1–10. [CrossRef]
6. Scott, L.J. Gadobutrol: A review of its use for contrast-enhanced magnetic resonance imaging in adults and children. *Clin. Drug Investig.* **2013**, *33*, 303–314. [CrossRef]
7. FDA. *DOTAREM (Gadoterate Meglumine) Injection for Intravenous Use*; FDA: Silver Spring, MD, USA, 2013; pp. 1–13.
8. Tombach, B.; Bremer, C.; Reimer, P.; Schaefer, R.M.; Ebert, W.; Geens, V.; Heindel, W. Pharmacokinetics of 1M Gadobutrol in Patients with Chronic Renal Failure. *Investig. Radiol.* **2000**, *35*, 35. [CrossRef]
9. European Medicines Agency. *Gadovist 1.0 mmol/Ml Solution for Injection*; European Medicines Agency: Amsterdam, The Netherlands, 2012; pp. 1–9.
10. *European Medicines Agency (EMA) Gadolinium-Containing Contrast Agents*; European Medicines Agency: Amsterdam, The Netherlands, 2010.
11. Aime, S.; Caravan, P. Biodistribution of gadolinium-based contrast agents, including gadolinium deposition. *J. Magn. Reson. Imaging* **2009**, *30*, 1259–1267. [CrossRef]
12. Clough, T.J.; Jiang, L.; Wong, K.-L.; Long, N.J. Ligand design strategies to increase stability of gadolinium-based magnetic resonance imaging contrast agents. *Nat. Commun.* **2019**, *10*, 1420. [CrossRef]
13. Sieber, M.A.; Lengsfeld, P.; Frenzel, T.; Golfier, S.; Schmitt-Willich, H.; Siegmund, F.; Walter, J.; Weinmann, H.-J.; Pietsch, H. Preclinical investigation to compare different gadolinium-based contrast agents regarding their propensity to release gadolinium in vivo and to trigger nephrogenic systemic fibrosis-like lesions. *Eur. Radiol.* **2008**, *18*, 2164–2173. [CrossRef]
14. FDA. *Gadovist (Gadobutrol) Injection, for Intravenous Use: US Prescribing Information*; FDA: Silver Spring, MD, USA, 2011.

15. Bussi, S.; Coppo, A.; Celeste, R.; Fanizzi, A.; Fringuello Mingo, A.; Ferraris, A.; Botteron, C.; Kirchin, M.A.; Tedoldi, F.; Maisano, F. Macrocyclic MR contrast agents: Evaluation of multiple-organ gadolinium retention in healthy rats. *Insights Into Imaging* **2020**, *11*, 11. [CrossRef]

16. Hazelton, J.M.; Chiu, M.K.; Abujudeh, H.H. Nephrogenic systemic fibrosis: A review of history, pathophysiology, and current guidelines. *Curr. Radiol. Rep.* **2019**, *7*, 5. [CrossRef]

17. Tóth, É.; Helm, L.; Merbach, A.E. *Relaxivity of MRI Contrast Agents*; Springer: Berlin/Heidelberg, Germany, 2002; pp. 61–101.

18. Brücher, E. *Kinetic Stabilities of Gadolinium(III) Chelates Used as MRI Contrast Agents*; Springer: Berlin/Heidelberg, Germany, 2002; pp. 103–122.

19. Woolen, S.A.; Shankar, P.R.; Gagnier, J.J.; MacEachern, M.P.; Singer, L.; Davenport, M.S. Risk of nephrogenic systemic fibrosis in patients with stage 4 or 5 chronic kidney disease receiving a group II gadolinium-based contrast agent. *JAMA Intern. Med.* **2020**, *180*, 223. [CrossRef]

20. Tombach, B.; Bremer, C.; Reimer, P.; Matzkies, F.; Schaefer, R.M.; Ebert, W.; Geens, V.; Eisele, J.; Heindel, W. Using highly concentrated gadobutrol as an MR contrast agent in patients also requiring hemodialysis. *Am. J. Roentgenol.* **2002**, *178*, 105–109. [CrossRef]

21. Von Knobelsdorff-Brenkenhoff, F.; Schüler, J.; Dogangüzel, S.; Dieringer, M.A.; Rudolph, A.; Greiser, A.; Kellman, P.; Schulz-Menger, J. Detection and monitoring of acute myocarditis applying quantitative cardiovascular magnetic resonance. *Circ. Cardiovasc. Imaging* **2017**, *10*, e005242. [CrossRef]

22. Minutoli, F.; Bella, G.D.; Mazzeo, A.; Donato, R.; Russo, M.; Scribano, E.; Baldari, S. Comparison between 99m Tc-Diphosphonate Imaging and MRI with late gadolinium enhancement in evaluating cardiac involvement in patients with transthyretin familial amyloid polyneuropathy. *Am. J. Roentgenol.* **2013**, *200*, W256–W265. [CrossRef]

23. Hussain, S.T.; Paul, M.; Plein, S.; McCann, G.P.; Shah, A.M.; Marber, M.S.; Chiribiri, A.; Morton, G.; Redwood, S.; MacCarthy, P.; et al. Design and rationale of the MR-INFORM study: Stress perfusion cardiovascular magnetic resonance imaging to guide the management of patients with stable coronary artery disease. *J. Cardiovasc. Magn. Reson.* **2012**, *14*, 65. [CrossRef]

24. Nacif, M.S.; Arai, A.A.; Lima, J.A.; Bluemke, D.A. Gadolinium-enhanced cardiovascular magnetic resonance: Administered dose in relationship to United States Food and Drug Administration (FDA) guidelines. *J. Cardiovasc. Magn. Reson.* **2012**, *14*, 18. [CrossRef]

25. ACR Committee on Drugs and Contrast Media. *ACR Manual on Contrast Media Version*; ACR: Silver Spring, MD, USA, 2022; ISBN 978-1-55903-012-0.

26. Cowper, S.E.; Bucala, R.; Leboit, P.E. Nephrogenic fibrosing dermopathy/nephrogenic systemic fibrosis—Setting the record straight. *Semin. Arthritis Rheum.* **2006**, *35*, 208–210. [CrossRef]

27. Cowper, S.E. Nephrogenic Systemic Fibrosis: An Overview. *J. Am. Coll Radiol.* **2008**, *5*, 23–28. [CrossRef]

28. Reiter, T.; Ritter, O.; Prince, M.R.; Nordbeck, P.; Wanner, C.; Nagel, E.; Bauer, W.R. Minimizing risk of nephrogenic systemic fibrosis in cardiovascular magnetic resonance. *J. Cardiovasc. Magn. Reson.* **2012**, *14*, 31. [CrossRef]

29. Zou, Z.; Zhang, H.L.; Roditi, G.H.; Leiner, T.; Kucharczyk, W.; Prince, M.R. Nephrogenic systemic fibrosis. *JACC Cardiovasc. Imaging* **2011**, *4*, 1206–1216. [CrossRef]

30. Prince, M.R.; Zhang, H.; Morris, M.; MacGregor, J.L.; Grossman, M.E.; Silberzweig, J.; DeLapaz, R.L.; Lee, H.J.; Magro, C.M.; Valeri, A.M. Incidence of nephrogenic systemic fibrosis at two large medical centers. *Radiology* **2008**, *248*, 807–816. [CrossRef]

31. Cowper, S.E.; Robin, H.S.; Steinberg, S.M.; Su, L.D.; Gupta, S.; LeBoit, P.E. Scleromyxoedema-like cutaneous diseases in renal-dialysis patients. *Lancet* **2000**, *356*, 1000–1001. [CrossRef]

32. Cowper, S.E. Nephrogenic Systemic Fibrosis: The nosological and conceptual evolution of nephrogenic fibrosing dermopathy. *Am. J. Kidney Dis.* **2005**, *46*, 763–765. [CrossRef]

33. Grobner, T. Gadolinium—A specific trigger for the development of nephrogenic fibrosing dermopathy and nephrogenic systemic fibrosis? *Nephrol. Dial. Transplant.* **2006**, *21*, 1104–1108. [CrossRef]

34. Girardi, M.; Kay, J.; Elston, D.M.; LeBoit, P.E.; Abu-Alfa, A.; Cowper, S.E. Nephrogenic systemic fibrosis: Clinicopathological definition and workup recommendations. *J. Am. Acad. Dermatol.* **2011**, *65*, 1095–1106.e7. [CrossRef]

35. Yee, J. Prophylactic hemodialysis for protection against gadolinium-induced nephrogenic systemic fibrosis: A doll's house. *Adv. Chronic Kidney Dis.* **2017**, *24*, 133–135. [CrossRef] [PubMed]

36. Kaewlai, R.; Abujudeh, H. Nephrogenic systemic fibrosis. *Am. J. Roentgenol.* **2012**, *199*, W17–W23. [CrossRef] [PubMed]

37. Mathur, M.; Jones, J.R.; Weinreb, J.C. Gadolinium deposition and nephrogenic systemic fibrosis: A radiologist's primer. *Radio-Graphics* **2020**, *40*, 153–162. [CrossRef] [PubMed]

38. Thomsen, H.S. Nephrogenic systemic fibrosis: A serious adverse reaction to Gadolinium—1997–2006–2016. Part 1. *Acta Radiol.* **2016**, *57*, 515–520. [CrossRef]

39. Deo, A.; Fogel, M.; Cowper, S.E. Nephrogenic systemic fibrosis: A population study examining the relationship of disease development to gadolinium exposure. *Clin. J. Am. Soc. Nephrol.* **2007**, *2*, 264–267. [CrossRef]

40. Wang, Y.; Alkasab, T.K.; Narin, O.; Nazarian, R.M.; Kaewlai, R.; Kay, J.; Abujudeh, H.H. Incidence of nephrogenic systemic fibrosis after adoption of restrictive gadolinium-based contrast agent guidelines. *Radiology* **2011**, *260*, 105–111. [CrossRef]

41. Prince, M.R.; Lee, H.G.; Lee, C.-H.; Youn, S.W.; Lee, I.H.; Yoon, W.; Yang, B.; Wang, H.; Wang, J.; Shih, T.T.; et al. Safety of Gadobutrol in over 23,000 patients: The GARDIAN Study, a global multicentre, prospective, non-interventional study. *Eur. Radiol.* **2017**, *27*, 286–295. [CrossRef]

42. Soyer, P.; Dohan, A.; Patkar, D.; Gottschalk, A. Observational study on the safety profile of gadoterate meglumine in 35,499 patients: The SECURE Study. *J. Magn. Reson. Imaging* **2017**, *45*, 988–997. [CrossRef]
43. Michaely, H.J.; Aschauer, M.; Deutschmann, H.; Bongartz, G.; Gutberlet, M.; Woitek, R.; Ertl-Wagner, B.; Kucharczyk, W.; Hammerstingl, R.; de Cobelli, F.; et al. Gadobutrol in renally impaired patients. *Investig. Radiol.* **2017**, *52*, 55–60. [CrossRef]
44. Tsushima, Y.; Awai, K.; Shinoda, G.; Miyoshi, H.; Chosa, M.; Sunaya, T.; Endrikat, J. Post-marketing surveillance of Gadobutrol for contrast-enhanced magnetic resonance imaging in Japan. *Jpn. J. Radiol.* **2018**, *36*, 676–685. [CrossRef]
45. Young, L.K.; Matthew, S.Z.; Houston, J.G. Absence of potential gadolinium toxicity symptoms following 22,897 Gadoteric Acid (Dotarem®) examinations, including 3,209 performed on renally insufficient individuals. *Eur. Radiol.* **2019**, *29*, 1922–1930. [CrossRef]
46. McWilliams, R.G.; Frabizzio, J.V.; de Backer, A.I.; Grinberg, A.; Maes, B.D.; Zobel, B.B.; Gottschalk, A. Observational study on the incidence of nephrogenic systemic fibrosis in patients with renal impairment following gadoterate meglumine administration: The NSsaFe Study. *J. Magn. Reson. Imaging* **2020**, *51*, 607–614. [CrossRef]
47. Penfield, J.G.; Reilly, R.F. NSF: What we know and what we need to know: Nephrogenic systemic fibrosis risk: Is there a difference between gadolinium-based contrast agents? *Semin. Dial.* **2008**, *21*, 129–134. [CrossRef]
48. Rudnick, M.R.; Wahba, I.M.; Leonberg-Yoo, A.K.; Miskulin, D.; Litt, H.I. Risks and options with gadolinium-based contrast agents in patients With CKD: A review. *Am. J. Kidney Dis.* **2020**, *77*, 517–528. [CrossRef]
49. Schieda, N.; Blaichman, J.I.; Costa, A.F.; Glikstein, R.; Hurrell, C.; James, M.; Jabehdar Maralani, P.; Shabana, W.; Tang, A.; Tsampalieros, A.; et al. Gadolinium-based contrast agents in kidney disease: Comprehensive review and clinical practice guideline issued by the Canadian Association of Radiologists. *Can. Assoc. Radiol. J.* **2018**, *69*, 136–150. [CrossRef]
50. Murashima, M.; Drott, H.R.; Carlow, D.; Shaw, L.M.; Milone, M.; Bachman, M.; Tsai, D.E.; Yang, S.L.; Bloom, R.D. Removal of Gadolinium by peritoneal dialysis. *Clin. Nephrol.* **2008**, *69*, 368–373. [CrossRef]
51. Basak, P.; Jesmajian, S. Nephrogenic systemic fibrosis: Current concepts. *Indian J. Dermatol.* **2011**, *56*, 59. [CrossRef]
52. Guo, B.J.; Yang, Z.L.; Zhang, L.J. Gadolinium deposition in brain: Current scientific evidence and future perspectives. *Front. Mol. Neurosci.* **2018**, *11*, 335. [CrossRef]
53. Gulani, V.; Calamante, F.; Shellock, F.G.; Kanal, E.; Reeder, S.B. Gadolinium deposition in the brain: Summary of evidence and recommendations. *Lancet Neurol.* **2017**, *16*, 564–570. [CrossRef]
54. Stoumpos, S.; Hennessy, M.; Vesey, A.T.; Radjenovic, A.; Kasthuri, R.; Kingsmore, D.B.; Mark, P.B.; Roditi, G. Ferumoxytol magnetic resonance angiography: A dose-finding study in patients with chronic kidney disease. *Eur. Radiol.* **2019**, *29*, 3543–3552. [CrossRef]

diagnostics

Interesting Images

Loeffler Endocarditis Causing Heart Failure with Preserved Ejection Fraction (HFpEF): Characteristic Images and Diagnostic Pathway

Silvia Lupu [1,2,*,†]**, Marian Pop** [3,4,†] **and Adriana Mitre** [1,2]

1   M3, Medicala V, "George Emil Palade" University of Medicine, Pharmacy, Science and Technology of Targu Mures, 540142 Targu Mures, Romania
2   Cardiology 1 Department, Emergency Institute for Cardiovascular Disease and Heart Transplant of Targu Mures, 540136 Targu Mures, Romania
3   ME1, "George Emil Palade" University of Medicine, Pharmacy, Science and Technology of Targu Mures, 540142 Targu Mures, Romania
4   Radiology and Medical Imaging Department, Emergency Institute for Cardiovascular Disease and Heart Transplant of Targu Mures, 540136 Targu Mures, Romania
*   Correspondence: silvia.lupu@umfst.ro
†   These authors contributed equally to this work.

Abstract: We report the case of a 69-year-old female patient in which echocardiography and cardiac magnetic resonance imaging were used to diagnose a patient presenting with heart failure with preserved ejection fraction (HFpEF) due to Loeffler endocarditis. Loeffler endocarditis is an uncommon cause of heart failure with preserved ejection fraction, triggered by eosinophil and lymphocyte infiltration of the endomyocardium, followed by the formation of thrombus in the afflicted area, and eventually fibrosis. This condition is due to an increased number of eosinophils associated with allergies, infections, systemic conditions, as well as malignancies and hypereosinophilic syndrome. Loeffler endocarditis can lead to serious complications, such as progressive heart failure, systemic thromboembolic events, or arrhythmias (including sudden cardiac death).

Keywords: Loeffler endocarditis; imaging; cardiac MRI; echocardiography

Citation: Lupu, S.; Pop, M.; Mitre, A. Loeffler Endocarditis Causing Heart Failure with Preserved Ejection Fraction (HFpEF): Characteristic Images and Diagnostic Pathway. *Diagnostics* **2022**, *12*, 2157. https://doi.org/10.3390/diagnostics12092157

Academic Editors: Minjie Lu and Arlene Sirajuddin

Received: 27 July 2022
Accepted: 3 September 2022
Published: 5 September 2022

Publisher's Note: MDPI stays neutral with regard to jurisdictional claims in published maps and institutional affiliations.

A 69-year-old female patient presented with dyspnea on moderate exertion, progressively aggravated within the last two years. The patient had no history of angina or syncope. Her medical records showed moderate hypereosinophilia, first documented 5 years before.

After ruling out a couple of common parasite infections by repeated stool sample analysis, no other causes were investigated. The lady had no history of overt allergy and nor of travel to any exotic destinations. She was treated for depression with escitalopram and lorazepam; both started after the hypereosinophilia was documented.

On clinical examination, the patient had a blood pressure of 110/80 mmHg, a heart rate of 90 bpm, fine crackles at the bases of both lungs, and bilateral edema of the inferior limbs.

The initial blood tests showed increased troponin I levels– 155 ng/L (>29 ng/L, Pathfast assay, Medscience Corporation Mitsubishi Chemical Europe), and a blood smear test was performed, yielding moderate hypereosinophilia with no dysmorphic features 4.95×10^9/L (55%). The number of leukocytes (7900/mm^3), hemoglobin levels (14.3 g/dL), and creatinine were normal (between 0.70 and 0.78 md/dL).

Echocardiography revealed a normal size left ventricle, with apical obliteration (Figure 2A,B), thickening of the posterior mitral valve and chordae, restricted movement and subsequent mild mitral regurgitation, as well as features of restrictive cardiomyopathy. Based on these characteristics, the diagnosis of Loeffler endocarditis was suspected [1].

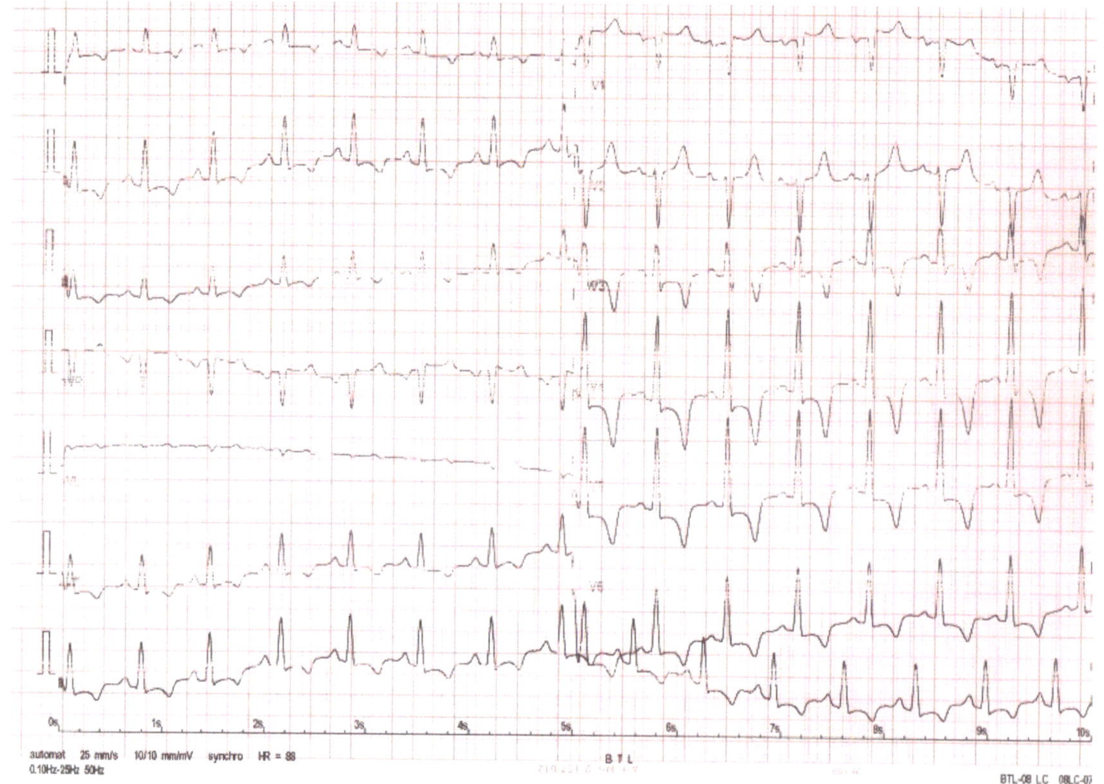

Figure 1. Twelve-lead electrocardiogram—sinus rhythm and deep inverted T waves in most leads (excepting V1, V2, and aVR), and a corrected QT interval of 533 msec.

Cardiac magnetic resonance imaging was further performed. Native and post-contrast (9 mmol gadolinium) images were acquired. bSSFP Cine imaging showed a non-dilated left ventricle with preserved ejection fraction and apical obliteration (Figure 2C and Supplementary Videos S1 and S2). Inversion recovery images 10 min after contrast injection showed sub-endocardial late gadolinium enhancement (LGE) pointing toward the apex and left ventricular apical non-enhancing mass-thrombus (Figure 2D)".

The imaging features and the presence of hypereosinophilia were consistent with the diagnosis of Loeffler endocarditis [1,2]. A coronary angio-computed tomography was performed, showing no significant coronary stenosis, a myocardial bridge in the mid-segment of the left anterior descendant artery, and a coronary calcium score of 0. The increase in troponin levels suggested active myocardial inflammation, prompting the initiation of treatment with corticosteroids (methylprednisolone 32 mg/day, starting dose), antihistamines (loratadine 10 mg), as well as loop and antialdosteronic diuretics, an angiotensin-converting enzyme inhibitor, and anticoagulants (acenocumarol and enoxaparin, which was discontinued when therapeutic INR was achieved) [3].

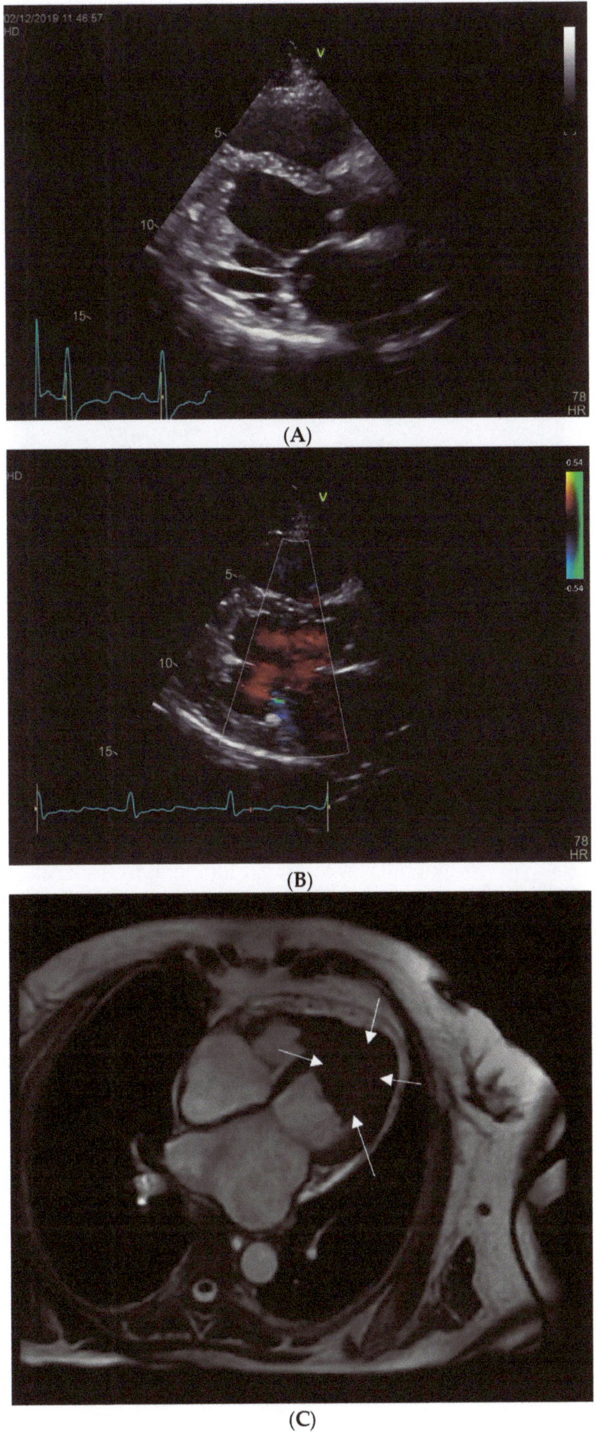

Figure 2. *Cont.*

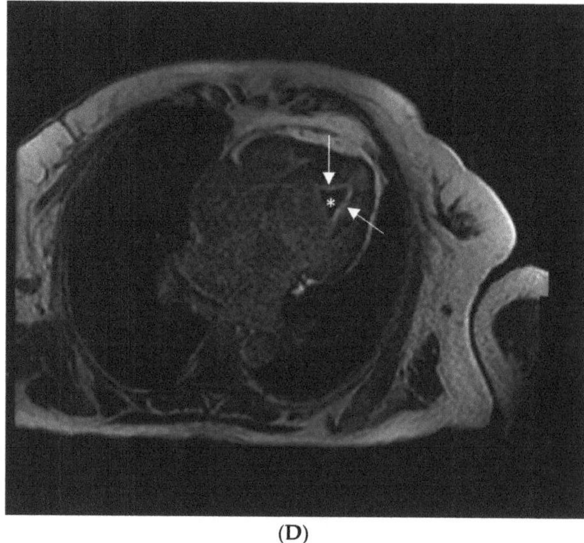

(D)

Figure 2. (**A**) Echo Parasternal long axis view showing obliteration of the left ventricular apex. (**B**) Echo Parasternal long axis view: normal-size left ventricle with normal systolic function, increased thickness of the posterior mitral leaflet and chordae, apical obliteration. (**C**) Cardiac MRI: 4 chamber view showing a normal-size left ventricle, with preserved ejection fraction and an apical mass engulfing the papillary muscles. (**D**) Cardiac MRI: inversion recovery images 10 min after contrast injection showing left ventricular apical thrombus (white arrows) and sub-endocardial enhancement (star).

Following treatment, the signs and symptoms of heart failure subsided, and the troponin levels quickly returned to normal levels (19.4 ng/L at discharge). The eosinophil count was normal after a week of treatment. We further attempted to identify the cause of hypereosinophilia. A computer tomography scan ruled out solid malignancy; an area of panacinary emphysema and mild fibrosis at the base of the right lung were visible. Repeated stool samples were negative for parasites. The patient was referred to the internal medicine department to complete the investigations panel for eosinophilia. We considered parasite infection, allergy, rheumatological and hematological disorders as potential causes [3] and scheduled the patient for an appointment with the hematologist to test for JAK2, FIP1L1, CAL-R, and CMPL. However, the patient was non-compliant with recommendations and followed her treatment inconsistently. At the time of discharge, the echocardiography showed the remanence of thrombus, with little change in size. While it is accepted that upon follow-up CMR can show both the decrease in cardiac chamber volumes, the resolution of apical thrombus, and the receding of subendocardial fibrosis, our patient skipped the follow-up visits and was, eventually, lost to follow-up.

Loeffler endocarditis is a rather rare cause of heart failure, which requires specific treatment and is associated with a very poor prognosis if left untreated. HFpEF in Loeffler endocarditis is due to a change in ventricular compliance, and it shares similar pathophysiology with other restrictive cardiomyopathies; however, its pathophysiology is due to endocardial fibrosis from eosinophilic injuries instead of deposits of amyloid (in amyloidosis), iron (hemochromatosis) or non-caseating granulomas (sarcoidosis).

Cardiovascular imaging plays a pivotal role in the diagnosis [2,4,5]. While echocardiography features are usually most obvious in the late stages of the disease and may provide insufficient details, cardiac magnetic resonance imaging may be used to identify early changes and to better characterize the endocardium and the presence of thrombus in the late stages [6–8]. Although endomyocardial biopsy remains the gold standard for diagnosing the disease, it is an invasive procedure and may sometimes yield falsely negative results [3]. However, cardiac MRI provides compelling evidence of the disease in this patient, such as the presence of subendocardial LGE, and apical thrombus, features that do not emerge in other cardiomyopathies associated with a restrictive pattern, such as sarcoidosis, amyloidosis, or Fabry disease. Although apical hypertrophic cardiomyopathy may exhibit a similar electrocardiographic pattern, the presence of apical thrombus and subendocardial LGE lining the hypertrophied area is unlikely in this condition [4].

Every effort should be made to identify the cause of hypereosinophilia and provide targeted treatment.

Supplementary Materials: The following supporting information can be downloaded at: https://www.mdpi.com/article/10.3390/diagnostics12092157/s1, Video S1: CMR 4C view, Video S2: CMR 2C view.

Author Contributions: Conceptualization, S.L. and M.P.; methodology, S.L.; software, M.P.; validation, A.M; formal analysis, S.L.; investigation, S.L. and A.M.; resources, S.L.; data curation, A.M.; writing—original draft preparation, S.L.; writing—review and editing, S.L and M.P.; visualization, S.L and M.P.; supervision, A.M.; project administration, S.L.; funding acquisition, S.L. and M.P. All authors have read and agreed to the published version of the manuscript.

Funding: This work was supported by the University of Medicine and Pharmacy of Tîrgu Mureș Research Grant number 15609/4 of 29 December 2017.

Institutional Review Board Statement: The study was conducted in accordance with the Declaration of Helsinki and approved by the Ethics Committee of the Emergency Institute for Cardiovascular Diseases and Heart Transplant of Tirgu Mures (protocol code 785/2018).

Informed Consent Statement: Written informed consent has been obtained from the patient(s) to publish this paper.

Data Availability Statement: Not applicable.

Conflicts of Interest: The authors declare no conflict of interest.

References

1. Kleinfeldt, T.; Nienaber, C.A.; Kische, S.; Akin, I.; Turan, R.G.; Körber, T.; Schneider, H.; Ince, H. Cardiac Manifestation of the Hypereosinophilic Syndrome: New Insights. *Clin. Res. Cardiol. Off. J. Ger. Card. Soc.* **2010**, *99*, 419–427. [CrossRef]
2. Park, J.; Hemu, M.; Kalra, D. Loeffler's Endocarditis: A Diagnosis Made With Cardiac Magnetic Resonance (CMR) Imaging. *J. Am. Coll. Cardiol.* **2019**, *73*, 2267. [CrossRef]
3. Roufosse, F.; Weller, P.F. Practical Approach to the Patient with Hypereosinophilia. *J. Allergy Clin. Immunol.* **2010**, *126*, 39–44. [CrossRef]
4. Habib, G.; Bucciarelli-Ducci, C.; Caforio, A.L.P.; Cardim, N.; Charron, P.; Cosyns, B.; Dehaene, A.; Derumeaux, G.; Donal, E.; Dweck, M.; et al. Multimodality Imaging in Restrictive Cardiomyopathies: An EACVI expert consensus document In collaboration with the "Working Group on myocardial and pericardial diseases" of the European Society of Cardiology Endorsed by The Indian Academy of Echocardiography. *Eur. Heart J. -Cardiovasc. Imaging* **2017**, *18*, 1090–1121. [CrossRef] [PubMed]
5. Gastl, M.; Behm, P.; Jacoby, C.; Kelm, M.; Bönner, F. Multiparametric Cardiac Magnetic Resonance Imaging (CMR) for the Diagnosis of Loeffler's Endocarditis: A Case Report. *BMC Cardiovasc. Disord.* **2017**, *17*, 74. [CrossRef] [PubMed]
6. Radovanovic, M.; Jevtic, D.; Calvin, A.D.; Petrovic, M.; Paulson, M.; Rueda Prada, L.; Sprecher, L.; Savic, I.; Dumic, I. "Heart in DRESS": Cardiac Manifestations, Treatment and Outcome of Patients with Drug Reaction with Eosinophilia and Systemic Symptoms Syndrome: A Systematic Review. *J. Clin. Med.* **2022**, *11*, 704. [CrossRef] [PubMed]
7. Schreiber, K.; Zuern, C.S.; Gawaz, M. Loeffler Endocarditis: Findings on Magnetic Resonance Imaging. *Heart* **2007**, *93*, 354. [CrossRef] [PubMed]
8. Allderdice, C.; Marcu, C.; Kabirdas, D. Intracardiac Thrombus in Leukemia: Role of Cardiac Magnetic Resonance Imaging in Eosinophilic Myocarditis. *CASE Cardiovasc. Imaging Case Rep.* **2018**, *2*, 114–117. [CrossRef]

MDPI AG
Grosspeteranlage 5
4052 Basel
Switzerland
Tel.: +41 61 683 77 34

Diagnostics Editorial Office
E-mail: diagnostics@mdpi.com
www.mdpi.com/journal/diagnostics